third edition

MASONRY
simplified

Tools, Materials, Practice

Bricklaying, Concrete Block
and Cement Masonry

J. Ralph Dalzell
Gilbert Townsend

 American Technical Society • Chicago, 60637

COPYRIGHT © 1958, 1956, 1973
BY AMERICAN TECHNICAL SOCIETY

Library of Congress Catalog Number: 72-93804

ISBN 0-8269-0627-3

First Edition

1st Printing 1948
2d Printing 1948
3d Printing 1949
4th Printing 1950
5th Printing 1952
6th Printing 1955

Second Edition (Revised)

7th Printing 1956
8th Printing 1958
9th Printing 1962
10th Printing 1964
11th Printing 1966
12th Printing 1967
13th Printing 1970
14th Printing 1971
15th Printing 1972

Third Edition (Revised)

16th Printing 1973
17th Printing 1974
18th Printing 1976
19th Printing 1977

No portion of this publication may be reproduced by any process such as photocopying, recording, storage in a retrieval system or transmitted by any means without permission of the publisher.

Illustrations on the cover are by courtesy of the Brick Institute of America.

PRINTED IN THE UNITED STATES OF AMERICA

Preface

The revised Third Edition of *Masonry Simplified* is designed to provide a step-by-step learning sequence in the bricklaying and cement masonry trades. *Masonry Simplified* provides a useful, thorough-going learning tool for the vocational student or apprentice as well as an invaluable reference work for the professional tradesman.

Tools and practices in the masonry trades have changed very little. However, the materials and manufacturing processes for tools and structural materials have been greatly refined over the last few years. As a result, modern codes and standards have been greatly changed and simplified. The standards and practices described in this book have been revised to conform to the latest standards and codes.

Safety, as always, is of the utmost importance on any construction site. The Third Edition includes a new chapter on safety which serves as an introduction and guideline to on the job safety. It also serves as a refresher course to supervising personnel.

The new edition also includes a complete chapter on tools for both brick masonry and cement masonry. In addition, advice is given on how to select the proper tools for a basic tool kit.

With the vast amount of new construction, workmen of different trades commonly work side-by-side on the job. The new edition includes, therefore, material on the closely related trades. The chapter on concrete construction has been greatly expanded. The carpenter who builds the formwork for footings and foundations, the cement mason who places them, and the brick mason who builds upon these foundations must understand how the other craftsman works in order to assure a sound structure.

The various formulas, designations and classifications of materials, and construction practices described in the text are not intended as hard and fast rules. They do, however, conform to the latest guidelines furnished by the federal and state governments and by interested private agencies. They are intended as typical examples of building codes and accepted construction practices.

Appendix A gives a review of blueprint reading for the professional while providing a basic introduction for the student or apprentice. Appendix B provides an outline of practical mathematics for craftsmen of all trades. A "Glossary of Trade Terms" is included at the end of the book; it has been greatly expanded and updated.

The text material of the revised Third Edition has been reviewed by Francis J. Miller, P.E., Coordinator of Building Construction Technology, Thornton Community College, South Holland, Illinois. His comments have been most helpful. Gratitude is expressed to the American Concrete Institute for their review and thoughtful comments on Chapters 6, 7 and 8. Thanks should be made to the National Concrete Masonry Association for valuable suggestions on Chapter 6. Appreciation is also due to John V. Burnett, Technical Editor, for his careful and precise editing, and contributions to the text.   THE PUBLISHERS

Contents

Chapter

1 Masonry as a Trade, 1
2 Safety on the Job, 13
3 Tools and Equipment, 29
4 Mortar: Basic Materials and Mixing, 46
5 Brick Masonry: Units and Building, 61
6 Concrete Block Masonry: Units and Building, 122
7 Concrete: Basic Materials and Mixing, 185
8 Concrete Construction, 204

Appendix

A Review of Blueprints, Conventions and Symbols for Masonry, 254
B Math for Masons, 274
   Glossary of Trade Terms, 286
   Index, 402

*Masonry is an ancient and proud craft.
There is pride and dignity in being a safe and
competent workman.*

Masonry as a Trade

Chapter **1**

The term *masonry* has traditionally meant the craft of building with bricks. Modern building trades practices, however, have given the term two different meanings: *brick masonry* and *cement masonry*. For the purposes of this book, and in concord with accepted trade terminology, the term brick masonry will refer to vertical (upright) construction made of masonry units (brick, concrete block, stone, glass brick) as bonded together by mortar. Cement masonry will refer to formed solid concrete structure such as slabs, floors, footings and foundations, driveways and sidewalks.

Structure of the Trade

The masonry trades are structured into various levels of craftsmen: apprentice, journeyman, foreman, superintendent.

Apprenticeship or learning usually lasts three years. From man's earliest history apprentices in the various crafts were indentured (a contract binding one person to work for another for a given length of time to learn a trade) to a master craftsman (a contractor) for a number of years to learn the trade.

In the past, in many cases, the

apprentice's father had to pay the master a fee to get him to teach his son the trade. From the medieval days down through most of the 19th century, the apprentice would live with the master and would get room and board plus some clothing. However, he was a virtual slave to the master, subject to his every wish. Both old and new certificates of apprentice indentures are reproduced on pages 3 and 4.

Note that today the apprentices are protected by Federal, State, and the Local J.A.C. (Joint Apprenticeship Committee) in regard to hours of work, wages, and conditions of employment, and there is no control over the apprentices outside of working hours. Also, apprentices are now selected from applicants who meet the standards of the local J.A.C. In most cases the apprentice is indentured to the J.A.C. and they assign him to a contractor. If the contractor runs out of work, the J.A.C. will place him with another contractor. This permits the J.A.C. to control the training and handle the federal and veteran's paperwork.

The *journeyman* or experienced craftsman is one who has completed an apprenticeship in the trade. He is now a free agent and can work for any contractor he pleases. He may travel from place to place, going where the work is to be found.

The *foreman* is a journeyman who has been placed in the job of supervising a group of men. He is given this job because of his ability as a craftsman and his knowledge of how to supervise other craftsmen.

The *superintendent* is usually a foreman who has been promoted to this important position. He is in charge of all the work in the field for his contractor and supervises the work of the foreman. Some large contractors have a *general superintendent* to oversee the *job superintendents*.

In the construction industry the *foreman* and *superintendent* keep their union membership. Many of the smaller contractors are permitted to retain their union cards in some unions.

Large contractors will employ an *estimator* who works in their offices to estimate the cost of the jobs the contractor wants to bid on. Working from blueprints and specifications, he performs a quantity take off, that is, measures areas to be constructed of masonry and determines the amount of materials required. He must be skilled in mathematics, blueprint reading, trade practices and the cost of labor and materials.

The last person in this team of workers, supervisors and estimator is the *contractor*. He must be knowledgeable in all phases of the business. He must know all the regulations governing the construction industry plus how to provide the money for payrolls and materials.

Masonry as a Trade

This Indenture witnesseth, That

Jacob Peterson, with the consent of his mother, Mary Ann Griffith, hath put himself, and by these presents, and for other good causes, doth voluntarily and by his own free will and accord, put himself apprentice to Hiram Miller to learn the art, trade, and mystery of a House Carpenter and, after the manner of an apprentice, to serve the said Hiram Miller from the day of the date hereof, for and during, and to the full end and term of Three Years Ten Months and Fifteen Days next ensuing. During all which time the said apprentice doth covenant and promise, that he will serve his master faithfully, keep his secrets, and obey his lawful commands—that he will do him no damage himself, nor see it done by others, without giving him notice thereof—that he will not waste his goods, nor lend them unlawfully—that he will not play at cards, dice, or any other unlawful game, whereby his master may be injured—that he will neither buy nor sell, with his own goods nor the goods of others, without license from his master—and that he will not absent himself day nor night from his master's service, without his leave—nor haunt ale-houses, taverns, nor play-houses; but in all things behave as a faithful apprentice ought to do, during the said term. And the said master, on his part, doth covenant and promise, that he will use the utmost of his endeavors to teach or cause to be taught or instructed, the said apprentice in the art, trade or mystery of a House Carpenter and that he will procure and provide for him sufficient meat, drink & lodging fitting for an apprentice during the said term—and that he will give him Forty dollars per Year payable Quarterly in lieu of Clothing and two Quarters Night Schooling—And for the true performance of all and singular the covenants and agreements aforesaid, the said parties bind themselves each unto the other, firmly, by these presents.

    In witness whereof, the said parties have interchangeably set their hands and seals hereunto. Dated the Twentieth day of July, Anno Domini 1844.

THE ABOVE IS AN EXACT TRANSCRIPT OF ORIGINAL INDENTURE WRITTEN IN THE YEAR 1844

Masonry Simplified

BRICKLAYERS AND STONE MASONS APPRENTICE AGREEMENT

THIS AGREEMENT Made this _____ day of _____, 19____, between _____, herein referred to as the "Apprentice", and (if a minor) _____, herein referred to as the "Guardian", and the BRICKLAYERS LOCAL 21 OF ILLINOIS APPRENTICESHIP AND TRAINING PROGRAM, a common law trust created by an agreement dated November 23, 1965, and administered by a Board of Trustees, herein called the "BOARD OF TRUSTEES."

WITNESSETH:

1. THE APPRENTICE AGREES:

(a) To work under the assignments and direction of the Board of Trustees for a period of 3 years as defined under the Rules and Regulations and Standards as promulgated by the Board of Trustees;

(b) To accept assignments by the Board of Trustees to such contractor or contractors as the Board of Trustees may from time to time select;

(c) To accept the wages, hours and working conditions established by the collective bargaining agreements between the UNITED ORDER OF AMERICAN BRICKLAYERS AND STONE MASONS' UNION NO. 21, herein called the "Union," and the BUILDERS' ASSOCIATION OF CHICAGO, and the ASSOCIATED MASONRY CONTRACTORS OF GREATER CHICAGO, in force from time to time during the period of this agreement, as the wages he shall receive from, and the working conditions under which he will be employed by such contractor or contractors;

(d) While attending the trade school established by the Board of Trustees, he will abide by the Rules and Regulations and Standards adopted by the Board of Trustees and any amendments thereto adopted during the period of this agreement;

(e) Upon the signing of this agreement and in accordance with the requirements of the collective bargaining agreements, to join the Union; and

(f) To accept, and be governed by, the decisions of the Coordinator as to his work assignments throughout the period of apprenticeship, subject only to the Apprentice's right of appeal herein provided for.

2. The Apprentice FURTHER AGREES that, if he fails to comply faithfully with all rules and regulations governing his attendance at the trade school established by the Board of Trustees, and all terms and conditions imposed on his on-the-job training by the collective bargaining agreements, he shall be subject to such reasonable penalties as may be imposed by the Coordinator.

3. The Apprentice FURTHER AGREES that the first three months of employment following the signing of this agreement shall be a probationary period; that during this period this agreement may be cancelled by either party without the formality of a hearing; that at the end of each six-month period of employment he will take such examination as may be prescribed by the Coordinator, and that if he does not successfully pass any one of such examinations, his indenture may be cancelled or his probationary period extended. The Apprentice shall have the right to appeal any ruling of the Coordinator to the Board of Trustees.

4. The Apprentice FURTHER AGREES that the Board of Trustees may cancel the apprenticeship agreement and remove the Apprentice from the Apprenticeship Program for cause. Such removal by the Board of Trustees shall cancel his classification of Apprentice and his opportunity to complete his training.

5. THE BOARD OF TRUSTEES AGREES:

(a) During the period of this agreement it will use every reasonable effort to place the Apprentice with contractors for on-the-job training so that the Apprentice will be employed as continuously as possible and receive training in all phases of the trade specified in the Rules and Regulations and Standards governing the Apprenticeship and Training Program in force from time to time during the period of this agreement. The Board of Trustees does not guarantee full time employment to the Apprentice, if the Board is unable to find contractors ready to accept the Apprentice for on-the-job training.

(b) Upon successful completion of the apprenticeship term fixed by this agreement, to cause to be issued to the Apprentice a certificate showing him to be a qualified journeyman bricklayer.

6. BOTH PARTIES AGREE that if any differences arise between the Apprentice and the contractor by whom he is employed concerning matters of apprenticeship, they shall be adjusted by the Coordinator, subject to review by the Board of Trustees. If, at any time, the Apprentice feels himself aggrieved by any decision of the Coordinator affecting his apprenticeship, he may appeal to the Board of Trustees for a review of the Coordinator's decision. Such appeal shall be in writing and filed within a reasonable time after the decision. It need not be in any particular form, but shall state clearly the reason for the Apprentice's dissatisfaction with the decision of the Coordinator.

7. This agreement may be cancelled by mutual consent of the parties hereto.

IN WITNESS WHEREOF, the parties hereto have signed this instrument the date and year first above written.

_____
(Signature of Apprentice)

BRICKLAYERS LOCAL 21 OF ILLINOIS
APPRENTICESHIP AND TRAINING PROGRAM

_____
(Parent or Guardian)

By _____
(Apprentice Coordinator)

ATTEST: For the United Order of American Bricklayers and Stone Masons. Local 21 of Illinois of the Bricklayers, Masons and Plasterers' Interntional Union of America.

_____
(President)

_____
(Secretary-Treasurer)

Registered by _____
(Name of registration agency)

Date _____ By _____
(Signature and title of authorized official)

Masonry as a Trade

On-the-Job Contracting Practices

It is common in the trade for a man to work his way up the ladder of responsibility from apprentice to contractor staying at each level until he is proficient and conditions, plus drive, enable him to move ahead. The majority of the contractors in the various skilled trades have worked their way up this ladder. They are craftsmen as well as businessmen.

When a craftsman has worked his way up the ladder of experience and aspires to become a contractor, he must prepare himself for operating under a whole new set of rules and conditions. As a contractor he will be risking his own money on the gamble that his knowledge and experience in estimating work, handling men and keeping up with the paperwork that is required today will pay a profit.

With these factors understood, the new contractor stands ready to bid against his fellow contractors on the work that is available. When there is a good supply of work ready to be bid and the new contractor has a crew of men that he knows can produce the work required, he will have a good chance of success. By starting small and bidding only on work that is fully understood, there is a chance to make money and progress to bigger and better jobs. Contracting is a rewarding endeavor for those who are suited for it by ability, temperament and aggressiveness to stick it out through good and bad times.

Therefore, to be successful the new contractor should consider the following rules, conditions, laws and facts which he will have to know and work with.

1. Develop a good knowledge of estimating material and labor costs to do a given job. Remember that no two jobs are ever alike. Conditions of weather, availability of men, changing costs for materials and labor, plus the differences in architects and general contractors can together or individually change the cost of doing a given job.

2. The contractor must know the laws and regulations he must follow, use and invoke at times to protect himself and his work force. These include the lien laws (used to protect against loss due to failure of the owner or general contractor to pay his bills), local permits and inspections as called for, the collection and payment of federal, state and local taxes on the men employed, sales taxes, use taxes, property and inventory taxes on office and storage yard.

Also there are a number of different types of insurances that must be carried. These include compensation, unemployment, accident, property damage, fire, theft, car and truck insurances, business loss, etc. Each one of these items costs quite a bit of money and must be included in the cost of doing business. Too many beginning contractors fail to consider these costs in their bidding of a job and end up having to take it out of the profit (if any) on the job—which is really taking these costs out of their own wages.

3. Most construction will involve tradesmen other than masons: such as carpenters, painters, electricians, plumbers, etc. Each trade, of course, has its own union and working rules. A masonry contractor must be fully aware of the rules of other trade unions and abide by them in order to successfully complete a job.

4. As the contractor advances in work he will find that each architect, general contractor and owner that he works for will have different conditions and rules governing the jobs they design and build. It is therefore very important to study and understand these variations so that the cost of working under these conditions can be added into the bid for the job. The plans and specifications must be gone over carefully to pick out each condition, rule and method set down in both items.

Anything that is not clear should be questioned in writing to the architect before the bid is sent in. No verbal agreement should ever be accepted or given. Everything should be in writing and signed by the architect or the general contractor. Written and signed work orders should be issued for any changes, additions or deletions in the work.

It is the responsibility of each contractor on a job to check the blueprints, specifications, and job site for omissions or errors in his particular field and call them to the attention of the architect. The architect, in turn, will provide the contractor with written or drawn corrections. If the contractor makes any errors in preparation for his work and attempts to overcome them on his own, he assumes responsibility for any resulting flaw or failure in his own work.

All changes in work to be done on the job site should be given in writing as work orders by the architect and general contractor. Temporary lighting, power, cleanup and removal of rubbish, use of scaffolds, use of hoist for lifting material and equipment to various floors, availability of work area in reasonable time to get the work done, storage of material at the job site and on various floors, use of water and many other items must all be agreed on by job conferences between the architect, general contractor and the masonry contractor. The other trades should be consulted on the cost of their services for anything that they will be asked to provide to the contractor.

5. The masonry contractor must

also have available for his men the required equipment which will enable them to do the job in the best and shortest time possible. Poor equipment or trying to do the job with insufficient equipment is a good way to lose money. Scaffolding is a good example of this rule; if the job is short of scaffolding, then there will be a lot of unnecessary moving of scaffolds, resulting in high labor costs.

Not having the masonry materials in the required amounts and in the right place at the right time will increase the labor cost due to high-priced skilled men waiting for material to arrive or to be moved.

6. Labor relations and cost accounting tied in with job financing all combine to make or break a contractor. The contractor must be aware of his labor and overall operating costs under varying conditions. He must be able to keep accurate cost accounts or have an accountant do it for him. Labor costs can vary considerably depending on how many men are available on the labor market and what the contractor's reputation is as a person to work for. As an overall statement, poor cost accounting is one of the most common causes of a contractor's failure.

7. The beginning contractor must also remember that on most commercial jobs, the general contractor retains ten percent of the total contract cost as a protection for him to make sure the masonry contractor will repair any defects or other problems for a period of one year. This means that if the contractor does a number of jobs in a year each one will have this ten percent withheld for one year. This will tie up a lot of his working capital and he may have to borrow additional money to stay in business. The cost of this borrowed money must be included in the cost of the job to enable the contractor to stay in business.

Joint Apprenticeship Committee and Apprenticeship Standards

The building industry has, in co-operation with the U.S. Department of Labor, Bureau of Apprenticeship and Training, set up National Standards of Apprenticeship. These standards define what the term *apprentice* in the trade shall mean. The standards set forth age limits, educational requirements, length of apprenticeship, ratio of apprentices to journeymen, hours of work and wages.

The Joint Apprenticeship Com-

Certificate of Completion of Apprenticeship

United States Department of Labor
Bureau of Apprenticeship and Training

This is to certify that

has completed an apprenticeship in the trade of

under sponsorship of

in accordance with the standards recommended by the Federal Committee on Apprenticeship

PRESENTED

mittee, commonly known as the J.A.C., is composed of equal representation from labor and management, with consultants from the Bureau of Apprenticeship and the State or local Board of Education attending to act as advisors without a vote.

The J.A.C. has the delegated power to set the local standards consistent with the basic requirements established by the National Committee. The apprentice, when he signs the indenture agreement, agrees to live up to all its provisions and in turn is protected by its rules and regulations.

The J.A.C. also establishes the curriculum for the related classroom work plus supervising the on-the-job training the apprentice receives. When an apprentice completes his training, the J.A.C. notifies the Bureau of Apprenticeship, and this agency issues a completion certificate which is recognized throughout the United States and Canada as proof of reaching journeyman status. See page 8.

Blueprints and Specifications

In addition to practical training with tools and materials, the masonry apprentice will devote much time learning to read and understand blueprints and specifications. This knowledge is essential to the journeyman as well as the foreman and contractor because it is the only way he will know the architectural requirements of a structure. Blueprints or working drawings are the architect's means of communicating instructions for the erection of a structure. They show the location, dimensions, and necessary materials for building the structure. Appendix A, at the end of the book, gives a review of blueprint reading.

Specifications are a written set of instructions as to kinds of material, quality of workmanship, etc., that accompany a complete set of working drawings for a structure.

Specifications amplify and supplement the set of working drawings. They also give much of the information that cannot adequately be presented on each sheet of the set of drawings. Specifications also include information such as the legal responsibilities, methods of purchasing equipment, and the insurance requirements on the building.

The blueprints and the specifications function together as a whole. What is in either is considered to be in both. All items which are necessary for the completion of the struc-

Masonry Simplified

FHA Form 2005
VA Form 26 1852
Rev. 3/68

Plan 329 (includes 2 car garage)
☐ Proposed Construction
☐ Under Construction

U. S. DEPARTMENT OF HOUSING AND URBAN DEVELOPMENT
FEDERAL HOUSING ADMINISTRATION

For accurate register of carbon copies, form may be separated along above fold. Staple completed sheets together in original order.

DESCRIPTION OF MATERIALS   No. _____
(To be inserted by FHA or VA)

Form Approved
Budget Bureau No. 63-R0055

Property address _____ City _____ State _____
Mortgagor or Sponsor _____
                              (Name)   (Address)
Contractor or Builder _____
                              (Name)   (Address)

INSTRUCTIONS

1. For additional information on how this form is to be submitted, number of copies, etc., see the instructions applicable to the FHA Application for Mortgage Insurance or VA Request for Determination of Reasonable Value, as the case may be.
2. Describe all materials and equipment to be used, whether or not shown on the drawings, by marking an X in each appropriate check-box and entering the information called for in each space. If space is inadequate, enter "See misc." and describe under item 27 or on an attached sheet.
3. Work not specifically described or shown will not be considered unless required, then the minimum acceptable will be assumed. Work exceeding minimum requirements cannot be considered unless specifically described.
4. Include no alternates, "or equal" phrases, or contradictory items. (Consideration of a request for acceptance of substitute materials or equipment is not thereby precluded.)
5. Include signatures required at the end of this form.
6. The construction shall be completed in compliance with the related drawings and specifications, as amended during processing. The specifications include this Description of Materials and the applicable Minimum Property Standards.

1. **EXCAVATION:**
   Bearing soil, type __Sand and Gravel__

2. **FOUNDATIONS:**
   Footings: concrete mix __2500 #__; strength psi _____ Reinforcing _____
   Foundation wall: material __Poured Conc. 8"__   Reinforcing _____
   Interior foundation wall: material _____   Party foundation wall _____
   Columns: material and sizes __3" adj. Post__   Piers: material and reinforcing _____
   Girders: material and sizes __1 Beam 7" at 15.3 #__   Sills: material _____
   Basement entrance areaway _____   Window areaways _____
   Waterproofing __Tar__   Footing drains _____
   Termite protection __Shield at Brick__
   Basementless space: ground cover _____; insulation _____; foundation vents _____
   Special foundations _____
   Additional information: _____

3. **CHIMNEYS:**
   Material __Face Brick__   Prefabricated (make and size) _____
   Flue lining: material __Vitrified Clay__   Heater flue size __8 x 12__   Fireplace flue size __12 x 12__
   Vents (material and size): gas or oil heater __5" G. I.__; water heater __3" G. I.__
   Additional information: _____

4. **FIREPLACES:** OPTIONAL
   Type: ☒ solid fuel; ☐ gas-burning; ☐ circulator (make and size) _____   Ash dump and clean-out __10__
   Fireplace: facing __Brick__; lining __Fire Brick__; hearth __Ceramic__; mantel __None__
   Additional information: _____

5. **EXTERIOR WALLS:**
   Wood frame: wood grade and species __Cedar #2__   ☒ Corner bracing. Building paper or felt __15# Felt__
   Sheathing __Asphalt impregnated Fiberboard__; thickness __½"__; width __4 x 8__; ☒ solid; ☐ spaced _____ o. c.; ☐ diagonal _____
   Siding __Aluminum__; grade _____; type _____; size _____; exposure __8"__; fastening __Nailed__
   Shingles _____; grade _____; type _____; size _____; exposure _____; fastening __Per Mfg.__
   Stucco _____; thickness _____"; Lath _____; weight _____ lb.
   Masonry veneer __Face Brick $60/M__ Sills __Lime Stone__ Lintels _____ Base flashing _____
   Masonry: ☐ solid ☐ faced ☐ stuccoed; total wall thickness _____"; facing thickness _____"; facing material _____
   Backup material _____; thickness _____"; bonding _____
   Door sills _____ Window sills _____ Lintels _____ Base flashing _____
   Interior surfaces: dampproofing, _____ coats of _____; furring _____
   Additional information: _____
   Exterior painting: material __Exterior lead and oil__; number of coats __3__
   Gable wall construction: ☐ same as main walls; ☐ other construction __Aluminum siding in gable__

6. **FLOOR FRAMING:** 2 x 10 – 16" o.c.
   Joists: wood, grade, and species __#2 Fir__; other _____; bridging __1 x 3__; anchors _____
   Concrete slab: ☒ basement floor; ☐ first floor; ☐ ground supported; ☐ self-supporting; mix __5 sk.__; thickness __3__";
   reinforcing _____; insulation _____; membrane _____
   Fill under slab: material __Sand__; thickness __4__". Additional information: _____

7. **SUBFLOORING:** (Describe underflooring for special floors under item 21.)
   Material: grade and species __½" plyscore__; size __4 x 8__; type __Plyscore__
   Laid: ☒ first floor; ☐ second floor; ☐ attic _____ sq. ft.; ☐ diagonal; ☐ right angles. Additional information: __Solid__

Fig. 1-1. The beginning of a typical specifications sheet.

Masonry as a Trade

ture, even though not mentioned in one or the other, are considered to be included. If the specifications and the working drawings are in conflict, normally the specifications take precedence. The specifications are binding on all parties, including the sub-contractors.

Fig. 1-1 shows part of a typical specifications sheet used by the Veterans Administration and the Federal Housing Authority.

The Construction Specifications Institute (CSI) Format

Advances in technology and methods of processing data have often made the reading of specifications cumbersome and confusing. The Construction Specifications Institute in Washington, D.C., has made an attempt to remedy this situation with the publication of the *CSI Format*. The *Format's* value lies in its potential for unifying and universalizing the specifications. It offers a logical framework as well as a standard for the many persons who must read the specifications. The *Format* is comprised of four major groupings:

Bidding Requirements
Contract Forms
General Conditions
Specifications (Technical)

Within this last grouping of *Specifications*, 16 permanent *divisions* are found. These divisions are constant in sequence and short in name. "Divisions" do not name units of work but rather relationships of units of work. The units of work are the "sections" within the divisions.

These divisions were established by considering construction relationships: materials, trades, functions and locations of specified work. The 16 divisions of Specifications in the *Format* are:

Division 1—General Requirements
Division 2—Site Work
Division 3—Concrete
Division 4—Masonry
Division 5—Metals
Division 6—Wood and Plastics
Division 7—Thermal and Moisture Protection
Division 8—Doors and Windows
Division 9—Finishes
Division 10—Specialties
Division 11—Equipment
Division 12—Furnishings
Division 13—Special Construction
Division 14—Conveying Systems
Division 15—Mechanical
Division 16—Electrical

Checking On Your Knowledge

The following questions give you the opportunity to check up on yourself. If you have read the chapter carefully, you should be able to answer the questions. If you have any difficulty, read the chapter over once more so that you have the information well in mind before you go on with your reading.

DO YOU KNOW

1. What is the difference between brick masonry and cement masonry?
2. How many years does the typical term of apprenticeship last?
3. What is the J.A.C.?
4. Who sets up national standards of apprenticeship?
5. What is an indenture agreement?
6. What happens if the specifications and working drawings are in conflict?
7. What is the CSI format?
8. How the 16 parts of the CSI Specifications are related?

Safety on the Job

Chapter **2**

Before beginning actual work on a building, the mason should carefully consider the safety measures necessary to protect himself and his fellow workers against accidents. Every building mechanic should be aware of the particular hazards of his own trade as well as those of associated trades. The accident rate is comparatively high in the building industry. Accidents often result in partial or total disability and are sometimes fatal. In addition to these serious accidents there is the possibility of sustaining innumerable minor cuts and bruises that are not only painful but temporarily handicap the workman. To reduce this accident rate to a minimum, the mason must become safety conscious; he must learn to think of the safety of his fellow workers as well as his own. Every man on the job must know how to prevent accidents, and must have a keen sense of responsibility toward his fellow workmen.

Safety education today has become an important phase of every training program. Under the 1970 Federal *Occupational Safety and Health Act* (OSHA), the employer is required to furnish a place of employment free of known hazards likely to cause death or injury. The *employer* has the specific duty of complying with safety and health standards as set forth under the 1970 act. At the same time, *employees* also have the duty to comply with these standards.

Of first importance in a "Safety First" campaign is the education of the worker. This education must become a part of his daily training

Masonry Simplified

National Safety Council

National Safety Council

as he learns the technical and manipulative skills of his job. Generally, a person becomes injured because of his own carelessness or the carelessness of some other person. To prevent accidents and injuries, observe all safety regulations, use all safety devices and guards when working with machines, and learn to control your work and actions so as to avoid danger. Training for safety is every bit as important as learning to be a skillful craftsman and should be a part of the worker's education.

In the performance of his work, the mason handles materials, manipulates hand tools, and operates machines which if improperly handled or used may result in serious injury. If an injury should occur, seek first aid no matter how slight the injury. Blood poisoning may result from an insignificant scratch. It is advisable to take a first aid course at the first opportunity.

General Safety on the Job

Safety is a combination of knowledge and awareness: *knowledge* and *skill* in the use and care of your tools and *awareness* on the job of the particular hazards and safety procedures involved. Tool skills may be learned; awareness, however, depends on attitude. An attitude of care and concern while on the job will help prevent injuries not only to yourself, but also to your fellow workers. Always be alert while on the job and follow recommended safety procedures. If in doubt, ask questions.

1. Wrestling, throwing objects, and other forms of horseplay should be avoided. Serious injuries may be the result.

2. Provide a place for everything, and keep everything in its place.

3. Keep the arms and body as

Safety on the Job

National Safety Council

nearly straight as possible when lifting heavy objects. Place your feet close to the object. Bend your knees, squat, and keep your back as straight as possible. Lift with the legs—not with the back. If the object is too heavy or too bulky, get help.

4. Never place articles on window sills, stepladders, or other high places where they may fall and cause injuries. Check scaffolds and ladders for articles before they are moved.

5. Oil, water, and other slippery substances left on the floor may cause a serious accident.

6. Keep all work spaces clear of tools and material. Things left scattered on the floor may cause stumbling and result in serious injury from a fall.

7. Notify your immediate superior of any known violations of safety rules or of conditions you think may be dangerous.

8. Replace faulty tools and equipment.

9. Immediately report all accidents, no matter how slight, to your superior, and report for first aid treatment.

10. Don't take chances.

National Safety Council

First Aid

A mason's equipment is not complete without a few essential first-aid supplies. These supplies should include at least antiseptics, bandages, and first aid for burns. Such provisions will help to prevent infection after minor skin injuries, bruises, and burns. Provision should be made also to avoid excessive loss of blood, heat exhaustion, and eye injuries. A minimum supply of first-aid provisions should include the following materials:

Antiseptic. A small bottle of an antiseptic with applicator of either mercurochrome or metaphan.

Bandages. Four compresses, three inches square, sealed in wax paper;

a small roll of 1-inch gauze; a five-yard roll of ½-inch adhesive tape; and a package of band aids.

First Aid for Burns. One tube of Butesin Picrate or Sulfa Diazine.

Heat Exhaustion. One bottle of 50 sodium chloride (common table salt) tablets. There are two kinds of sodium chloride tablets — the plain and the enteric coated. The enteric coated do not dissolve until they reach the intestines. This prevents stomach disturbances.

Eye Protection. Every tool kit should contain a good pair of goggles for eye protection when drilling or cutting stone or concrete.

Instruction. The physical well-being of the mason is of equal or greater importance than his skill or knowledge of the trade. The mason must be physically fit in order to do his work properly. He must not only keep himself physically fit but he must be safety conscious also in order to protect himself and his fellow workers against accidents and the consequent loss of time on the job. Although the mason may be safety conscious and take every precaution possible to prevent accidents, nevertheless, he should be able to administer simple first aid to an injured worker when accidents occur. A reliable instruction book on first aid should have a place in his tool kit. The *American Red Cross First Aid Book* is recommended.

Accidents are all too frequent in the construction industry, yet the severity of these accidents is not so great as in many other industries. Large construction organizations have their safety engineers, doctors, nurses, and hospital facilities. However, the small organizations, unfortunately, cannot provide these aids. Therefore, the mason in the small organization must be his own safety engineer and be prepared to administer first aid to an injured worker. It is of prime importance then that every mechanic become safety conscious, thinking in terms of safety for himself and others while performing every operation in the process of erecting a building. Since safety instruction becomes most effective when given as the situation or need arises, such instruction is given throughout this book in connection with the various construction operations.

It is not within the scope of this text to deal with first aid, hence this information must be obtained from another source such as the textbook issued by the American Red Cross. However, a few suggestions are given here.

1. The mason should develop safety consciousness, since "an ounce of prevention is worth a pound of cure".

2. He should protect his eyes with goggles when working near flying objects.

Safety on the Job

National Safety Council

3. Slight cuts, bruises, or skin breaks should be treated immediately with an antiseptic and protected with a bandage to prevent infection. *Note:* Never put adhesive tape directly on the wound.

4. Air, dust, and dirt should be excluded from burns with butesin picrate or sulfa diazine, then covered with a bandage immediately.

5. To avoid heat exhaustion, a construction worker should drink plenty of water and take salt tablets to replace the salt lost from the body through perspiration.

6. Before moving an injured worker, always examine him for broken bones. This precaution may prevent compound fractures.

7. In case of serious injuries, always call or see a doctor as quickly as possible.

Clothing

Safety, comfort, and convenience are the watchwords for selecting clothing to be worn on any construction job. The mason should avoid clothing with unnecessary protrusions such as loose cuffs, loops, flaps, etc., which may catch on tools, materials or scaffolding. If the mason is working with any sort of caustic material, he must be careful not to expose his skin to it. Long sleeves with snug fitting cuffs, rubber gloves, and safety glasses are recommended.

Good, sturdy, leather shoes with hard toe-caps should be worn to protect the feet from falling objects and from dampness. Never wear soft-soled shoes as you might step on a nail which would puncture the soft sole and enter the foot.

The hard hat is a must on many construction jobs. Any building, except in the house field, requires the protection of an approved metal or plastic hard hat. There is the constant danger of some workman dropping a tool or piece of material from one of the floors above or from a scaffold. A hard hat should be worn if there is any danger of falling materials or if there is a crane being used. Wear safety goggles and a dust mask when the work requires it.

Tools and Equipment

The greatest care must be taken in handling tools and equipment. Careless handling is very likely to cause serious accidents. Remember that trowels may have sharp edges. Be especially careful with them when working near others. Do not leave

Masonry Simplified

trowels and other hand tools where you or another worker might trip over or fall upon them. Also, never go off and leave them on scaffolding, walls, superstructure, etc., because they may fall or be knocked off and injure someone below.

Most jobs will employ mechanical equipment. Motors, gears, and other moving parts are always hazardous unless treated correctly and with care. Fig. 2-1 shows a modern mixer with the motor and gears enclosed in a protective housing. This housing should be removed *only* for servicing —never while the machine is in operation. The guard over the filler opening should be closed whenever the machine is running, except when loading. When loading, hands must be kept away from the mixer blades.

Power saws are also very dangerous when handled carelessly and must be treated with great respect. Fig. 2-2 illustrates the safe use of a power saw.

One very important rule is "know your machine". Machinery must be operated by someone who is thoroughly familiar with it. Always fol-

Fig. 2-1. A mechanical mixer must provide protection from moving parts. (Gilson Bros. Co.)

Safety on the Job

Fig. 2-2. Safety glasses should be worn when operating a power saw.

metal improves the grounding condition, permitting a greater flow of current to pass through the body. This causes severe burns and possible death.

Never attempt to touch a person who has live current flowing through him or you too may be killed. Try to remove the wire or equipment creating the problem by pushing or lifting it off using a dry piece of wood. Shut off the electricity immediately, if possible; this is the safest method.

National Safety Council

low the manufacturer's instructions or suggestions.

Electricity

Electricity is the power supply for much of the mechanical equipment used on construction jobs today. Temporary lighting wires are found strung throughout the jobs during construction. Electricity is a very dangerous power. There is nothing to show that the power is flowing through the wires except when a light is burning or a motor is turning. The electricity is there whether it is used or not. Great care must be taken not to touch any bare wire, or to create any condition where the current can flow through your body to a ground. Wet scaffolds, metal lath ceilings and wet ground are dangerous, as the wetness or mass of

Check all wires and equipment for bare spots, poor connections and for proper grounding before using them. Keep wire up off the ground and never operate electrical equipment in wet locations without the proper grounding conditions or instantaneous overload-shutoff devices.

Make sure all equipment has three-wire, grounded-type cords, using adequate size wire to carry the required load. Undersized wire may cause a fire due to overheating or may cause a motor burnout when operating under a heavy load.

Never connect electrical equipment to a power source unless the switch is in an OFF position. When work is completed, shut off the power. Report defective power tools and remove from the job site.

Chemicals

The modern mason in the course of his work, may come into contact with chemicals. Some of them when used without the proper precautions can cause serious burns or loss of sight. Never use any material without first determining what dangers it may present and how to use it safely. Read all the manufacturer's directions and check with previous users to find out what problems may be encountered.

Wear rubber gloves and safety glasses plus the proper body covering to prevent chemicals from coming into direct contact with the skin. Keep a pail of clean water available to wash off any chemicals that might accidentally come in contact with the skin.

Scaffolds

Since a great deal of the mason's work is done on scaffolding of various kinds, there is an ever present possibility of a serious injury — to yourself and others. The three main hazards while working on or under scaffolds are falling, dropping tools or material, and faulty scaffolding. Always watch your step, keep your balance, and handle your equipment carefully.

Scaffolds must be built to support the load they are to carry. According to the National Safety Council, they should be designed to support at least four times the anticipated load of men and materials. This is necessary for safety because sometimes unexpected additional loads are placed upon the scaffold.

It is essential that those who erect scaffolding be familiar with the requirements of the safety codes or statutes of his own state as well as the national standards. The *Standard Safety Code for Building Construction* gives detailed requirements for the materials to be used and the manner of erection of scaffolds.

The mason uses four basic types of scaffolds and each of these types can be constructed of either wood or metal and sometimes a combination of the two materials. The basic types are as follows:

1. *Trestle scaffolds:* planks laid across trestles make safe scaffolds up to 10 feet high.
2. *Built-up scaffolds:* either wooden pole scaffolds or steel sectional scaffolds.
3. *Rolling scaffolds:* scaffolds on wheels that permit them to be moved.

Safety on the Job

4. *Hanging scaffolds:* suspended scaffolds supported by cables or metal straps. (Some large scaffolds in high ceiling areas are hung from cables to beams above to keep the floor area clear.)

A *swing stage scaffold* is also used at times for some exterior work, but it is not as common as the others. A swing stage scaffold, such as the one shown in Fig. 2-3, is supported from above, using ropes and pulleys which permit it to be raised or lowered as needed. This type of scaffold is difficult to use, but is sometimes necessary for veneer and other exterior work, such as tuckpointing.

For *wooden scaffolds,* all supports and planks must be of sound lumber, free of large knots, cracks or split ends. All uprights must be cross-braced; ledgers and bearer planks must not be spaced more than 8 feet apart for use with 2″ x 10″ planks. New lumber standards will now reduce this plank to 1½″ x 9½″ net size. Allowances will have to be made for this by reducing ledger spacing.

On all scaffold platforms, the planks must be laid tight together to keep the materials from dropping

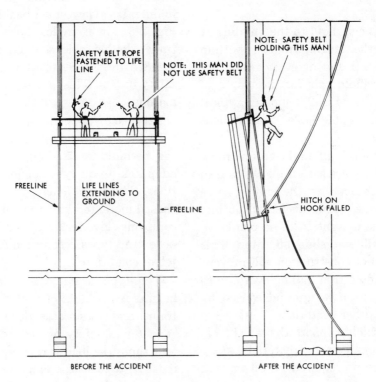

Fig. 2-3. Swing stage scaffold safety devices. (National Safety Council.)

on the workers below. Plank ends should extend at least 6 inches but not more than one foot beyond the bearer planks so they will not slip off. On steel scaffolds planks should have cleats nailed across their ends underneath so they cannot slide off the metal cross bars.

Scaffolds for masonry must be at least four feet wide so as to permit a safe passage between the mortar board and the building.

Guard rails and toe boards should be used on all scaffolds over one stage in height. Not more than 4 to 5 inches of open space should be allowed between the scaffold and the walls.

Tie the scaffold to the building at every other staging from the bottom to the top and at every other upright scaffold pole for the length of the scaffold. Fig. 2-4 shows a typical wooden double pole, independent scaffold.

Scaffolds built on the ground must have the poles set on planks laid in solid contact with the ground so as to provide a firm unsinking footing. Nail the upright to these planks so they will not slide off later. This method of construction will prevent the scaffold from falling over due to sinking into wet ground caused by rain or other conditions.

To safely support the weight of the planks, men and materials, 4" x 4" poles are required for all wooden mason scaffolds over one stage in height. Poles of 4" x 6" are recommended for the first 32 feet of scaffolds exceeding 32 feet in height.

Most contractors stock only certain basic scaffold material in their storage yards. It is cheaper to use a 4" x 4" pole for a simple scaffold when you have it on hand anyway. It is useable for all scaffolds and because of its size can be nailed repeatedly without splitting.

Wooden scaffolds should not exceed 40 feet in height. Anything above this height should be built of steel for fire safety and strength.

State and National Safety Regulations vary considerably for scaffolding. Local laws must be followed. Fig. 2-5, for example, shows heavy trade, double pole scaffold recommended by the State of California safety orders. This is acceptable for masons.

Steel scaffolds are now widely used for overall adaptability. They can be built to any height and are adaptable to all types of job conditions. These scaffolds can be purchased or rented and the supplying companies give technical services on needs and types best suited for each job or condition.

The type usually used for low heights is made up of prefabricated frames and cross braces. See Fig. 2-6. A factor of safety of not less than four times the load is required. Care must be used in the erection of the scaffold so that it rests on a firm base

Safety on the Job

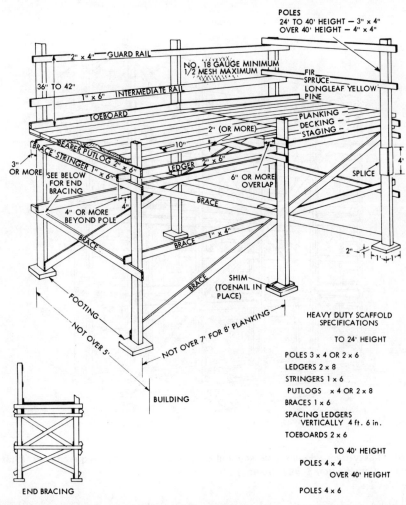

Fig. 2-4. Typical double pole wood scaffold, showing safe construction for a single stage height. Multiple stage scaffolds require 4″ x 4″ poles and 2″ x 8″ bearer planks. (National Safety Council)

and is kept plumb and level as it is assembled. It must be inspected daily. Care must be taken to keep the frames from injury or from rusting so that they do not lose part of their design strength. Safety rules for metal scaffolding, recommended by the Steel Scaffolding and Shoring Institute, are shown on page 26.

Rolling scaffolds should have large strong wheels provided with locking devices. Never move a rolling scaffold while men are on the scaffold. Always clean the floor ahead of

Masonry Simplified

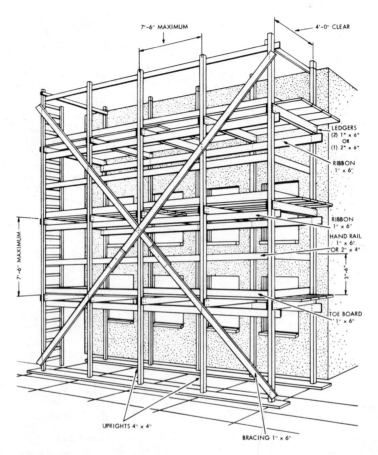

Fig. 2-5. A heavy trade, double pole scaffold must meet state code requirements. (State of California, "Construction Safety Orders")

the move to be made so that the wheels will not be blocked by an obstruction which might cause the scaffold to tip over. All planks on rolling scaffolds should be securely fastened down so they cannot slide off while the scaffolds are moved.

Hanging and *swing stage scaffolds* depend upon the cables or block and tackle to support them. Never use frayed cable or cable clamps that are worn out. Inspect all these items before using the scaffold; your life depends upon it. Old ropes and worn blocks are perhaps the greatest danger in using swing stage scaffolds. Insist on good equipment, make sure the supports are securely fastened and tied off so they cannot slip or work loose.

Safety on the Job

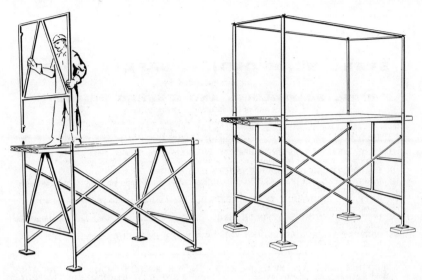

Fig. 2-6. Prefabricated metal frames and diagonal braces are assembled quickly to provide safe scaffolds.

When the mason has to work on a swing stage scaffold, it must be tied to the building at each end so as not to swing away from the building.

Safety belts fastened to life lines should be used at all times. See Fig. 2-3. Use only safety approved equipment with the proper guard rails installed. Never overload the scaffold; keep it under its rated load capacity at all times.

Building Enclosure

With year-round construction now an established practice, the enclosure of buildings to permit both interior and exterior work to continue even in freezing weather often creates a dangerous condition for the worker. Temporary heaters, if they are not properly vented, may give off obnoxious or deadly fumes. Salamanders, gas or gasoline heaters are very dangerous, as the carbon monoxide gases given off are odorless but lethal.

The mason usually works on scaffolds and may be up near a ceiling; therefore he will get the full force of these gases first because the gas is lighter than air and will rise to the ceiling. All temporary heaters must be constructed with positive venting of the combustion chamber, and the resulting gases must be piped to the outside. The vent pipes must be of the proper size and have gastight points. "You only live once —don't make your life a short one."

Masonry Simplified

STEEL SCAFFOLDING SAFETY RULES
as Recommended by
STEEL SCAFFOLDING AND SHORING INSTITUTE
(SEE SEPARATE SHORING SAFETY RULES)

Following are some common sense rules designed to promote safety in the use of steel scaffolding. These rules are illustrative and suggestive only, and are intended to deal only with some of the many practices and conditions encountered in the use of scaffolding. The rules do not purport to be all-inclusive or to supplant or replace other additional safety and precautionary measures to cover usual or unusual conditions.

I. **POST THESE SCAFFOLDING SAFETY RULES** in a conspicuous place and be sure that all persons who erect, dismantle or use scaffolding are aware of them.
II. **FOLLOW LOCAL CODES, ORDINANCES** and regulations pertaining to scaffolding.
III. **INSPECT ALL EQUIPMENT BEFORE USING** — Never use any equipment that is damaged or deteriorated in any way.
IV. **KEEP ALL EQUIPMENT IN GOOD REPAIR.** Avoid using rusted equipment — the strength of rusted equipment is not known.
V. **INSPECT ERECTED SCAFFOLDS REGULARLY** to be sure that they are maintained in safe condition.
VI. **CONSULT YOUR SCAFFOLDING SUPPLIER WHEN IN DOUBT**—scaffolding is his business, **NEVER TAKE CHANCES.**

A. **PROVIDE ADEQUATE SILLS** for scaffold posts and use base plates.
B. **USE ADJUSTING SCREWS** instead of blocking to adjust to uneven grade conditions.
C. **PLUMB AND LEVEL ALL SCAFFOLDS** as the erection proceeds. Do not force braces to fit — level the scaffold until proper fit can be made easily.
D. **FASTEN ALL BRACES SECURELY.**
E. **DO NOT CLIMB CROSS BRACES.**
F. **ON WALL SCAFFOLDS PLACE AND MAINTAIN ANCHORS** securely between structure and scaffold at least every 30' of length and 25' of height.
G. **FREE STANDING SCAFFOLD TOWERS MUST BE RESTRAINED FROM TIPPING** by guying or other means.
H. **EQUIP ALL PLANKED OR STAGED AREAS** with proper guard rails, and add toeboards when required.
I. **POWER LINES NEAR SCAFFOLDS** are dangerous—use caution and consult the power service company for advice.
J. **DO NOT USE** ladders or makeshift devices on top of scaffolds to increase the height.
K. **DO NOT OVERLOAD SCAFFOLDS.**
L. **PLANKING:**
  1. Use only lumber that is properly inspected and graded as scaffold plank.
  2. Planking shall have at least 12" of overlap and extend 6" beyond center of support, or be cleated at both ends to prevent sliding off supports.
  3. Do not allow unsupported ends of plank to extend an unsafe distance beyond supports.
  4. Secure plank to scaffold when necessary.

M. **FOR ROLLING SCAFFOLD THE FOLLOWING ADDITIONAL RULES APPLY:**
  1. **DO NOT RIDE ROLLING SCAFFOLDS.**
  2. **REMOVE ALL MATERIAL AND EQUIPMENT** from platform before moving scaffold.
  3. **CASTER BRAKES MUST BE APPLIED** at all times when scaffolds are not being moved.
  4. **DO NOT ATTEMPT TO MOVE A ROLLING SCAFFOLD WITHOUT SUFFICIENT HELP** — watch out for holes in floor and overhead obstructions.
  5. **DO NOT EXTEND ADJUSTING SCREWS ON ROLLING SCAFFOLDS MORE THAN 12".**
  6. **USE HORIZONTAL DIAGONAL BRACING** near the bottom, top and at intermediate levels of 30'.
  7. **DO NOT USE BRACKETS ON ROLLING SCAFFOLDS** without consideration of overturning effect.
  8. **THE WORKING PLATFORM HEIGHT OF A ROLLING SCAFFOLD** must not exceed four times the smallest base dimension unless guyed or otherwise stabilized.

N. For "PUTLOGS" and "TRUSSES" the following additional rules apply:
  1. **DO NOT CANTILEVER OR EXTEND PUTLOGS/TRUSSES** as side brackets without thorough consideration for loads to be applied.
  2. **PUTLOGS/TRUSSES SHOULD EXTEND AT LEAST 6"** beyond point of support.
  3. **PLACE PROPER BRACING BETWEEN PUTLOGS/TRUSSES** when the span of putlog/truss is more than 12'.

Safety on the Job

Fig. 2-7. Building enclosures must be properly ventilated.

Fig. 2-7 shows a typical building enclosure.

Housekeeping

Good housekeeping is not just another extra chore for the workman. It is an important element in accident prevention and efficiency on the job. Good housekeeping begins with planning ahead. Storage areas should be planned for ease of access to materials but not in the way of traffic and construction. Materials should be neatly stockpiled. Access areas and walkways should be kept clear of loose materials and tools. Containers for trash and pits or bins for waste materials should be provided and conscientiously used.

It is the responsibility of each and every man on the job to maintain his area in good working order. A neat and orderly work area is a reflection of a proper attitude toward safety by all concerned.

Checking On Your Knowledge

The following questions give you the opportunity to check up on yourself. If you have read the chapter carefully, you should be able to answer the questions. If you have any difficulty, read the chapter over once more so that you have the information well in mind before you go on with your reading.

DO YOU KNOW

1. What is necessary to do a safe, competent job?
2. How to lift a heavy object?
3. What to do if safety violations are discovered on the job?
4. The safe use of a scaffold?
5. What kind of clothing a man should wear?
6. Three types of protective equipment a mason may use?
7. When the motor housing on a mechanical mixer may be removed?
8. Three things to check for before using power tools?
9. What to do when you have finished using a power tool?
10. Why good housekeeping is important?

Tools and Equipment

Chapter 3

Learning the proper use and care of tools is an essential part of apprenticeship in any trade. The tools for brick and cement masonry are highly specialized in that there is usually only one type of tool that is suitable for each step in brick or cement masonry construction, and also, that these tools are used *only* in the masonry trades. A piece of finished masonry construction may be considered a "hand-tooled" product. For this reason, the brick or cement mason may take particular pride in the individuality of his profession.

This chapter will cover the kinds of tools used in both brick and cement masonry, how to select the best quality tools, and the specific purposes for which they are intended.

Tools for Brick Masonry

Compared with most other building trades, the brick mason carries relatively few hand tools to the job. The mason's basic personal tool kit which he will be expected to bring to the job consists of assorted trowels and jointers, a plumb rule (level), brick hammer and brick set, steel square, line, retractable steel tape, and folding rule. (Fig. 3-1)

Masonry Simplified

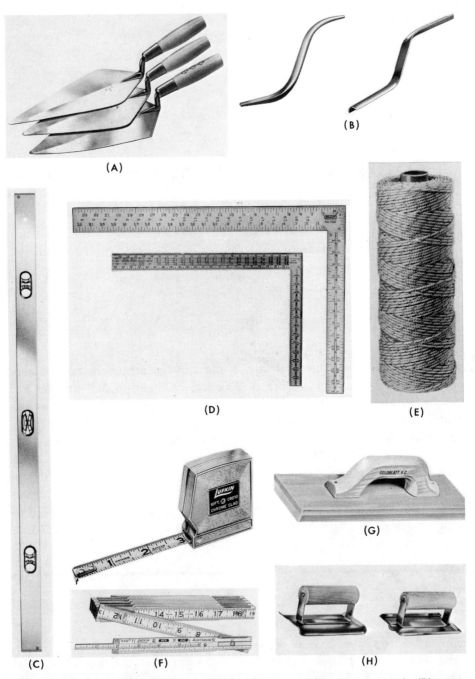

Fig. 3-1. Some basic masonry tools: (A) trowels, (B) jointers, (C) level or plumb rule, (D) square, (E) line, (F) rules, (G) float, (H) edger and groover.

Tools and Equipment

The most important things to consider in selecting personal tools are strength of material, design, and comfort. Some types of tools, such as trowels and jointers, will vary from different manufacturers in size, shape, and weight. The first consideration should be to obtain the tool made of the best possible materials. This not only assures long life but also makes for speed and efficiency on the job. The sizes or shapes will be determined by the nature of the job and personal preference. As the mason gains more experience, he will develop an individual "feel" for particular sizes, shapes, and weights of tools. In other words, whichever tool "feels" the best is probably the best tool for the job.

Brick Trowels

Trowels are the most important and most used tools in brick masonry, and great care should be taken in selecting them. As with all tools, the mechanic should buy the best that he can afford. Better trowels are hand ground to the proper shape, taper and balance out of one-piece forgings of high grade steel alloys. Stamped, welded trowels are cheaper, but will readily warp, wear down, and break. The mason will be money ahead in the long run by buying trowels that will give long and dependable service.

Trowels are available in lengths from about 9" to 14" and widths from about 4½" to 7" and may be either wide heeled or round heeled. Again, the type of job and personal preference will determine what sizes, shapes and lift to buy. Fig. 3-2 shows some typical brick masons' trowels.

Fig. 3-2. Brick trowel patterns: (top) Philadelphia, (middle) wide heeled London, (bottom) narrow heeled London. (Goldblatt Tool Co.)

Jointers

A jointer (also called joint tool or finishing tool) is used to "finish" the mortar joints between the units in a masonry structure. Finishing the joint smooths and compresses the face of the joint, giving a pleasing appearance and, as will be discussed later, assures watertightness of the joint. Jointers are usually either cast or forged metal rods, or stamped split tubes with the ends rounded or angled to provide a particular shape of convex joint. For long horizontal joints, a joint runner (sometimes called a "sled" runner) is recommended. Fig. 3-3 shows some typical jointers. Fig. 3-4 shows the joint or sled runner being used in laying concrete block.

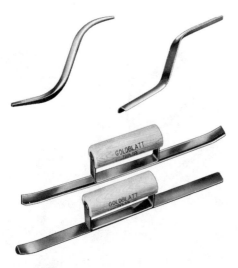

Fig. 3-3. Jointers. Regular jointers (top) are used to finish mortar joints in most brickwork. Longer jointers, called sled runners (bottom) are used to finish long horizontal joints. (Goldblatt Tool Co.)

Fig. 3-4. Sled runner being used in concrete block construction. (Goldblatt Tool Co.)

Tools and Equipment

Levels

Professional masonry is not possible without the combination level and plumb rule commonly called *level*. (Fig. 3-5) It is used constantly to check the wall as it goes up to assure that it is absolutely true vertically (plumb) and that each course of brick is level horizontally. The mason knows that the level and trowel "go hand in hand" as he will often have one in one hand and one in the other (Fig. 3-6). Levels suitable for masonry are available in lengths from 42" to 48". (The mason may keep a shorter level, 14" to 24", for smaller work or inside spaces.) A good level has two bubble gages in the center for leveling and two at either end for plumbing. This allows the mason to pick up the level from any position and immediately apply and read it.

Brick Hammer

The brick hammer is a tool designed and built especially for the brick mason. The head is square and made of very hard tempered steel. The edges of the head are kept very sharp. The particular purpose of the brick hammer is for splitting brick. The sharp edge of the hammer head is used to make a cutting line where the brick is to be split. This is done by making light blows in a line all the way around the brick. Then one sharp blow with the hammer head on the line at about the center of the long face will split the brick (Fig. 3-7, left). The rough spots on the split face are then trimmed with the blade part of the hammer (Fig. 3-7, right). This is a skill which requires some practice which may be done on scrap or waste bricks.

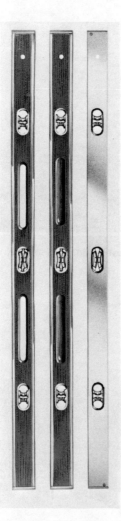

Fig. 3-5. Levels are available in either wood or metal clad. (Goldblatt Tool Co.)

Masonry Simplified

Fig. 3-6. Plumbing a corner with the plumb rule (level). (Goldblatt Tool Co.)

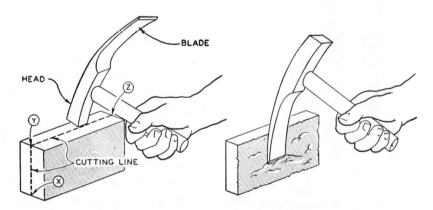

Fig. 3-7. The brick hammer is used to cut or split bricks.

Tools and Equipment

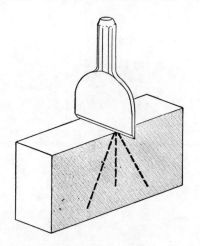

Fig. 3-8. The brick set is used to cut bricks to exact dimensions.

Brick Set

The brick set is a special chisel-like tool used to cut bricks when exact dimensions or surfaces are required. The bricklayer's hammer is used to strike the brick set, thereby cracking the brick. See Fig. 3-8.

Steel Square

A steel square (Fig. 3-9) is used by masons when *laying up* (constructing) corners to assure a true 90 degree corner. It is also used to set the corner bricks or blocks in the dry layout of the first course in a masonry structure. The steel square is the same tool as the carpenter's framing square.

Line

The mason's line is important in keeping each course of brick or block level and the wall true and *out-of-wind* (free of hollows or bulges) as the units are being laid up between the leads (corners). The line is usually a strong cord of nylon or similar material. It is available in different

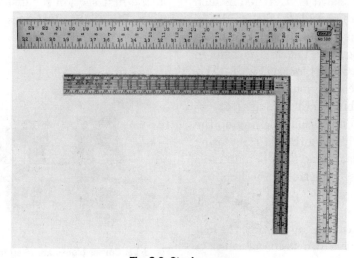

Fig. 3-9. Steel squares

Masonry Simplified

Fig. 3-10. Mason's line is used to align the top edge of each course.

sizes and colors, and in lengths up to 1000 feet. Fig. 3-10 shows a line in use. Note that it is stretched in such a way as to provide a level guideline for the top edge of the brick or block in each course. The line is held in place by corner blocks (Fig. 3-11).

If the corner must be kept clear for plumbing and truing (for instance, if one mason is laying up the corners while another is completing the courses) line pins or cut nails may be used to stretch the line.

Rules

For measuring and layout, the mason will have two kinds of rules: a retractable steel tape, usually ten to twelve feet long, and a six foot folding rule with a six inch sliding rule on the first section for inside measuring. Fig. 3-12. illustrates the two

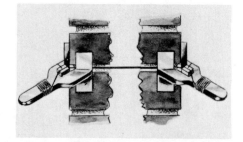

Fig. 3-11. Corner blocks are used to hold the line securely in place. (Goldblatt Tool Co.)

Tools and Equipment

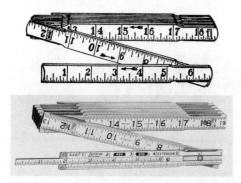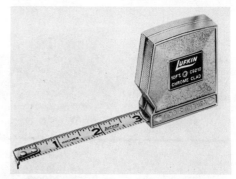

Fig. 3-12. Folding rule and tape.

commonly used rules. These types of rules are also available with special scales and markings for masonry.

Tool Bag

The bricklayer usually carries his personal tools in a tool bag. The 18" canvas bag is the most popular and will usually contain the following items: large trowel, pointing trowel, round and V jointer, brick hammer, brick set, mason's line, corner blocks, line pins, cut nails, folding rule, steel tape, and short plumb rule. The long plumb rule may be carried in the leather straps on the outside of the tool bag.

In addition to the tools just described, the mason will usually carry such items as chalk, pencils, and knives in his tool kit.

Power Tools and Equipment

Besides his own hand tools, there will be other larger tools and equipment available to the mason. Most of the heavy time-consuming operations such as transporting, lifting and mixing which were formerly done by hand are now done by mechanical means. Although the mason himself will not be operating this equipment, he will benefit directly from it and should be familiar with it. He should know what it does, how it does it, what its capacities and limitations are in order that he and others may use it to full advantage. Fig. 3-13 shows a modern gasoline engine driven mortar mixer. Note the safety devices: the engine and gears are enclosed; the mixing drum is completely covered by a safety guard. The only time this guard is opened is when the machine is stopped.

Electric masonry saws (Fig. 3-14) are used when a large number of brick, block or glazed tile must be cut to a special size. As with all machinery, these tools should be oper-

Masonry Simplified

Fig. 3-13. Gasoline engine powered mortar mixer.

Fig. 3-14. Electric saws must be used with caution.

Tools and Equipment

Fig. 3-15. A modern hoisting device. (Insley Manufacturing Corp.)

ated only by experienced and cautious persons. Fig. 2-2 shows a table-mounted masonry saw being used to cut a concrete block.

Fig. 3-15 shows one type of modern hoisting device, a great time and labor saver.

Tools for Cement Masonry

Almost all of the hand tools used in cement masonry are those involved in finishing *concrete flatwork;* that is, such horizontal structures as slabs, floors, sidewalks and driveways, etc. The tools described below

39

are presented in the order in which they are usually employed in finishing concrete. A detailed description of how and where they are used is given in Chapter 8, "Concrete Construction."

Screeds

The screed is the first tool used as the concrete is placed. It is used to "strike off" the excess concrete and bring the surface to the proper grade (level or height). It is also referred to as a "straightedge" or "strike-off tool." It is made of either wood or metal. The two requirements for a screed are that it have one long edge that is perfectly straight and that it be long enough to straddle the tops

Fig. 3-17. Mechanical screed with a power vibrator. (Stow Manufacturing Company)

of the forms on either side of the structure. See Fig. 3-16. Fig. 3-17 shows a mechanical screed with a power vibrator striking off newly placed concrete. The vibrator helps to compact the concrete.

NOTE: The term *screed* also applies to the grade level forms in concrete flatwork. This is discussed in detail in Chapter 8, "Concrete Construction."

Tampers

Tampers, commonly called "jitterbugs," are sometimes used in heavy concrete flatwork to compact the aggregate below the surface. Jitterbugs are not commonly used in average construction as they tend to segregate the aggregate, leaving only very fine aggregate and mortar on the surface. See Fig. 3-18.

Fig. 3-16. Screeding freshly placed concrete. (Portland Cement Association)

Tools and Equipment

Fig. 3-18. Tamper. (Goldblatt Tool Company)

Darby

A darby is a long, flat, rectangular piece of wood, aluminum, or magnesium with a raised handle. It may be from 30 to 80 inches long and 3 to 4 inches wide. It is used directly after screeding. It further levels the concrete, smoothing high spots, and filling depressions left by straightedging. See Fig. 3-19.

Bull Floats

The bull float is sometimes used

Fig. 3-19. Typical darbies. (Goldblatt Tool Co.)

Masonry Simplified

Fig. 3-20. Magnesium bull float. (Goldblatt Tool Co.)

instead of a darby and is used for the same purpose. It is a wooden or metal float with a long handle. It can cover a wide area by an operator standing in one position, therefore it is usually substituted for the darby when larger areas are to be finished. See Fig. 3-20.

Edger

The edger is a trowel-like tool with one edge curved down to form an arc with a ⅛ to 1½ inch radius. See Fig. 3-21. Its purpose is to round off the sharp edges of concrete slabs such as sidewalks and driveways thereby improving appearance and reducing edge damage.

Fig. 3-21. Edger. (Goldblatt Tool Co.)

Groover

The groover is a trowel-like tool with a longitudinal projection called

Fig. 3-22. Groover. (Goldblatt Tool Co.)

42

Tools and Equipment

a *bit* down the middle of the tooling surface. The bits may be from 3/16 to 1 inch deep. It is used to cut control or contraction joints in the surface of the fresh concrete. See Fig. 3-22.

Floats

Floats are used to prepare the concrete for final finishing. They are used to level any small lumps or hollows left after darbying or bull floating, and to imbed any exposed aggregate. Floats are made of wood, aluminum or magnesium and are made in lengths of 12 to 20 inches long and 3½ to 4½ inches wide. See Fig. 3-23.

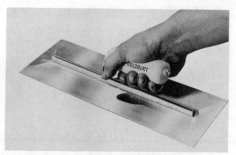

Fig. 3-24. Trowel. (Goldblatt Tool Co.)

Power Trowels

The power trowel with 3 or 4 rotating steel trowel blades, driven by a gasoline or electric engine, is widely used where there are large flat areas to finish. See Fig. 3-25.

Fig. 3-23. Float. (Goldblatt Tool Co.)

Trowels

Trowels for cement masonry are usually flat steel rectangles. They are available in several sizes to be used according to the size of the job. Trowels perform the final finishing of the surface of the concrete providing a dense, smooth surface. See Fig. 3-24.

Fig. 3-25. Power trowel. (Stow Manufacturing Company)

Masonry Simplified

Fig. 3-26. Power joint cutter. (Stow Manufacturing Company)

Power Joint Cutters

Power joint cutters are used on larger jobs to cut control joints. They are actually power saws with special diamond or shatterproof abrasive blades. For large jobs, there are mobile saws equipped with wheels and guideline markers. On smaller jobs, there are portable hand carried saws. Also, the regular portable power saw may be used if equipped with the special concrete blades. See Fig. 3-26.

Fig. 3-27. Small portable concrete mixer. (Gilson Bros.)

Concrete Mixers

Portable concrete mixers for on-the-job mixing are available in many sizes; from capacities of ½ cubic foot up to as much as 7 cubic yards. Fig. 3-27 shows a typical type and size of mixer. More commonly, on the job concrete is delivered in ready-mix trucks. Fig. 3-28 illustrates a ready-mix concrete truck.

Fig. 3-28. Ready Mix concrete truck.

Tools and Equipment

Checking On Your Knowledge

The following questions give you the opportunity to check up on yourself. If you have read the chapter carefully, you should be able to answer the questions. If you have any difficulty, read the chapter over once more so that you have the information well in mind before you go on with your reading.

DO YOU KNOW

1. What should be the first consideration in buying a tool?
2. What is the most important and most used tool in brick masonry?
3. Why the plumb rule is necessary in masonry?
4. The main purpose of the brick hammer?
5. How to assure a true 90 degree corner in masonry work?
6. Two other names for a screed?
7. When a bull float is used instead of a darby?
8. What is the final step in finishing concrete?
9. Where power trowels and joint cutters are used?
10. What is the most common method of providing concrete for a job?

Chapter 4

Mortar: Basic Materials and Mixing

Any masonry structure, such as a wall, a foundation, a fireplace, although made up of bricks, stone or concrete blocks, is a solid, one-piece unit. This is because the individual units are bonded together by a strong, durable material called *mortar*. This chapter will cover the materials used in making mortar, the various types of mortars, their properties and where they are used, and the different methods of mixing and delivery.

Mortar Materials

Mortar for unit masonry may be defined as a compound of cementitious materials and sand with sufficient water to reduce the mixture to a workable consistency. The cementitious materials are cement and hydrated lime. The cement is the main binding agent and supplies most of the strength of the mortar. The most common type of cement is portland cement. In fact, over 95 percent of all the cement manufactured in the United States is portland cement. See Fig. 4-1.

Portland Cement

In 1824, Joseph Aspdin, a plasterer and bricklayer of Leeds, England, was experimenting to produce a mortar that would harden under water. He achieved this by burning limestone and clay together in his

Mortar: Basic Materials and Mixing

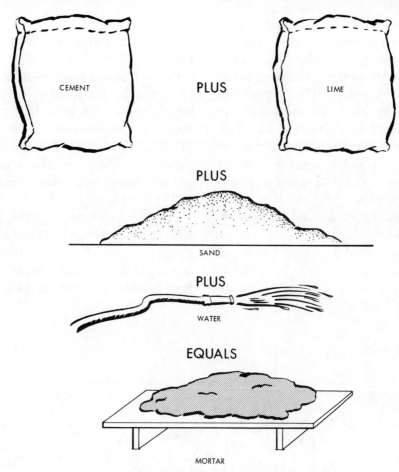

Fig. 4-1. The ingredients of mortar.

kitchen stove. The gray powder was called portland cement because of its resemblance to stone quarried on the Isle of Portland.

Since its invention, portland cement has replaced almost all other cements, both natural and artificial. This is due to low cost and superiority. Portland cement has become a standardized product of high quality and uniformity, regardless of where it is made. It is manufactured all over the United States and Canada and in most other countries.

Modern cements are manufactured in the same basic manner that Joseph Aspdin used. Some form of lime, which may be obtained from

Masonry Simplified

limestone, marble, chalk, slag, oyster shells or coral, is mixed with certain kinds of clay; this is ground and then calcined at temperatures around 2,700°F. The ingredients lime, silica, aluminum oxide, and iron oxide combine chemically and form lumps or clinkers. The clinkers are pulverized and a small amount of gypsum is added to control the setting properties. Fig. 4-2 illustrates the process of manufacturing portland cement.

This cement is a combination of calcium silicates and aluminates. Water starts a complex reaction yielding crystalline substances in an amorphous gel which sets as a hard mass.

Cement hardens as a result of hydration of the materials in it. While setting takes place in a short time, the cement requires several days for the hydration to become complete. As a result, cement continues to increase in hardness for about a month and in some cases for years.

Portland cements are quite uniform commercial products and when the term *cement* is used without qualification, portland is usually the type to which reference is made. To

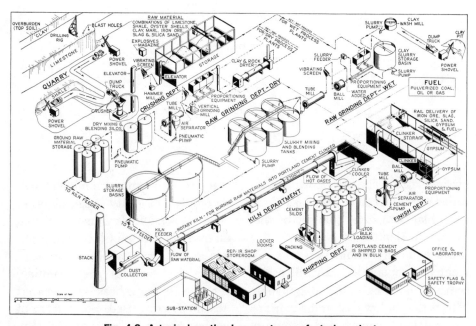

Fig. 4-2. A typical portland cement manufacturing plant.

avoid misunderstandings, it should always be called portland cement.

The specifications of the American Society for Testing and Materials (ASTM) cover five types of portland cement and establish requirements relative to the physical properties and chemical composition for each type. Cement of any one of these types is satisfactory for use in mortars, but Types I and III are those generally used in mortar. The other types are used mainly in concrete. *Type I* is a general purpose cement. It is the one the mason will use under most conditions. *Type III* is "high early strength" cement. Although it takes as long as Type I to set, Type III achieves its full strength much sooner. It is sometimes specified for cold weather masonry because it requires shorter protection time.

Air-Entraining Cement. This product contains small quantities of a chemical which increases the amount of air held by the mortar or concrete made from the cement. The additional air usually improves the workability of the mortar and concrete and increases the resistance to freezing and thawing. The strength, however, is usually less than that obtained with the average untreated cement. Such cements are not required for average construction but may be specified for special jobs where the specific use requirements would best be met by a material of these characteristics. Air-entraining portland cements are designated *Type I-A* and *Type III-A*, and may be used interchangeably with Types I and III.

Lime

While cement provides the strength of mortar, if it were used by itself in the mortar, the mixture would be stiff and unworkable. For this reason, lime is introduced into the mixture as a plasticizing agent. Plasticity means the degree of smoothness and workability of the mixture. Lime also increases the water retentivity (water-holding capacity) of the mortar. This decreases the tendency of the mortar to lose water, called bleeding, and reduces separation or segregation of the sand.

Lime, calcium oxide (CaO) or hydrated calcium oxide, $(Ca(OH)_2)$ is one of the most common minerals in the world. It is very active chemically and combines easily with other elements. In some forms it is extremely caustic and in all forms it is highly alkaline. While the mineral lime is present in many combinations, only the combination with carbon dioxide which forms limestone (calcium carbonate, $CaCO_3$) is important as a source of lime.

Lime for mortar or other use is obtained by quarrying limestone, which is a rock made up mostly of calcium carbonate. In order to change limestone into a lime suitable

Masonry Simplified

Fig. 4-3. Dolomitic lime plant and quarry. (Ohio Lime Co.)

Fig. 4-4. Rotary kilns in which lime is burned. (Marblehead Lime Co.)

Mortar: Basic Materials and Mixing

for mortar it is crushed, screened, selected, washed, and graded. Fig. 4-3 shows how lime is removed from an open pit quarry. The selected stone is then placed in kilns where it is heated up to 2,500°F. This process drives off the moisture and also removes certain gasses from the stone, especially carbon dioxide. This is a calcining process similar to that used for gypsum. It is also referred to as *lime burning*. Fig. 4-4 shows rotary kilns in which lime is burned.

The product of the calcining or burning is *quicklime*, chemically calcium oxide (CaO). Calcining lime is a process that is nearly as old as the use of fire. It is likely that it was discovered by ancient man when he built a fire on limestone rock.

Quicklime is a very caustic material. When it comes in contact with water a violent reaction occurs that is hot enough to boil the water. It can also cause severe burns on the skin. In the past quicklime was the material the mason received on the job and he had to use it to make mortar. The first step in using quicklime is to slake or hydrate it. This is done by adding enough water to it so that the oxide becomes a hydroxide $(Ca(OH)_2)$. During the slaking process the caustic quicklime becomes very hot and is hazardous to work with.

Today the lime manufacturers slake the lime as part of the process of producing lime for mortar. The slaking is done in large tanks where water is added to convert the quicklime to *hydrated lime* without saturating it with water. The hydrated lime is a dry powder with just enough water added to supply the chemical reaction. Hydration is usually a continuous process and is done in equipment similar to that used in calcining. After the hydrating process the lime is pulverized and bagged.

The ASTM designates lime for clay masonry as *Type S*.

Sand

Sand is the aggregate component of mortar. It is commercially available almost anywhere and is quite inexpensive, but it is still just as important in mortar as the cementitious materials. For good strong mortar, the sand must be clean and well graded.

Cleanness. Sand purchased from a reputable commercial supplier will almost always be well washed and ready to use. However, a mason or contractor will sometimes have to obtain sand from a questionable source. If there is any possibility that the sand may contain impurities, it should be tested. The two chief impurities that make sand unfit for use in mortar are silt and organic matter. There are simple tests to determine the presence and amount of each in a given batch of sand. Fig. 4-5 shows the test for silt content. Simply put two inches of

Masonry Simplified

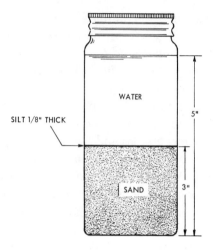

Fig. 4-5. Test for silt content in sand.

the sand to be used in a clear glass jar. An ordinary quart mason jar will work perfectly for this test. Then fill the jar partly full with clean water and shake vigorously. Let it stand for an hour. The sand will settle to the bottom leaving any silt in the sample lying on top of the sand. If the silt layer measures more than 1/8 of an inch, the sand is unfit for use in mortar and must be washed before using.

Organic matter in sand will cause a weak and non-durable mortar. If the amount of organic matter is not known, the *colorimetric test* will easily determine it. First, make a three percent solution of sodium hydroxide by dissolving one ounce of sodium hydroxide (available at any drug store) in one quart of distilled water. (*CAUTION:* Sodium hydroxide is highly caustic and can cause serious burns and ruin clothing so take great care while handling it.) Next, fill a pint jar that is graduated in ounces to the 4 1/2 ounce mark with a dry sample of the sand. Then add the sodium hydroxide solution to the seven ounce mark. Replace the lid tightly and shake the mixture thoroughly. Then let it stand 24 hours. At the end of this time, the color of the liquid will indicate the amount of organic matter in the sand. If the liquid is clear, the sand is free of organic matter. If it is slightly straw-colored, there is a slight amount of organic matter but not enough to be seriously objectionable. Any darker color indicates too much organic matter and the sand should not be used. See Fig. 4-6.

Fig. 4-6. Colorimetric test for organic matter in aggregate.

Mortar: Basic Materials and Mixing

Grading. Grading means the difference in size of sand particles. A well graded sand contains particles of several different sizes. See Figure 4-7.

Sand must be well graded as to particle size if a satisfactory mortar is to be obtained. Owing to the great differences in the sands of various localities, the standard specifications permit a rather wide range in sizes but a better graded sand, if available, will yield a more workable and better mortar in almost every respect.

The ASTM requirements for grading represent the extreme limits and it is specified that "the gradation of the material from any one source shall be reasonably uniform and shall not be permitted to vary over the extreme range" and that it "shall be so graded that neither the proportion finer than a No. 16 sieve and coarser than a No. 30 sieve nor the proportion finer than a No. 30 sieve and coarser than a No. 50 sieve exceeds 50 percent." For purposes of comparison, Table 4-1 gives the ASTM specification ranges of sieve analyses together with the analysis of an average commercial sand and that of one of more nearly ideal grading.

In a good mortar, all the sand particles are completely coated with cementitious material (paste) and hence a sand containing a high proportion of fine particles may require much more paste than either a properly graded material or one made up chiefly of larger particles. However, sufficient fine material should be

Fig. 4-7. A well graded sample of sand; mixed (above) and separated (below).

Masonry Simplified

TABLE 4—1 SIEVE ANALYSES FOR MORTAR SAND

SIEVE NUMBER	PERCENTAGES PASSING EACH SIEVE		
	A.S.T.M. SPECIFICATIONS	COMMERCIAL	MORE NEARLY IDEAL
4	100	100	100
8	95-100	98	97
16	60-100	88	84
30	35-70	64	50
50	15-35	26	27
100	0-16	5	6

present to separate the larger particles and to fill the spaces (voids) between them, thus yielding an easily workable mortar. When all the sand particles are coated and lubricated with the paste of cementitious material, the smaller particles act more or less as ball bearings, thus permitting the particles of aggregates to roll over each other and producing a plastic workable mortar which will serve as a continuous uniform bedding for the brick or other structural units.

Water

The water used in mortar should be clean and practically free from acids, alkalis, salts, and organic matter. As a general rule, water that is used for drinking (potable) will be suitable for making mortar.

Admixtures

Standard mortar consists of only four ingredients: cement, lime, sand, and water. Any other material added to the basic mortar is called an admixture. The most common admixtures are accelerants and retardants. An accelerant is an agent which speeds up the curing time of mortar. This type of admixture is sometimes specified when the working conditions are very cold and there is the danger of mortar freezing before it is completely set. Calcium chloride is the admixture usually specified for this purpose. Calcium chloride present in a mortar will corrode metal. Therefore, its use in reinforced masonry is discouraged by most building codes. Studies have shown that calcium chloride in excess of two percent of the total volume of mortar will eventually weaken it. Building codes and specifications on the allowable amount of calcium chloride vary from zero to two percent. For cold weather work (temperatures below 40°F.) most codes recommend warming the materials, the

Mortar: Basic Materials and Mixing

use of heating devices and Type III (high early strength) cement instead of admixtures.

Retardants are admixtures which slow down the curing time of mortar. They may be specified when the conditions are very hot and dry, causing the mortar to set too rapidly and never attain its strength. Again, most codes discourage the use of retardants where other means of retarding, such as insulation, may be used.

Mortar Colors

Mortar may be colored in order to match or contrast with the color of the masonry units. There are several commercial coloring additives available on the market. These are usually pure mineral or metallic oxides that are colorfast and resistant to acids and alkalis. When they are used properly they do not affect the composition and workability of the mortar. The reputable manufacturer will have exact specifications for mixing printed on the container. As with any additive to the basic mortar mix, these specifications must be followed exactly.

Specifications for the amount of coloring additives to use vary, but the amount should not exceed 15 percent of the weight of the cement.

Carbon black may be used to darken mortar in amounts up to 3 percent of the weight of the cement.

Mortar Properties and Types

The American Society for Testing and Materials has classified different formulas for mortar into *types* according to proportions, properties and usage. Where the mortar is to be used, such as in interior or exterior walls, foundations, or columns and pilasters, will call for certain *properties* which, in turn, are created by the proportions of the materials in the mortar mix. Properties are the characteristics that determine which type of mortar is to be used for a specific job. These properties are strength, durability, workability, and water retentivity.

Although each type of mortar has all these qualities, the degrees of each will vary from type to type. The mason must remember that all mortar, according to the conditions to which it will be exposed, must be prepared and applied toward the goals of permanence and watertightness.

Strength and Durability

The terms *strength* and *durability* should not be confused. Strength means the amount of stress (tension and compression) that the mortar is able to withstand. Durability means the degree of resistance to external elements, such as weather and chemicals, and to aging. The two do not go hand in hand. In fact, it will be seen that a very high strength mortar, because of its composition, will lose a small amount of durability. On the other hand, a mortar proportioned for high durability will have to sacrifice a small amount of strength.

Workability

Workability means the ease or difficulty with which the mortar works under the trowel. As with all properties, workability is determined by the proportion of materials. Workability is increased by including more of the plasticizing agent, lime, or by reducing the amount of sand. Again, to increase workability, some strength and durability may have to be sacrificed. It must be emphasized that *all* good mortars are strong, durable, and workable. There is no one type that is best for all conditions. The mason must be able to analyze the purpose and conditions of each job and know which type of mortar is best suited for it.

Water Retentivity

Water retentivity is the property of mortar which prevents the rapid loss of water to masonry units of high suction and prevents "bleeding" when the mortar is in contact with relatively impervious units. In both cases the result is incomplete curing which causes the mortar joint to be weak and permeable to moisture. High water retentivity also improves the workability of mortar.

Lime has the property of water retentivity; therefore, mortars with high lime content will have high water retentivity. This property is quite important where the working or weather conditions are very hot and dry, causing a high rate of evaporation.

Mortar Types

Most building codes and specifications call for ASTM mortar *Types M, S, N,* or *O*. Table 4-2 shows the proportion for each type.

Note that the formulas allow for variations of proportions within the types but that the volume of sand never exceeds three times the combined volume of cement and lime. Any more sand would make for a weaker and less workable mortar. Typical mortar formulas are: M—1 part cement, ¼ part lime, 3 parts sand; S—1 part cement, ½ part lime, 4½ parts sand; N—1 part cement, 1 part lime, 6 parts sand; O—1 part cement, 2 parts lime, 9 parts sand.

Type M. Note in Table 4-2 the high cement to lime proportion. This makes for a very strong mortar. Type M would be specified where high compressive stress occurs, such as in heavy load bearing walls. The rel-

Mortar: Basic Materials and Mixing

TABLE 4-2 PROPORTIONS OF MORTAR TYPES

TYPE	MATERIAL (parts by volume)			
	PORTLAND CEMENT	HYDRATED LIME		SAND
		MIN.	MAX.	
M	1		1/4	Not less than 2 1/4 and not more than 3 times the sum of the total volume of cement and lime
S	1	1/4	1/2	
N	1	1/2	1 1/4	
O	1	1 1/4	1 1/2	

atively high proportion of cement makes this type more prone to expansion and contraction, thus slightly less durable than Type S. For this reason, it is usually specified for exterior construction in moderate climates and for interiors. Type M mortar is also specified for below grade structures in contact with earth such as foundations and retaining walls.

Type S. This is a very good general-purpose mortar. It has good workability and strength and excellent durability. It is specified for above grade exteriors exposed to severe weathering. It is also used in interiors and in all load-bearing structures unless only Type M is specified. Types M and S are usually interchangeable.

Type N. Excellent workability is the characteristic of this mortar because of its high lime content. Although it does not have the strength of Types M or S, it may still be used in bearing walls above grade if the stress is not too great. It is widely used in veneers, partitions and some exterior walls where climatic conditions are negligible.

Type O. Type O is extremely plastic and workable but has relatively low strength. Some codes do not allow this type of mortar at all while others allow it for non-bearing partitions.

Table 4-3 shows a typical building code for allowable compressive stresses which indicates the comparative strengths of the different types of mortar.

Masonry Simplified

TABLE 4-3 ALLOWABLE COMPRESSIVE STRESSES IN UNREINFORCED UNIT MASONRY

CONSTRUCTION	COMPRESSIVE STRENGTH OF UNITS IN PSI	ALLOWABLE COMPRESSIVE STRESSES OVER GROSS CROSS-SECTIONAL AREA IN PSI			
		TYPE M MORTAR	TYPE S MORTAR	TYPE N MORTAR	TYPE O MORTAR
Solid Masonry of solid brick	8000 plus 4500 to 8000 2500 to 4500 1500 to 2500	400 250 175 125	350 225 160 115	300 200 140 100	250 150 110 75

Mortar Measurements

The proportions for mortar types described above are based on *volume* measurements, and mortar materials are packaged and delivered accordingly. Portland cement is packaged and delivered in bags containing 1 cubic foot, which weighs 94 pounds. Hydrated lime is packaged in 50 pound bags which contain approximately 1 cubic foot. Sand is delivered by the cubic foot or cubic yard. So a 1:1:6 mortar would call for one bag of cement, one bag of lime and six cubic feet of sand for six cubic feet of mortar.

Epoxy

Regular portland cement mortars, as previously described, if properly mixed and applied, are sufficiently strong for most masonry work. In a few instances, such as for prefabricated brick panels, a higher strength mortar may be necessary. This is accomplished by introducing a chemical additive to the mortar mix. The most common additive for this purpose is *epoxy*. Epoxies and the amount to be used vary with different manufacturers. As with any commercial product, the manufacturer's directions should be followed exactly.

Mortar: Basic Materials and Mixing

Mortar Mixing

Mortar for masonry is usually mixed on the job site in portable mechanical mixers of the type described in Chapter 3. On some very small jobs, it may be mixed by hand in a mortar box using a mortar hoe especially designed for this purpose.

For hand mixing, put one half the desired quantity of sand in a clean box. Over it spread the specified quantity of hydrated lime and portland cement. Then add the remaining half of the sand. This "sandwich" operation permits a more thorough mixing with less effort. While dry, turn the mixture twice with a hoe and then pull it to one end of the box. Add water and cut the dry mixture back into it. Continue to add water until the desired consistency is obtained.

For machine mixing first add a small amount of water to the drum. This will prevent the mixture from balling or caking up on the machine paddles. Next add one third the required amount of sand and then all the required amounts of lime and cement. As the paddles are turning, add sand and water until the desired consistency is obtained. After all the ingredients are in the drum, continue mixing for at least three more minutes. The drum must be completely emptied before the next batch is begun. The manufacturer's recommendations for maximum load should never be exceeded.

Portable mechanical mixers have several advantages over hand mixing. They assure a uniform mixture, are very speedy, and are easily operated by one man. See Fig. 3-13.

Checking On Your Knowledge

The following questions give you the opportunity to check up on yourself. If you have read the chapter carefully, you should be able to answer the questions. If you have any difficulty, read the chapter over once more so that you have the information well in mind before you go on with your reading.

DO YOU KNOW

1. What quantities of lime, cement, and sand are used in Type M mortar?
2. What the primary components of mortar are?
3. What the names are for the most commonly used cementitious materials?
4. What is meant by the term "bleeding"?
5. The names of four general types of mortar.
6. What benefits to mortars are experienced when air-entraining cements are used?
7. How natural cements are produced?
8. What would happen to mortar made only of cementitious materials?
9. Why it is that when sand composed of very fine particles is used in mortar, *more* cementitious materials are necessary?
10. What special treatment should be given to sand which contains clay?
11. How much sand can be safely used in a mortar?

Brick Masonry: Units and Building

Chapter

5

The art of building with bricks is thousands of years old. Modern archaeologists have discovered ancient brickwork in northern India which is believed to be well over 6000 years old. The first bricks to be inscribed, found in ruins of ancient cities in Mesopotamia (now Iraq), are positively dated as about 3800 years B.C. These bricks were made by hand by molding clay mud from the banks of the Tigris and Euphrates Rivers and allowing them to dry in the sun. The Chinese are known to have made and used bricks in ancient times, but exactly how long ago is not certain.

Brick making in ancient Egypt is evidenced by a painting on a wall in one of the royal tombs in Kurna. It depicts in detail a brickyard scene showing each step in the process the Egyptians used in making bricks. These bricks were inscribed and archaeologists have been able to date them as far back as about 1300 B.C.

The Romans probably learned the art from the Egyptians, thus introducing masonry to the Western World. As the Romans expanded their great empire, they carried the art of brickmaking and masonry construction with them throughout most of Europe, including England.

The early Spanish explorers in the New World built houses using brick which they had brought with them from Spain.

Brickmaking began in America in the English settlements in Virginia and Massachusetts in the 1630's. These bricks were, of course, hand made. They are still being made in a few brickyards, mainly in New Hampshire.

The term *brick masonry*, as already mentioned, should be applied only to that type of construction employing comparatively small building units made of burned clay or shale. There are bricks made of other materials but when such bricks are used in masonry work, the term *brick masonry* should not be employed. The standard physical properties for bricks used in brick masonry structural work are set forth by the United States Government and by the American Society for Testing and Materials in what are known as *Federal* and *Standard Specifications* for brick masonry. These specifications are available and are suggested for anyone interested in selecting and buying brick for use in masonry work.

Ordinary brick are economical in cost and when hard-burned and laid in good mortar are one of the most durable construction materials now in use. Bricks are comparatively small in size and are, therefore, easy to handle. Since they are thoroughly burned during their manufacture, they can be used in construction work intended to be fire-resistive as well as in other forms of construction that are intended to be absolutely permanent in character.

The purpose of this chapter is to explain the materials used and processes involved in the manufacture of bricks; and to describe many principles and construction applications which young or inexperienced masons should understand. It should be made clear that the applications of brick masonry principles explained herein are only those which are commonly encountered in average brick masonry work.

Brick Manufacture

Raw Material

The raw materials in the form of clays or shales from which burned bricks are made are found in all parts of the country. Clays are produced naturally by the weathering of rocks. Shales are produced naturally in practically the same way and from the same material, but differ from clays in that the shale has been compressed, and, in some cases heated, producing a material that is much more dense than clay, and consequently more difficult to remove from banks or pits. The chemical composition of clay or shale and the method of firing them give the various colors and textures to bricks. The colors vary from a light cream to very dark red, with red, buff, and cream predominating.

Clay and shale are dug or quarried (sometimes called "winning") and in some cases allowed to weather in the open for a period of time. See

Brick Masonry: Units and Building

Fig. 5-1. Quarrying or "winning" the raw material. (Brick Institute of America)

Fig. 5-1. The purpose of this weathering is to allow the material to break down naturally into a workable mass. Brick plants are usually located near the source of the raw material. Since the process of brick manufacturing is uninterrupted, the raw materials must be dug or quarried continuously so that delivery to the brick plant will be constant. If the clay or shale delivered to the plant contains the proper ingredients, it is a simple process to grind, mix, and mold the bricks preparatory to firing. However, it is necessary sometimes to mix two or more kinds of shale, clay, or other ingredients in order to produce a mixture of the proper consistency.

Methods of Molding Bricks

There are several methods of molding bricks. Among these are the stiff-mud, soft-mud, and dry-press processes. Each process requires slightly different molding equipment and treatment in mixing. The shale or clay must be ground and mixed to the proper consistency for molding and the correct amount of water added.

Stiff-Mud Process. In the stiff-mud process of forming and molding, the clay is delivered to an auger machine which forces the plastic mass through a die in a continuous stream called a column. The die molds the mass into the desired shape for brick and as the column is extruded, it passes through a machine which cuts it into the desired lengths. See Fig. 5-2.

Soft-Mud Process. In the soft-mud process, machines press the clay into forms rather than extrude it from a die. The final results are the same as in the stiff-mud process.

Dry-Press Process. The dry-press process permits the use of more or less nonplastic and relatively dry clays. The clay is put into molds and subjected to pressure of from 550 to 1500 pounds per square inch.

Drying

The wet molded shapes (called "green brick") as they come from the cutting or molding machines contain from 7 to 30 percent moisture, depending upon the forming process. In order to assure strength and uniformity, most of this water must be evaporated before the molded units are burned. This process is done in drying kilns at temperatures of 100 to 400 degrees. Drying time varies from 24 to 48 hours according to the amount of moisture to be evaporated. The process must not be too rapid.

Fig. 5-2. Forming and cutting by the stiff-mud process. (Brick Institute of America)

Brick Masonry: Units and Building

Burning

Burning is the most critical step in the manufacturing process. In the past, bricks were burned in oven-like kilns which were fired by wood or coal. These kilns could not distribute heat evenly and consequently, each batch of bricks varied greatly in hardness and strength. The soft underfired bricks (called "salmon brick" because of the predominant color) had little strength and therefore, their structural use was quite limited. Modern burning kilns, fired by natural gas or oil, however, allow for even distribution of heat, assuring even quality brick throughout each batch.

The main types of kilns used today are the *periodic* and *tunnel kilns*. In each type, the molded and dried units are stacked in such a way that the heat may circulate freely around each unit. See Fig. 5-3. In the periodic kilns, the bricks remain stationary while the temperature is changed for each stage of burning. In the tunnel kiln, the bricks are on moving cars which pass through different temperature zones. Fig. 5-4 shows a modern tunnel kiln.

The burning process requires from as little as 40 hours in tunnel kilns, or up to as much as 150 hours in periodic kilns. Again, the time required will vary according to the

Fig. 5-3. Molded units stacked and ready for the kiln. (Photo by C. J. Burnett)

Masonry Simplified

Fig. 5-4. A modern tunnel kiln. (General Shale Products Corp.)

molding process employed. The bricks are cooled gradually in the kiln. If the bricks are cooled too rapidly, cracking and checking of the surfaces may occur. Cooling in periodic kilns takes from about 48 to 72 hours. Tunnel kilns, however, seldom require over 48 hours for cooling.

Weight of Brick

The weight of brick varies because of the materials used in their manufacture, the amount of burning, and their sizes. Since every manufacturer produces brick of different weight, such information should be obtained directly from him. An approximate weight, especially of building brick, is about 4½ pounds each.

Quality of Brick

Brick should be uniform in shape and size; their edges should be fairly square, straight, and well defined; they should be free of cracks, pebbles, twists, and broken corners; and should be well burned but not vitrified or brittle. A good test of bricks is to strike two of them together. They should emit a metallic ring. Surfaces should not be too smooth because some roughness is required to assure good bonding with the mortar. A good brick should not absorb more than 10 to 15 percent of its

Brick Masonry: Units and Building

weight in water after having been soaked in water for 24 hours.

Colors and Surface Finishes of Brick

In general, the brick produced in the United States from natural clays and shales without special mixing are red in color. There are a few localities where the materials available produce bricks which tend to be yellow. The slight difference in the clays and shales and the manufacturing processes account for these various shades of red and yellow. Some difference in color also is possible between burnings which make it advisable to purchase enough bricks for a job all at one time to assure the same coloring.

Such minerals as iron, lime, and magnesia are responsible for the coloring in bricks. These minerals occur naturally in the clays and shales. For example, iron in clay will produce yellow, orange, red, and blue. Magnesia produces a brown color. When the manufacturer carefully controls the amounts of these minerals he can produce bricks of almost any desired color provided he also controls the amount of heat in the kiln. Kiln heat also plays an important part in the production of colors.

Many surface textures in face bricks are possible by various steps in the manufacturing. For example, rough textures can be obtained by mixing coarse materials with the other brick materials, or wires can be used to cut the bricks as they are extruded from a die. Wire cutting produces a type of finish or texture which is pleasing to the eye. There are many possible textures produced. Brick manufacturers will supply literature relative to their products free of charge. Such literature can be used in selecting texture.

Classification of Bricks

Bricks are classified in three ways: by size and shape, by where they are to be used, and by the properties of strength and durability.

Brick Sizes and Shapes

Bricks are manufactured in many sizes and shapes. There are two standards for quoting measurements of brick: non-modular and modular. The measurements for non-modular brick are the actual size as they come from the manufacturer. At the present, the most popular types of non-modular brick are *standard, three inch,* and *oversize.* Fig. 5-5 shows the shapes and dimensions of typical non-modular bricks.

Modular brick sizes were developed as a way to simplify estimating

Masonry Simplified

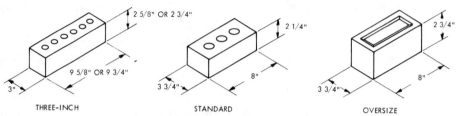

Fig. 5-5. Non-modular brick dimensions.

the number of bricks needed for a particular job. Modular dimensions are *nominal* as opposed to the actual size of the brick. A nominal dimension is the actual dimension of the brick *plus* the thickness of one mortar joint. This allows the estimator to simply measure the area to be laid up in brick and divide into it the nominal dimensions of the specified type of brick. Fig. 5-6 illustrates some of the more common modular brick sizes. Note that the given *nominal* dimensions, for instance, of "standard modular" brick are 4 x 2⅔ x 8 inches. If the building specifications call for a ⅜ inch mortar joint, the actual manufactured dimensions would be 3⅝ x 2¼ x 7⅝ inches. These are the dimensions used in ordering the brick.

Face and Building Brick

Bricks are also classified according to where they are to be used. The main classifications in this category are *face brick* and *building or common brick*. Face brick are those which are to be used on the exterior tiers (wythes) of walls and sometimes on interior walls for special decorative purposes. They are made of selected clays and given special treatments in the molding or burning processes to produce surfaces with special textures or colors.

Building brick (formerly called common brick) have very much the same properties but are given no special treatments as to appearance. Building brick are used mainly in backing tiers of multi-wythe walls, foundations, or partitions where appearance is not so important as strength and economy. However, attractive surfaces may be achieved by using building brick and many designers, for either economy or beauty, will specify building brick instead of face brick for exposed surfaces.

Types of Facing Brick. The ASTM divides facing brick into three *types* according to allowable variations in color, size, and texture of the finish. *Type FBS* allows for the normal moderate variation in size, color and face texture expected in

Brick Masonry: Units and Building

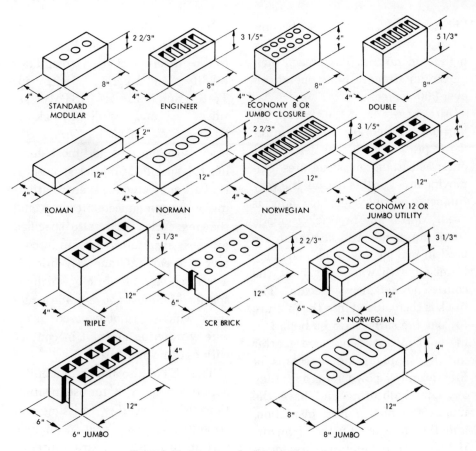

Fig. 5-6. Modular (nominal) brick dimensions.

regular manufacturing practice. *Type FBX* is a specification for brick with all units as nearly identical as possible with respect to size, color and texture. *Type FBA* allows for a wide range of variation in size, color, and texture of the individual units. Facing brick *types* pertain only to the design or decorative function of the brick.

Firebrick

Firebrick are the most specialized of all masonry units. They are manufactured from a special clay called fire clay which is found only in certain areas of the country. Firebrick can withstand extremely high temperatures and are used exclusively where this property is necessary, such as in fireplaces and furnace

69

linings. In fact, firebrick become stronger when exposed to extremely high temperatures. They reach their maximum strength at temperatures over 900 degrees.

Airtightness is of the utmost importance in firebrick masonry in order to protect the less heat-resistant outer sheathing of structures employing fire. For this reason, the very thinnest possible mortar joint is used. Mortar for firebrick is made from the same fire clays that are used in making the brick. The clay is mixed with water to about the consistency of heavy cream. The brick is then dipped into the mixture to coat the surfaces to be bound. A mason may make his own mortar mix by simply grinding scraps of firebrick into a powder. Because they are not loadbearing and have the unique function of heat insulation, and, therefore, must make up an airtight structure, firebrick are sometimes laid up with no mortar at all. In this case, the surfaces exposed to the heat source may be "painted" with the mortar mixture to assure airtightness.

Brick Properties

Brick are also classified according to the properties of strength and durability. The American Society for Testing and Materials (ASTM) classifies face and building brick as Grades SW and MW. *Grade SW* brick is the hardest and most durable. It is specified for brick structures that are subjected to severe weathering and pronounced freezing and thawing cycles. It is also specified for most structures that are in contact with earth. *Grade MW* brick is a strong durable brick which is specified for most above-grade structures which are not exposed to severe weathering and not in contact with freezing moisture.

The ASTM also allows for a building brick *Grade NW* for use in situations where weather resisting and strength requirements are negligible, such as in non-load bearing interior partitions.

Table 5-1 shows the comparative strength properties of face and building brick Grades SW and MW and building brick Grade NW.

TABLE 5-1

DESIGNATION	MINIMUM COMPRESSIVE STRENGTH (BRICK FLATWISE), PSI, GROSS AREA	
	Average of 5 Brick	Individual
Grade SW	3000	2500
Grade MW	2500	2200
Grade NW	1500	1250

Brick Masonry: Units and Building

Courses, Joints, and Bonds

Courses

Each horizontal layer of brick in a masonry structure is called a *course*. The brick laid flat with the long face parallel to the wall is called a *stretcher*. When all the brick in the course are laid in this manner, it is called a *stretcher course*. Brick laid flat and perpendicular to the face of the wall are called *headers*. A *header course* consists entirely of headers. Header and stretcher courses are the most commonly used in a brick wall. Fig. 5-7 shows a partly completed eight course wall. The first and eighth courses are headers with six stretcher courses in between.

Sometimes a special type of course is used for either structural or decorative purposes. The *rowlock course* (frequently spelled *rolok*) is similar to the header course except that the brick are laid on narrow or face edge. This type of course is often used as the top course, or *cap* of garden walls and as window and door sills. See Fig. 5-8.

A *soldier course* is one in which brick are laid standing on end. This type of course is sometimes used for decorative effects over door and window openings and in fireplace facings. See Fig. 5-9.

Closures. In some courses, particularly at or near the ends or corners, there will be a space to be filled by a shorter length of brick. These pieces are called *closures*. On small

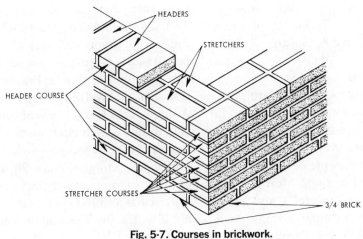

Fig. 5-7. Courses in brickwork.

Masonry Simplified

Fig. 5-8. Rowlock course.

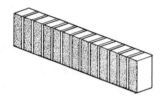

Fig. 5-9. Soldier course.

jobs, the mason may make closures out of whole brick by using the brick hammer or brickset as was shown in Chapter 3. On big jobs, the closures will probably be cut in advance with a power saw. Note in Fig. 5-7 that the corner of the header course is started with a ¾ length of brick which is called a ¾ closure.

Mortar Joints

As bricklayers are laying up walls, for example, they should keep in mind that unless the mortar joints are made *properly*, the mortar will fail to some extent, in one or all of its required functions. Mortar joints must be made *carefully* and properly. The importance of this cannot be overemphasized. Excellent bricks and excellent mortar cannot produce good walls or any other structural items unless the mortar is properly applied. Much of the brick masonry done in the past has been poor only because of carelessness in the use and application of the mortar in the horizontal and the vertical joints in brickwork.

Strength of Bond in Mortar Joints. While good workmanship is of the utmost importance in assuring a good mortar joint, certain physical characteristics of the materials have a direct bearing on the strength of the bond.

For maximum strength, the water retentivity of the mortar should be high and the flow of the mortar should be the maximum consistent with workability (use the maximum amount of water possible). The suction rate of the brick (the amount of water it absorbs in a given length of time) should be low. The maximum rate of suction allowed under ASTM standards and most building codes is 20 grams (0.7 oz.) of water per 30 sq. in. of surface area per minute. Any brick with a higher suction rate should be wetted prior to laying. A simple field test for rate of suction is to draw a 1 inch circle on the bedding surface of a sample brick. With a medicine dropper, place 20 drops of water in the circle. If the water in the circle is absorbed in less than 1½ minutes, the brick should be wetted. The brick should be thoroughly

Brick Masonry: Units and Building

wetted and allowed to stand from 3 to 15 hours before using, to allow for even absorption of the water.

Bed Joints in Stretcher Courses. A bed joint, as illustrated in Fig. 5-10, is a horizontal joint upon which bricks rest. Bed joints are important from the standpoint of bonding brick together, creating equal pressure throughout a wall, and making the wall moistureproof and airproof.

Mortar for bed joints should be spread thick and uniform as indicated by the thick bed of mortar under the trowel in Fig. 5-10. The mortar can be one inch or more in thickness. It is customary to run the point of the trowel along the mortar, as shown in the illustration, to make a furrow. This furrow should be along the middle of the mortar bed and should be shallow, not deep. The furrow provides a double track of mortar which makes it possible for the brick to be laid with the least pressure by the hand. The pressure of the brick being laid fills in the furrow, making a solid mortar joint.

The mortar should be spread a distance of only four or five brick lengths in advance of laying. This is especially important during hot and dry weather. By spreading the bed not more than a few brick lengths in advance of laying, the mortar remains soft and plastic and allows bricks to be laid (bedded) easily and properly. Another advantage is that the mortar does not have a chance to dry out much before the bricks are bedded and thus sticks to the bricks as they are placed.

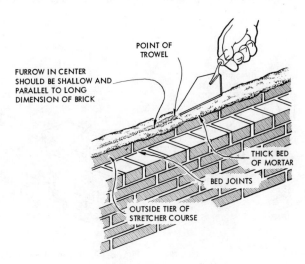

Fig. 5-10. Proper bed joints in walls.

Masonry Simplified

Head Joints in Stretcher Courses. A head joint, as illustrated in Fig. 5-11, is a vertical joint which joins bricks together at their ends. Head joints are also important, especially from the standpoint of preventing cracks in the wall and making the wall moistureproof and airproof.

Head joints, like bed joints, must be *completely filled with mortar*. There are different ways in which this can be accomplished. Plenty of mortar, as thick as will stick, should be thrown on the end of each brick to be placed. This should be done in such a way that the mortar will be scraped off the trowel by the bottom edge of the end of each brick. The bricks can then be placed on the mortar bed and pushed into place, as shown in Fig. 5-11 so that the excess mortar squeezes out at the head joint and at the side of the wall. Or, a dab of mortar may first be spotted on the corner of the brick, such as brick X in Fig. 5-11 already in place. This is followed by placing additional mortar on the end of the brick to be laid. The brick to be laid is then pushed into position as previously explained.

Both of the methods described in the previous paragraph succeed in making a full head joint. These methods are the *only* methods a good bricklayer should use.

Bed and Head Joints for Backing in Stretcher Courses. Fig. 5-12 shows part of a wall in which the face tier has been laid up ahead of the backing tier. The bed and head joints here are also important.

The best method for making good bed and head joints as backing is laid is as follows: A large trowelful of mortar should be thrown at the place where backing bricks are to be laid. Plenty of mortar should be

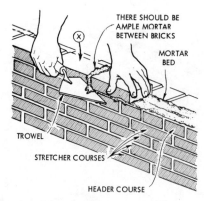

Fig. 5-11. Proper head joints in walls.

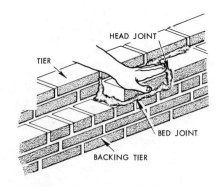

Fig. 5-12. Proper bed and head joints for backing.

used—all that can be carried by the large-sized trowel. Then the bricks should be shoved into this deep mortar so that it oozes out from the bed and head joints as shown in Fig. 5-12. This method makes absolutely certain that joints are full of mortar at every point.

Cross Joints in Header Courses. Cross joints also should be carefully made to assure their being *absolutely filled with mortar*.

First, plenty of mortar should be spread, several brick widths in advance, to form the bed joint. This mortar should be up to an inch thick and evenly spread. Before each header brick is laid, the edge shown in Fig. 5-13 should be entirely covered with all the mortar that will stick to it. Then, as shown in Fig. 5-13, the header should be shoved into place so that mortar oozes out above the cross joint as well as at the bed joint. The excess mortar is scraped off with the trowel.

Closure Joints in Stretcher Courses. The last brick to be placed in a stretcher course must be laid so that both head joints are completely filled with mortar.

With the bed joint mortar already in place, the first step in laying a closure is to apply plenty of mortar to the ends of bricks X and Y in Fig. 5-14, which are already in place. Also, ample mortar should be thrown on the two ends of the brick to be placed. The mortar should entirely cover the two ends. Finally, the closure should be laid as indicated in the illustration, without disturbing the brick already in place.

Closure Joints in Header Courses. The laying of a closure brick in a header course is illustrated in Fig. 5-15. Before laying the closure brick, plenty of mortar should be placed on the sides of both bricks already in place. Also, mortar should be carefully and amply spread on both edges of the closure brick so that the edges

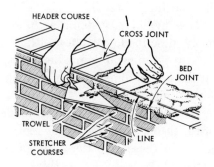

Fig. 5-13. Proper cross joints in walls.

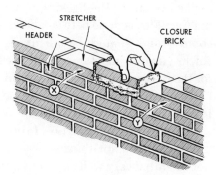

Fig. 5-14. Proper closure joints in walls.

Masonry Simplified

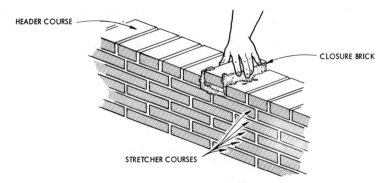

Fig. 5-15. Proper closure joints in walls.

are completely covered to a thickness up to an inch. Then the closure should be laid without disturbing the bricks already in place.

No matter what kind of brickwork is being laid — thick joints or thin — the joints should be completely filled. Any deviation from this rule is poor bricklaying.

Parging. Parging, or the plastering of brick tiers, is a good means of making walls more moistureproof and airproof. Fig. 5-16 shows a wall being back-plastered. The plaster is applied to either the backing or facing tier. This is done between header courses and should be about $3/8''$ thick. Regular mortar can be used for the purpose. This back-plastering may be omitted and the backing tier laid as indicated in Fig. 5-12.

Thicknesses of Mortar Joints. The matter of joint thicknesses cannot be stated very well as a hard and fast rule because of the many possible variables which are encountered.

Bricks made by either the stiff or soft-mud process are likely to be somewhat irregular in shape and thus must be laid up using mortar joints approximately $1/2''$ thick. This varies according to the condition of the brick. The thinner mortar joints, such as $1/4''$, are the strongest and should be used whenever possible. Bricks manufactured by the pressed process are likely to be regular enough in shape to allow the use of $1/4''$ mortar joints.

The modular system, as previously described, calls for either a $3/8''$ or $1/2''$ mortar joint, depending on the actual dimensions of the masonry unit.

When walls have face tiers made of rough-textured bricks, the mortar joints in these face tiers are sometimes made up to $3/4''$ in thickness. In such cases, the joints in the

Brick Masonry: Units and Building

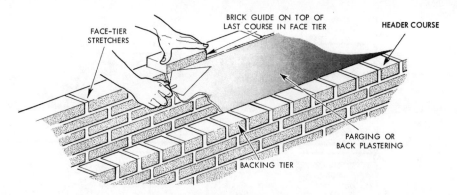

Fig. 5-16. Parging between facing and backing courses.

backing tiers are adjusted either by additional courses or thicker mortar joints than usual so as to bring the two tiers to the same level at the header courses.

When colored mortar is used in face tier joints, the thickness of the joints is frequently decided by the effect of the color in the wall surface. When light-colored bricks are used as facing, thin, colored mortar joints give the principal effect. Usually, however, the wider the colored mortar joint, the more colorful the wall will be.

Joint Finishing. Joint finishing (also called *tooling* or *jointing*) is a very important step in good masonry. Proper finishing not only improves the appearance of brickwork; it is absolutely necessary in making the joint waterproof. Tooling is done when the mortar has set long enough to be "thumbprint hard". Although waterproofing is the main reason for tooling joints, care must be taken not to seal the joint *too* tightly. At the time of tooling, the mortar still contains some water. If the surface is sealed airtight, the water is trapped and cannot evaporate. Thus, in freezing weather, this water will freeze and expand, causing cracking in the mortar joint which will eventually weaken the joint and cause leakage.

Fig. 5-17 shows some of the commonly used joint finishes. The concave, rodded and V joints are the best for waterproofing and therefore are the most often used. The raked joint is sometimes specified for decorative purposes as it provides for interesting light and shadow patterns, but does not have good weather resistance. It is also specified when the joint is to be tuckpointed with colored and/or textured mortar.

The diameter of the tool tip should be slightly larger than the thickness

Masonry Simplified

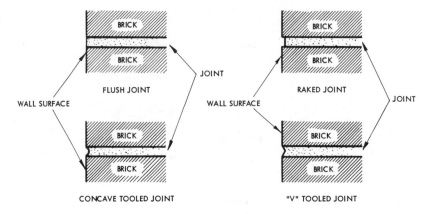

Fig. 5-17. Types of mortar joints.

of the mortar joint to insure complete sealing of both top and bottom edges of the joint. See Fig. 5-18.

Brick Bond

The strength of any masonry wall depends to a great extent upon the *bond* used in erecting the wall. Bond, in this sense, refers to the arrangement of brick or stone in the wall. The arrangements are designed to prevent the vertical joints between the masonry units from being directly above each other. There are many patterns for placing brick which will produce a structurally sound wall. The variation between bonds is brought about by the distribution of *stretchers* (the length of the brick laid parallel with the face of the wall) and *headers* (laid with the length at right angles to the face of the wall) laid in various *courses* (rows).

The following paragraphs list some of the basic bonds used in brick work.

Running or Stretcher Bond. (See Fig. 5-19A) This bond uses stretcher courses with the joints breaking at the center of each brick immediately above and below. Face, building, Roman, or SCR brick is used for this bond.

Common Bond. (See Fig. 5-19B.) The common bond, or American bond as it is sometimes called, is a variation of the running bond, with a header course every 5th, 6th, or 7th course. This ties the wall to the backing masonry material. The header courses are centered on each other. Face or building brick is usually used in the common bond.

English Bond. (See Fig. 5-19C) Alternate courses of headers and stretchers are laid so the joints between stretchers are centered on

Brick Masonry: Units and Building

Fig. 5-18. Left: V joint. Right: Concave joint. (Brick Institute of America)

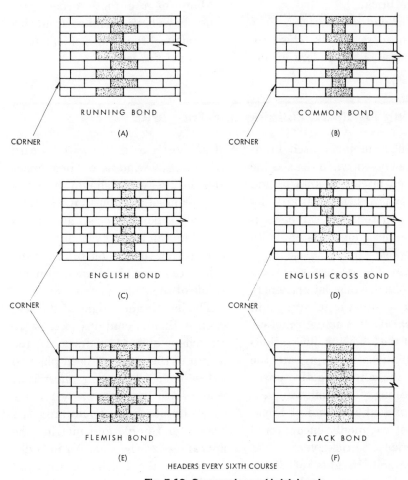

Fig. 5-19. Commonly used brick bonds.

Masonry Simplified

the headers. Stretchers are centered on stretchers; headers on headers. Face or building brick is usually used for the English bond.

English Cross or Dutch Bond. (See Fig. 5-19D) This bond uses alternate header and stretcher courses. The joints of the stretchers center on the stretchers two courses above and below; headers center on headers. This bond is usually building or face brick.

Flemish Bond. (See Fig. 5-19E) Alternate headers and stretchers are in each course. The headers in one course are centered above and below the stretchers in the other course. Face or building brick is used for this bond.

Stack Bond. (See Fig. 5-19F) All courses are stretchers and all joints are in line. This is used primarily for aesthetic purposes—it has relatively little structural value. The most effective brick for this type of bond is Roman.

Many of the more ornamental bonds have been excluded from this discussion — they are seldom used because of the cost.

Bricklaying Practices: Building and Face Brick

When the commonly used kinds of brick are known, when the specific uses of such brick are known, and when the typical brick masonry details are understood and can be visualized, the next logical step is to learn how this kind of masonry is actually laid. The processes or techniques involved, while not extremely difficult, do require careful study and then a great deal of actual practice. The aim of the following illustrations and explanations is to prepare the reader for the actual practice phase of his training. The following examples represent the kind of bricklaying most commonly encountered by inexperienced bricklayers.

Lengths and Heights of Brick Courses. The lengths and heights of all walls and the widths and heights of all window and door openings in the walls should be carefully planned in order that whatever brickwork bond is used, all courses and tiers can be laid without having to use other than whole bricks in the facing tier. Fig. 5-20 shows an example of this.

The length and height of the wall is such that it contains exactly 15 stretchers and 41 courses. If the length of the wall had been planned a few inches longer or shorter, it would have been necessary for one stretcher to be cut, causing the bricklayer trouble and marring the appearance of the wall. An inch difference in the height of the wall would have made it necessary to

Brick Masonry: Units and Building

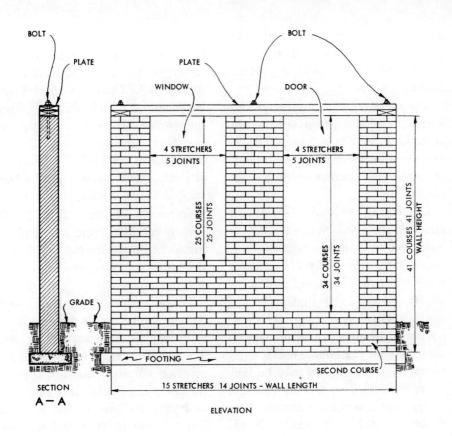

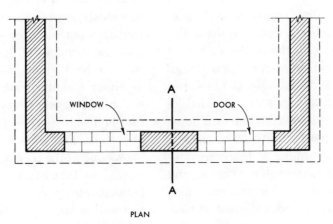

Fig. 5-20. Lengths and heights of brick courses.

make the last course out of split units or else, omit one stretcher course and top out with a rowlock course.

The window and door location and sizes were planned to avoid the use of small closures. Windows and doors can be purchased in many standard sizes which makes possible exact planning.

Laying an 8" Common Bond Brick Wall. This wall of building brick has ½" joints between the units. The bricks are laid in common bond which is illustrated in Fig. 5-19(B). Note in (B) that there are five courses of stretchers; then one course of headers. In other words, there is a header course every sixth course. Header courses may be laid every fifth, sixth, or seventh courses.

Fig. 5-21 shows more details of an 8" brick wall laid in common bond. The sketch at (C) illustrates a corner and shows three courses of the two tiers required. It can be seen that the first course is a header course, as is generally required. The sketch at (B) shows the details of this course. Along the lengths of walls, headers are easy to place, but at the corners, closures are used for proper bond between courses. The use of three-quarter and quarter closures permits the filling of the corners as shown in the sketch. The second course consists of stretchers which can be laid without the use of fractional closures. This can be seen in sketch (A). The third, fourth, fifth, and sixth courses, although not shown in Fig. 5-21, also consist of stretchers laid so as to break joints. For example, note in sketch (C) how the corner bricks for courses two and three are alternated. This alternating of stretchers, or breaking of joints, can be seen to good advantage in Fig. 5-19(B).

In many buildings, 12" brick walls laid in common bond are required. Fig. 5-22 illustrates the positions of the bricks and fractional closures in such a wall.

When masons erect brick walls, corners are laid up in advance of the wall lengths between the corners. This is called "raising the lead". This practice affords them the opportunity of using the corners as a guide in laying the rest of the walls. In Fig. 5-23 assume that the foundation shown in (A) has been erected, that it has been designed so as to fit an even number of bricks (as explained previously, relative to Fig. 5-21), and that it is perfectly level and square and ready for an 8" modular brick wall to be laid on it. The bricks at either end illustrate the corners which bricklayers always lay up before laying bricks the full length of the wall. The sketch at 5-23(B) shows one of the corners laid up and racked so that other bricks can be bonded into them when the rest of the wall is laid. Note that only the face tier is laid in these corners. It

Brick Masonry: Units and Building

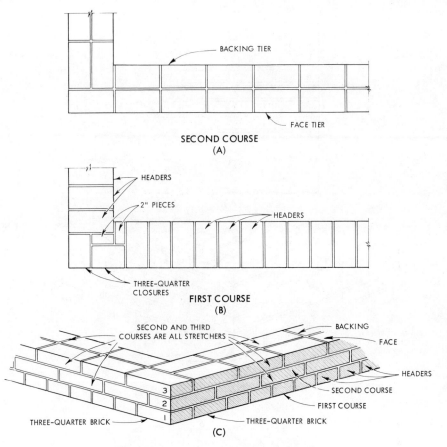

Fig. 5-21. Common bond in eight-inch wall.

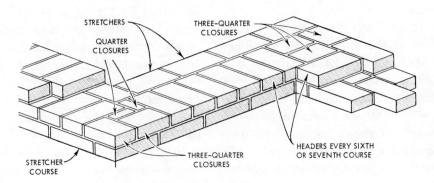

Fig. 5-22. Common bond in twelve-inch wall.

Masonry Simplified

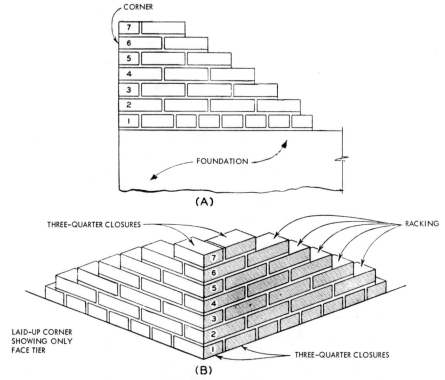

Fig. 5-23. Laying corners.

should also be noted that the corners are laid up to a height of the second header course or a distance of seven courses.

In Fig. 5-24(A) is shown a plan view of the same foundation illustrated in Fig. 5-23. Before the corners are started, bricklayers lay dry bricks (without mortar) along the wall to test its length in terms of bond. As shown in (A), the wall length is just right in terms of whole bricks. A stretcher course is used for such a test.

Note that the first stretcher course (course 2) in Fig. 5-23(B) has four stretchers on one side and three on the other for a total of seven. The sum of the stretchers used in this course always equals the number of courses the lead may be raised. Another example is that if there were five stretchers on one side and four on the other, the lead would be raised nine courses.

In Fig. 5-24(B) is shown the first step in laying up a corner. Mortar should be spread evenly, directly on

Brick Masonry: Units and Building

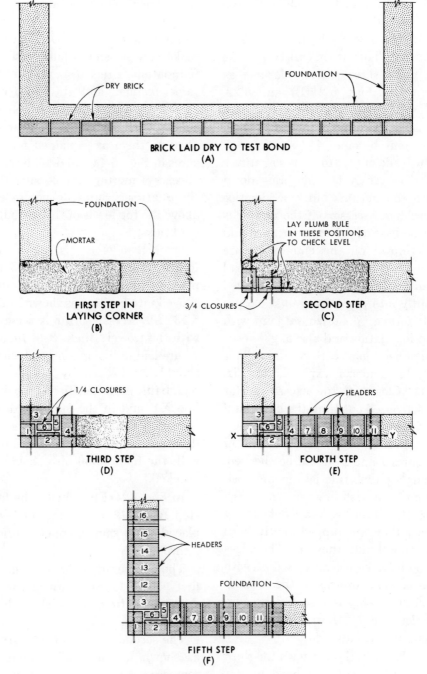

Fig. 5-24. Laying first course at corners.

the concrete of the foundation top and should be at least ¾" deep.

In Fig. 5-24(C) is shown the second step. The mason cuts two bricks to make three-quarter closures as shown in Figs. 5-21(B) and 5-22. Closure *1* is laid on the mortar and pressed down with the hand until the joint becomes ½" as required. The blade of the trowel is sometimes used to gently tap the brick down into the mortar, but this practice is seldom necessary if the mortar is properly spread. Next, mortar should be thrown on the end of closure *2* as previously explained for head joints. This closure is then shoved gently into place up against closure *1* to form a ½" horizontal joint with the foundation and also a ½" vertical or head joint with closure *1*. This should be done as explained for Fig. 5-11. Clean off the excess mortar which oozed out from the horizontal and head joints by holding the trowel as shown in Fig. 5-11. The level of closures *1* and *2* should be checked, using a plumb rule laid in the position of the dash lines shown in Fig. 5-24(C). If the bricks are not exactly level, they are tapped gently with the trowel until they are. The edges of both bricks also must be flush with the outside surfaces of the foundation. This step is essential because bricks *1* and *2* form the base for squaring the entire corner.

In Fig. 5-24(D) is shown the third step. Brick *3* should have mortar thrown on its edge and laid as shown. Check its level using the plumb rule in the position of the dash line and make sure its end is flush with the foundation. Brick *4* is laid in the same manner and its level, etc., checked. The quarter-closures should next be cut and mortar applied to them as explained for closures in Figs. 5-14 and 5-15. Remove all excess mortar and be sure that the quarter-closures do not extend above the top levels of the surrounding bricks.

In Fig. 5-24(E) is shown the fourth step. The edge of brick *7* has mortar thrown on it and is then shoved into position as shown in Fig. 5-13. All excess mortar is removed with the trowel. Bricks *8, 9, 10,* and *11* are laid in like manner. The level should be checked by placing the plumb rule in the position of the dash line *XY* and in the opposite direction, indicated by the second set of dash lines. The matter of flushness with the foundation also should be checked.

In Fig. 5-24(F) is shown the fifth step. Bricks *12, 13, 14, 15, and 16* are placed in the same manner as bricks *7* through *11*.

The number of bricks to lay in the first course for the corners can be determined from a sketch such as shown in Fig. 5-23(B).

The second course for the corner, as shown in Fig. 5-25, should be composed of stretchers, but only the

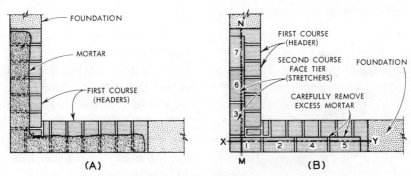

Fig. 5-25. Laying second course at corners.

facing tier should be laid at this stage of the wall construction. Fig. 5-25(A) shows the first step in laying the second course of stretchers. Mortar should be spread over the first course to a depth of at least ¾ inch. The furrow in the mortar, as explained for bed joints and illustrated in Fig. 5-10, should be shallow. Brick *1* should be laid on the mortar and gently shoved or tapped down until the mortar joint is ½ inch. Mortar should be thrown on the end of brick *2* and the brick shoved into position. Remove excess mortar as shown in Fig. 5-11. The joint should be checked for thickness. Bricks *3, 4, 5, 6,* and *7* should be laid in the same manner. Next, the level and plumbness should be checked. The check for level is made by placing the plumb rule in the positions of dash lines *XY* and *MN*. The plumb check is made by placing the plumb rule in a vertical position against the foundation and bricks at several points.

Note: The number of bricks necessary in the second course for the corners can be determined from the sketch shown in Fig. 5-23(B).

The rest of the courses, see Fig. 5-23(B), are laid in the same manner. Care must be exercised in constantly checking the level, and after the first few courses, the plumbness of the rising corner. This is especially important because the wall between corners is laid using the corners as a guide. If the courses are not plumb, the bricks must be moved in or out until the plumb check is perfect. It is not good practice to move bricks once they have been laid, so special care should be taken to lay the corner plumb as it progresses.

The joints should be well pointed before the mortar has set. The outside joints are finished using a pointing tool. The inside joints can be finished flush with the trowel.

The foregoing procedure should be followed in laying up all corners.

Masonry Simplified

Fig. 5-26. Laying to the line. (Brick Institute of America)

In order to assure a level course, the mason stretches a cord (line) from corner to corner at the top edge of the course and beds each brick with the top edge exactly even with it (Fig. 5-26).

The face joints should be tool pointed and the inner joints troweled smooth prior to the time the mortar sets.

When the face tier for two corners and the intervening wall have been laid up to the height of the second header course (six courses as shown in 5-27) the backing tier is laid. It is best to lay the backing bricks first for the two corners, as shown in Fig. 5-27, and to lay the rest of the backing after that. It is not necessary to use a line for the backing except for 12" walls.

Apply mortar to backing bricks, and lay them as explained for Fig. 5-12. Closures for backing tier courses should be laid as previously explained.

When the backing for the first wall is complete up to header height, another corner, adjacent to one of the corners already completed, can be laid up and the same procedure followed.

When all walls have been completed to the second header height, corners are again laid up to the third

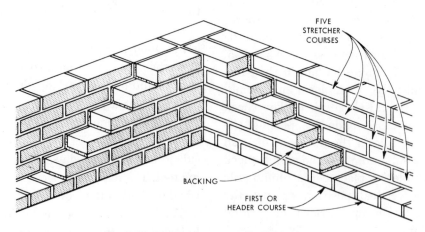

Fig. 5-27. Backing for corner of eight-inch wall.

88

Brick Masonry: Units and Building

header height and the same process repeated. This general procedure is followed until the proper total height of the wall has been reached. Provisions for window and door openings are discussed in advanced examples.

Laying an 8" Common Bond Wall with Building and Face Brick. This type of wall is laid following exactly the same procedure just explained for a common bond wall in building brick, except for the following differences:

First, the face tier is laid using face bricks.

Second, the headers are all laid using face brick.

Third, the backing stretcher courses are laid using building brick.

Fourth, face brick generally require more precise joint finishing for appearance.

Fifth, there may be some backing tier joint variation if the face brick tier must be laid using very thin joints. In such cases, bricklayers sometimes use one less course of bricks in the backing tier with much thicker joints. The main object is to make both tiers even at every header course. The joint thickness required can be determined by trial.

Bonding Face Brick. When face brick are laid as an exterior tier and backed by building brick, they must be bonded to the backing. This can be accomplished in two ways. If the facing bond has courses of headers, these headers can be face brick and bonded into the common bricks like any other header course. Or, if the face brick tier is laid in stretcher bond, such as in Fig. 5-28, metal ties must be used as anchors.

In Fig. 5-28 is shown a 12" brick wall having face brick laid in stretcher bond and anchored to the backing by means of galvanized metal ties. These ties are spaced two or three bricks apart horizontally as

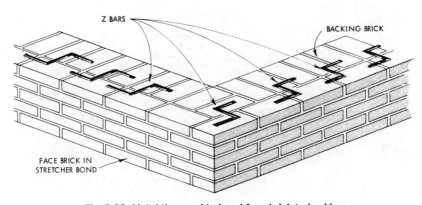

Fig. 5-28. Metal ties used to bond face brick to backing.

89

Masonry Simplified

part of the mortar joint. For best results they should be laid in at least every other course.

Cavity Walls. Cavity walls are made up of two tiers (wythes) of brick with an air space (usually about two inches) between them. They are employed where economy, rather than strength, is important. As air is a relatively low heat conductor, cavity walls have good insulating qualities. They are also quite watertight. If any moisture should happen to penetrate the exterior wythe, it is contained in the cavity and drains out through weep holes placed in the lower exterior mortar joints. For lateral stability, the two wythes must be tied together. This is achieved by using metal ties as shown in Fig. 5-29. Most specifications call for the ties to be embedded

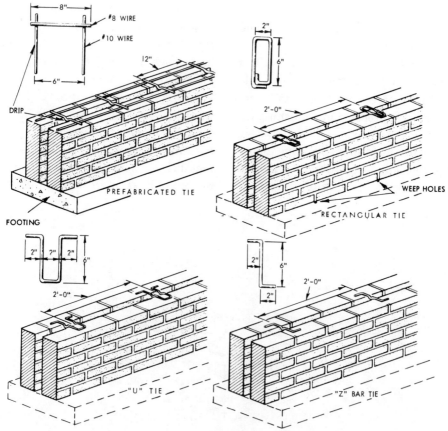

Fig. 5-29. Cavity walls showing metal ties and weepholes.

Brick Masonry: Units and Building

in the mortar joint as the brick are being laid in every sixth course on 24 inch centers.

The cavity wall is frequently called the barrier wall or dual masonry wall. This design of wall is becoming more and more popular in this country because of economical insulating qualities. The continuous air space or cavity forms a barrier to the penetration of moisture and heat or cold. It is suitable for use as exterior or interior bearing or non-bearing walls above grade. A two-wythe 10″ cavity wall has ample strength to support loads present in one and two story residential or small commercial buildings. Taller structures require an 8″ inner wythe to carry the load.

Cavity walls may also be bonded by brick units laid as header courses across the two wythes. A wall of this type is a *rolok-bak* wall, shown in Fig. 5-30. Note that the bricks in the interior wythe are laid on edge, while the exterior wythe is laid in stretcher bond. The brick size should be such that four interior courses are equal in height to six exterior courses. This enables laying the bonding headers at every seventh course. Tight, solid mortar joints are essential in the header course as the headers provide a moisture bridge to the inner wythe. Excess mortar should not be allowed to drop into the cavity as it may accidentally plug the weep holes. The best way to prevent this is to bevel the mortar in the bed joint away from the cavity. If plenty of mortar is used (which is always good practice), as the brick is pushed into place, the mortar will spread into a good solid joint with very little excess being squeezed out. See Fig. 5-31.

Reinforcing. Brick walls that will be subjected to greater than nor-

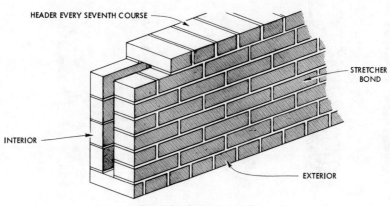

Fig. 5-30. Rolok-bak wall.

Masonry Simplified

Fig. 5-31. Left: A beveled mortar bed for a cavity wall. Right: If the bed is not beveled, excess mortar may fall into the cavity. (Brick Institute of America)

mal stresses are designed to include reinforcing in the form of deformed steel bars. Deformations are projections on the surface of the bar which cause a rigid, inflexible bond with mortar. See Fig. 5-32. These bars are imbedded vertically in the footing or foundation as it is being placed.

After the footing has hardened, the wall is laid up just as the cavity wall previously described. After the two wythes are laid, the cavity is then completely filled with *grout*. Grout is simply mortar with enough water added to allow it to be easily poured. After pouring, the grout should be

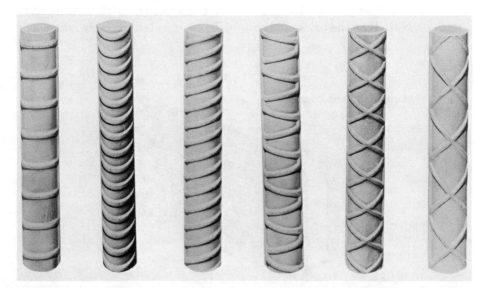

Fig. 5-32. Various types of deformed reinforcing steel bars. (Concrete Reinforcing Steel Institute)

Brick Masonry: Units and Building

Fig. 5-33. Puddling grout in a steel reinforced cavity wall. (Brick Institute of America)

puddled (worked with a stick) to assure that it is completely consolidated. See Fig. 5-33. Horizontal metal ties as shown in Figs. 5-28 and 5-29 are also usually employed in reinforced brick masonry. Fig. 5-33 shows a typical reinforced brick masonry wall.

Window and Door Details. Fig. 5-34 shows window and door details in brick walls. Note how the bonding is carried out around the window and door and how wood bricks, called "jamb blocks", are employed to secure the door frame.

Jamb blocks are the size of a brick and are usually laid as a stretcher in the backing tier.

Laying Window Openings in 8″ Common Bond Brick Walls. If one or more windows are required in a brick wall, the bricklayer must plan for them in advance in order that one course of bricks will be exactly the right height above the foundation for the window sills. Note the window in the wall shown in Fig. 5-34(A). Not counting the rowlock headers, there were exactly 12 courses up to the bottom level of the sill. By measuring the vertical distances window sills must be above foundations, bricklayers can determine how many courses that distance is equal to. This can be done by measuring such a distance on a wall already laid and then counting the courses. Architects generally consider brick courses when they determine window location dimensions, but it is wise for bricklayers to check for their own convenience.

For convenience, a story pole may be marked for locations of sill highs, tops of door and window openings, etc. Also, each course of brick may be marked on the pole by cutting saw kerfs to size for each mortar joint.

When a brick wall has been laid up to sill height, the rowlock sill, shown in Fig. 5-34(A), is laid. Note that this rowlock course should be sharply pitched and should take up vertical space equivalent to two courses. When laying the rowlock course, ample mortar should be used under and between bricks. The surface joints must be very carefully tooled to make them waterproof.

After the rowlock sill has set, the window frame, indicated in Fig. 5-34(A), is placed on the sill and temporarily supported by wood braces. Once the window frame is in place, the bricklayer's next concern is to

93

Masonry Simplified

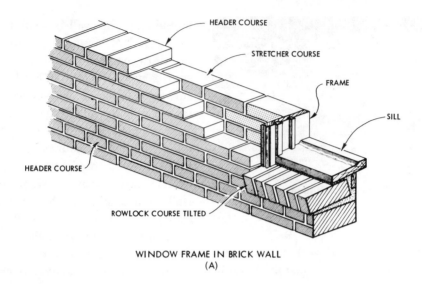

Fig. 5-34. Window and door details in an 8" masonry wall.

Brick Masonry: Units and Building

lay up the surrounding wall so that a course of brick near the top of the frame comes at a level not more than ¼″ higher than the frame. To accomplish this, he marks with a pencil the position of the top course on both sides of the frame. Below these marks, he makes other marks indicating where each header course should be. Sometimes mortar joints have to be varied a little in order to bring header courses to the marks on the frame.

Brick are laid on both sides of the frame using the line stretched across the frame opening as previously explained. If window locations have been well planned by the architect, the bond in the face tier will not be disturbed. In other cases, bricks of the face tier may have to be cut.

When the wall has been built up to the height of the frame, mortar is applied about ½″ deep to the top of the frame and to the wall on both sides of the frame. Steel lintels are then placed over the window opening and bedded in the mortar. Once the two pieces of the lintel are in place, the wall is continued above them. (Note in Fig. 5-35 how the bricks are cut so they will fit around the lintels.)

Laying Door Openings in 8″ Common Bond Brick Walls. Practically the same procedure is followed in laying bricks around door openings as explained for window openings.

Note the wood brick grounds shown in Fig. 5-34(B). These pieces of wood are cut to the size of a full brick and then laid into the various courses, using mortar the same as for the clay brick. When the wall is complete and has set, the door frame is secured to the wood brick by means of nails or screws.

Steel Lintel Details. When steel lintels in the forms of angle irons, etc., are used over window and door openings, the brick must be fitted around them carefully so as to maintain the bond and good appearance. Note in Fig. 5-35 that brick are cut so as to fit snugly around the steel members. Plenty of mortar should be thrown on around the steel as the brick are laid around it. In many cases, flashing may be required as a moisture barrier over the exterior side of the lintel.

Laying a 4″ Brick Veneer Wall on a Wood Frame. Brick veneer, as shown in Fig. 5-36, generally consists of one tier with the brick laid in stretcher bond.

The laying of brick veneer follows closely the procedure explained for 8″ walls except that no backing or headers are used and metal ties are employed to secure the veneer to the frame wall. Corners should be laid up seven or eight courses high and a line used to guide the laying of intervening courses. The window frames are already in place when the veneer is started so the bricklayer can judge easily where the last

Masonry Simplified

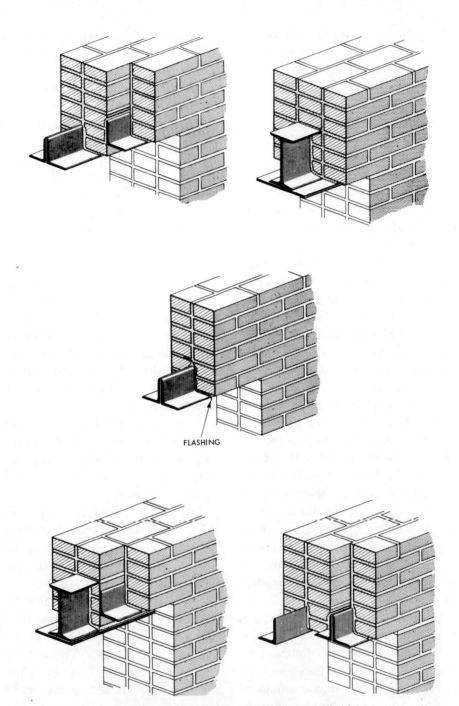

Fig. 5-35. Manner of laying brick around steel lintels.

Brick Masonry: Units and Building

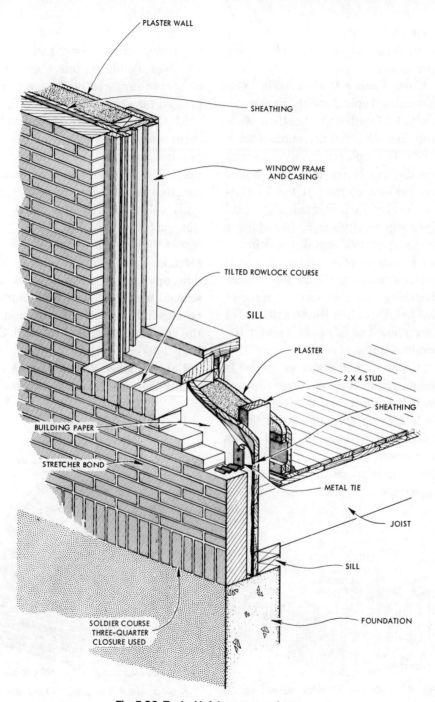

Fig. 5-36. Typical brick veneer on frame.

Masonry Simplified

course before the rowlock sill must be in terms of distance above the foundation.

Brick Veneer Details. Fig. 5-36 illustrates typical brick veneer on a wood-framed and sheathed wall. Note that the stretcher-bonded brick veneer is supported by the concrete foundation. Note, too, the soldier and rowlock courses at the foundation level and window sill. The metal ties are generally spaced two or three bricks apart horizontally and in every fourth or fifth course. There is a space between the paper-covered sheathing and the inside surface of the bricks. This illustration also shows how bricks are laid about the window jambs and sill.

In Fig. 5-36 the soldier course is composed of three-quarter closures or, if it is desirable, this course can be laid using specially made bricks about 5" or 6" long. Fig. 5-37 shows how queen closures can be used at the corners so as not to break the appearance.

The soldier course is laid in much the same manner as explained for Fig. 5-24. Following this, the courses of stretchers are laid as was explained for Fig. 5-25.

Metal ties must be installed as the corners and intervening wall courses are laid, spaced as previously explained. Care should be taken to nail the ties in such positions that the nails will penetrate the studs and not just the sheathing. This will insure their holding firmly. Laying brick around and over door and window openings is carried on as described for the 8" wall. Joints must be carefully tooled so as to add to the good appearance of the wall. See Fig. 5-38.

Cold Weather Practices. As was pointed out in Chapter 4, most build-

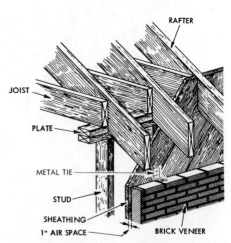

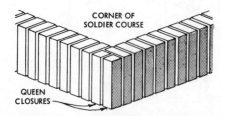

Fig. 5-37. Corner of brick veneer soldier course.

Fig. 5-38. Metal ties used to hold the masonry wall to the frame superstructure.

Brick Masonry: Units and Building

ing codes and standards either discourage or do not allow for retarding or "anti-freeze" admixtures in mortar. Instead, the practices of warming the materials and enclosing the structure are recommended. ASTM standards state that the brick units should be heated to at least 40 degrees and that the mortar may be warmed by heating the water and sand before mixing, not to exceed 160 degrees F. The minimum and maximum temperatures for mortar when laid are 70 and 120 degrees F. When work is not proceeding, the tops of unfinished walls should be covered and, if possible, kept warm.

If Type M or S mortar is used, the air temperature on both sides of the wall should be maintained at about 50 degrees F. for at least 48 hours. Type N requires 72 hours. If high-early cement is used, these periods may be reduced to 24 hours for Types M and S, and 48 hours for Type N.

Construction Details: Building and Face Brick

The art of bricklaying is not confined just to laying brick in courses. There is much fine detail work techniques that the apprentice must learn. Most detail work is concerned with openings in walls (doors, windows, etc.) and fireplaces. The following section describes some of the more common detail work that the mason will encounter.

Footing and Foundation Details. Although brick is now seldom used for footings and foundations, the following information is provided in case the need should arise. Fig. 5-39 shows how the brick for a footing and foundation can be bonded to provide strength and durability. The use of ample mortar to assure full mortar joints is important so that the pressure in the foundation and on the footing will be equalized.

If the soil is wet, the exterior surface of the footing and foundation can be plastered with regular mortar to a thickness of ½" or ¾" as a means of further waterproofing.

Laying Pilasters. When foundations must support a heavily loaded beam end, it is wise to strengthen the foundation by means of a brick pilaster laid up against it. Such a situation is illustrated in (A) of Fig. 5-40. Such supports or pilasters may be of any size depending upon the load, but those shown in (B) and (C) are typical. Pilasters must be supported by footings made of concrete or brick.

The pilaster at (B), being slightly greater than 8" x 12" in section, is bonded using three alternate courses as shown. The pilaster at (C), being slightly greater than 4" x 8" in

Masonry Simplified

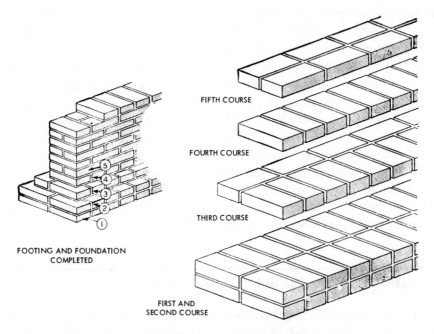

Fig. 5-39. Brick footing and foundation details.

section, has only two alternate bonding courses.

The first step in the laying up of a pilaster such as the one shown at (B) is to spread mortar evenly about ¾" deep on the footing. Then throw ample mortar on one edge of brick *1* and shove or gently tap it with the trowel into place, making a ½" joint between it and the foundation and between it and the footing. Lay halfbrick *2* in the same manner. Then throw mortar on one edge of halfbrick *3* and shove it into position. For brick *4*, it is necessary to throw mortar on one end and one edge before shoving it into position. Remove all excess mortar, and then test the level with the plumb rule.

The second course is laid in much the same manner but with the bond variations shown. It is essential to have an ample bed of mortar.

To lay the third or header course, spread mortar over the second course; then throw mortar on one end of brick *1* and shove it into place. For bricks *2* and *3*, throw mortar on one end and one edge before shoving them into place. Check their level with plumb rule.

Continue courses *1*, *2*, and *3* in alternate manner to the top of the pilaster. Use the plumb rule to plumb

Brick Masonry: Units and Building

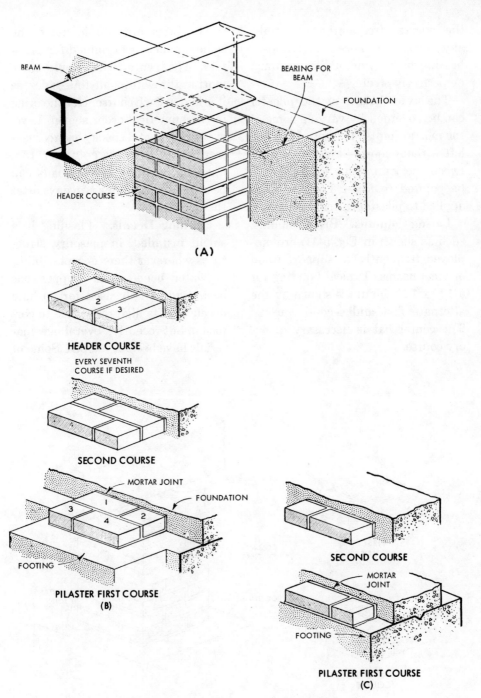

Fig. 5-40. Brick pilaster construction details.

Masonry Simplified

the pilaster frequently and make any corrections necessary. A load-bearing masonry detail like this must be perfectly erect.

The exact height of the pilaster can be governed by varying some of the mortar joints. The top should be left without mortar except that between the joints. The joints should be pointed to taste. Loads can be applied to pilasters after two days.

Laying Columns. Brick columns, such as shown in Fig. 5-41, are employed frequently to support wood or steel beams. Typical bonding for a 12″ x 12″ column is shown by the alternate first and second courses. The center bat is necessary in every course.

The brick are laid in much the same manner as explained for pilasters. The center bat should have mortar thrown on all four sides as it is laid in each course. Lay the brick in the numerical order shown. Level every course and check the erectness frequently because columns, too, must be perfectly erect. Loads can be applied to columns two days after they are laid.

Flashing Details. Flashing is a shield installed in masonry structures wherever there is a possibility of water penetration. Copper has been traditionally used for flashing because of its high resistance to corrosion, but recently, several new materials have been developed. Some of

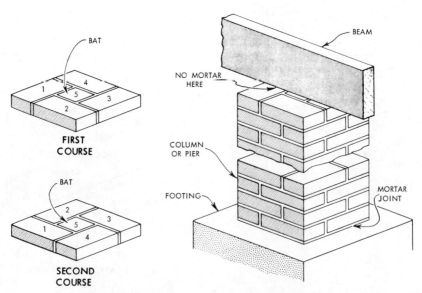

Fig. 5-41. Brick column or pier construction details.

Brick Masonry: Units and Building

these are various metal alloys, metal cloths and papers, plastics and laminates.

Flashing is most often employed at the grade line (foundation top), below and above windows and doors, around the intersection of chimneys and roofs, parapet walls, and other places where rain penetration may occur. Figs. 5-42 and 5-43 show some typical flashing installations.

Wall Ventilation. Some type of ventilation must be supplied to the solid masonry or masonry veneer wall to drain off any moisture which may form. Ventilation in a solid or veneer masonry wall (Fig. 5-43) may be produced by either: (1) weepholes placed every 2' O.C.; or (2) cavity ventilators spaced according to size and manufacturers' specifications. Either method of ventilating

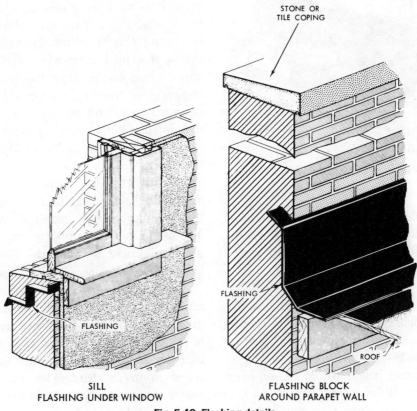

SILL
FLASHING UNDER WINDOW

FLASHING BLOCK
AROUND PARAPET WALL

Fig. 5-42. Flashing details.

Masonry Simplified

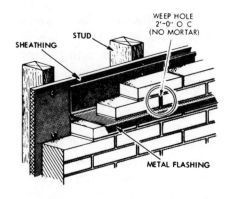

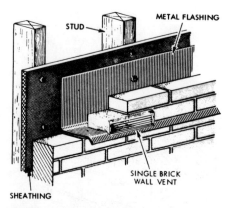

Fig. 5-43. Ventilation methods in a brick wall.

should be placed immediately above the flashing at the base of the wall, over flashing on the second story, and over openings.

Laying Bricks around Joist Bearings in Brick Walls. As brick walls are being laid, the bricklayer usually makes what is called a story pole to use in helping him lay up the wall to the exact height where second-floor joists will have their bearing in the wall. The story pole generally is equal in length to the distance from the rough flooring of the first floor to the height where joist bearing is necessary including the thickness of the finished ceiling. With the aid of such a pole and by varying mortar joints, bricklayers can easily bring certain courses to required heights. When brick walls have reached the joist-bearing heights (called "story high") carpenters usually place the joists in their proper position. The laying of the wall is then continued.

Joists extend at least 4" into a wall. Thus, bricks must be laid around them. This is accomplished by cutting the bricks but without changing the bonding. In other words, if a joist takes some of the space which would otherwise be occupied by a stretcher, enough of the stretcher is cut off so that the remaining part can be laid in the available space in the wall. When filling the brickwork around joists, care should be taken to fit the trimmed or cut bricks no more than enough to allow them to fit into position. The use of small, irregularly shaped pieces of bricks between the joists is not good practice. See Fig. 5-44.

Joist anchors are set by the brickmason and should be solidly bedded in mortar. The mason should carefully check their number and location according to the plans and specifications and lending institution requirements, or the provisions of the local building code.

Brick Masonry: Units and Building

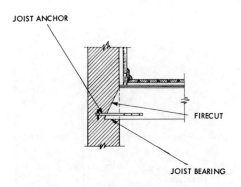

Fig. 5-44. Joist bearing in a solid masonry wall.

Laying a Simple Chimney. Fig. 5-45 shows the alternate courses for a simple, flue-lined, brick chimney.

When the exact location of the chimney has been determined, it is wise to lay out its exact outline on the footing with chalk or a pencil. The steel square should be used to make certain all corners are 90°.

To lay the first course, spread mortar evenly around the lines which indicate the chimney position on the footing. Shove brick *1* into place, making a ½" bed joint. Then throw mortar on the end of brick *2* and shove it into place. Gently tap it down with the blade of the trowel if necessary. Lay the other five bricks, being sure to throw mortar on properly. Clean off excess mortar and use the plumb rule to level the entire course. Apply bed mortar over the first course and, being sure to throw mortar on all bricks properly, lay the second or alternate course. This course also must be checked carefully for level and plumb. When several courses have been laid, the first section of flue lining can be lowered so that it rests on the footing.

The laying of courses is continued until the wall of the chimney is about two courses below the top of the first section of flue lining. From two courses below to two courses above the joint in flue-lining sections, mortar should be applied between the inside edges of the bricks and the sides of the flue lining. This practice helps make the chimney more fireproof.

When height of the chimney wall

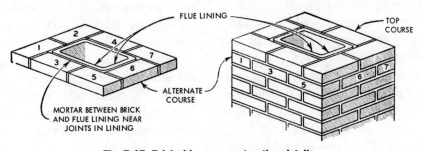

Fig. 5-45. Brick chimney construction details.

Masonry Simplified

reaches the top of the first section of flue lining, mortar is carefully applied around the edges of the flue-lining top and another flue lining carefully laid in place. The wall laying is then continued with care being taken to put mortar between the brick and the flue lining for two courses above and two courses below the flue-lining joint. This method of placing flue linings requires more care and is a little slower but makes for an absolutely fireproof chimney which is otherwise impossible.

The rest of the chimney is laid up in the same manner. Frequent checking to make certain the chimney is absolutely plumb is necessary. The formwork of the building in which the chimney is located will help to keep the chimney erect, too. The joints should be tooled before the mortar sets.

Corbeling Details. In many instances walls are corbeled out, that is, enlarged as a means of carrying some extra load or supporting another structural member. Sometimes the corbeling is entirely a means of adding pattern or beauty to a wall. Chimneys are frequently corbeled to increase their wall thicknesses where they are exposed to the weather.

Fig. 5-46 shows the details of typical wall corbeling. Note that headers are used to a great extent and that in order to give the corbeling strength, these headers extend into

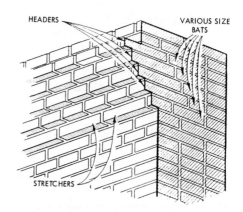

Fig. 5-46. Typical corbelling.

the wall farther than they project beyond it. The first continuous projecting course can be stretchers but all other projecting courses should not extend beyond the under courses more than 2″ and the total projection of the corbeling should not extend more than the thickness of the wall.

A pleasing and stable appearance may be accomplished by allowing only about ¾″ projections. The total projection of the corbeling should not exceed ½ the thickness of the wall, particularly if the corbeling is to support other structural members.

Extreme care must be taken in corbeling to see that all joints are completely filled with mortar and that all bricks are level and plumb. The various bats should be cut carefully and fitted into their places.

Brick Masonry: Units and Building

Fireplace Details

Fig. 5-47 shows a pictorial cross-sectional view of one type of fireplace. Note that it employs most of the bricklaying skills and practices used in any type of masonry construction, including face brick with

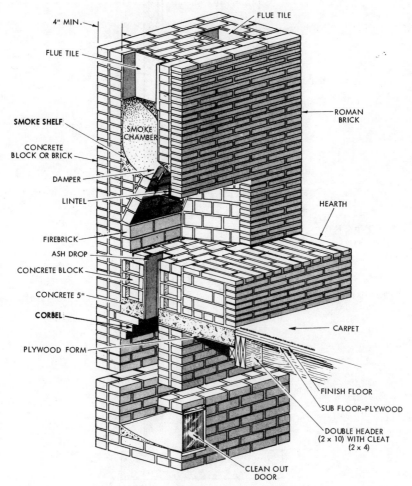

Fig. 5-47. Fireplace details and nomenclature.

107

Masonry Simplified

backing tiers, parging, corbeling, lintels, firebrick, etc. A well designed and well built fireplace is a masterpiece of the bricklayer's art.

Although the elements of a fireplace as shown in Fig. 5-47 are common to all fireplaces, individual designs will vary greatly. Fig. 5-48 is a section view of another style of fireplace, employing floor level hearth and wooden mantel.

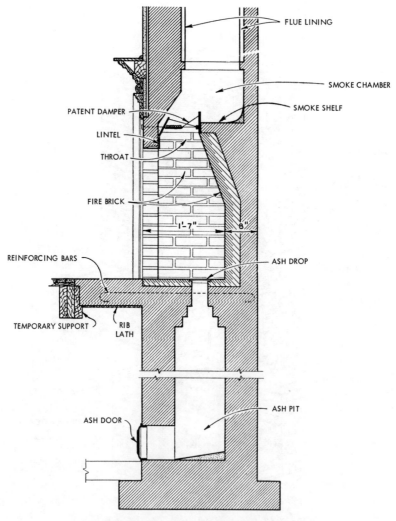

Fig. 5-48. Section view of fireplace and chimney.

Brick Masonry: Units and Building

SCR Brick

Most single-family homes being erected today have only one-story load-bearing exterior walls. Because of the sizes of standard brick, such houses are ordinarily built with 8″ walls. A wall of that thickness possesses sufficient strength for a three-story structure and is unnecessarily heavy for a dwelling of one story. This is emphasized by the fact that nationally recognized building codes approve 6″ masonry walls for one-story residences. (The wall height must not exceed 9′ to the eaves or 15′ to the peak of the gables.)

Many local building codes have long recognized the adequacy of 6″ masonry walls for one-story buildings. With the increasing popularity of this type of construction, additional local codes have extended their approval. A variety of 6″ clay building products have appeared for use in one-story residential and industrial buildings. Some of these are hollow units classed as tile, whereas others are considered as solid units, or brick. In this section, we shall treat in some detail a representative example, the SCR brick. The data presented can be regarded as fundamental and applicable to the use of similar products.

The SCR brick was developed by the Structural Clay Products Research Foundation. It was designed as a through-the-wall unit requiring no backup material; with it, a nominal 6″ wall can be constructed with a single wythe. The SCR brick is intended to adapt masonry construction to present housing design trends at a cost that can compete favorably with other building materials.

As can be seen in Fig. 5-49 the SCR brick is conventional in appear-

Fig. 5-49. The SCR brick.

Masonry Simplified

ance. It presents a face outline like that of the standard Norman brick. Its actual dimensions are $2\frac{1}{6}" \times 5\frac{1}{2}" \times 11\frac{1}{2}$ inches. When the $\frac{1}{2}"$ allowance for joints is added, the brick has, for construction purposes, a nominal size of $2\frac{2}{3}" \times 6" \times 12$ inches. Its weight, when it is of usual composition, is about eight pounds. The unit has ten vertical holes, each $1\frac{3}{8}"$ in diameter. The holes constitute less than 25 percent of the total volume, so the brick is regarded as a solid masonry unit. In one end of each brick, a $\frac{3}{4}" \times \frac{3}{4}"$ jamb slot is provided to facilitate construction around openings.

Except in size, the SCR brick does not differ from ordinary brick. It is made from the same materials, and by the same processes employed in manufacturing conventional units. The choice of colors will be the same as for standard brick produced by the manufacturer. Single-wythe walls of SCR brick have been subjected to the customary laboratory tests with satisfactory results. These included tests for strength, fire resistance, and moisture penetration.

Laying the SCR Brick

Working with SCR brick does not present any particular problems. The mason may span the entire unit with his hand, as demonstrated in Fig. 5-50, or, for easier handling, the core holes provide a convenient hold.

Fig. 5-50. Construction of an SCR brick wall.

Brick Masonry: Units and Building

These same holes help to reduce "floating" of the brick on very wet or plastic mortar. Tests have disclosed that where the maximum transverse strength is desired, mortar Type S is to be preferred. SCR brick must be laid with completely full mortar joints to insure maximum strength and weatherproofing.

The SCR brick is a modular unit, that is, its nominal dimensions can be taken as a unit of measurement in laying out work. It is most easily used with a stretcher (half) running bond, as shown in Fig. 5-51. Walls and corners can then be laid using only whole brick. If door and window widths and wall lengths are planned in multiples of 6" or 12",only half-units will be required to work around the openings with no attendant waste of material. Three courses of SCR brick make 8" of wall height; and 450 units are needed for 100 square feet of wall area.

Construction Details

Wall Construction. A 2"x 2" furring strip is recommended for use with SCR brick wall construction. Such furring provision allows ample space for a moisture barrier and installation of electrical equipment as well as insulation. The 2"x 2" stock offers sufficient rigidity to permit anchoring the strips only at three points. Since the strips can be nailed to the top wall plate, only two clips

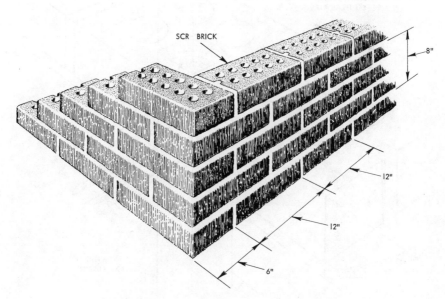

Fig. 5-51. Simple wall construction with SCR brick.

Masonry Simplified

will be required. The special clip shown in Fig. 5-52 has been devised for use with this construction arrangement. The furring strips are driven onto the staples after the wall is completed. The width of the clip allows the horizontal positions of the staples to be adjusted as needed to align them properly. The clips hold the furring approximately ¼″ away from the masonry. This feature allows air to circulate freely and permits moisture readily to drain down to the weep holes in the bottom course.

The space provided by the furring can be insulated in the same manner as in any wall. The cavity is adequate to carry most standard electrical installations and piping. It must be observed, however, that pipes or ducts cannot be cut into single wythe SCR brick walls. Building codes do not permit reducing the nominal thickness of a 6″ wall. Where pipes or ducts cannot be run through the furred-out space, they must be boxed in. For heating systems, special "out-of-wall" and baseboard registers are available.

Foundations. Laying SCR brick on a standard 8″ foundation requires no special instructions. With slab-on-the-ground building where no

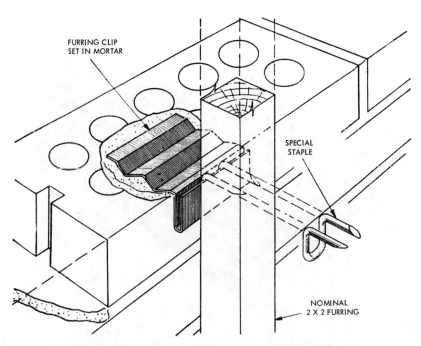

Fig. 5-52. Furring clip designed for use with **SCR** brick construction.

Brick Masonry: Units and Building

basement is to be provided, the wall is very simply erected on the foundation as shown in Fig. 5-53. In cold climates a rigid mineral insulation should be installed between the floor construction and the exterior foundation wall.

Floor joists cannot be framed into a 6″ masonry wall. If a basement or crawl-space is planned, metal hangers must be installed or the wall must must be corbeled out to provide a bearing surface for the joists. The latter arrangement is preferable and one method of doing this is illustrated in Fig. 5-54. It should be noticed that in all cases the furring strips extend only to within 2″ or 3″ of the floor level. This allows for a gradual curve of the metal flashing to improve drainage to the weep holes.

Because of unusual local conditions in some areas, building regulations may require 10″ or 12″ foundation walls. Single methods have been worked out to accommodate the SCR brick to such foundations. Fig. 5-55 shows how the wall can be erected on a 10″ foundation. In this case, the SCR brick are backed up from the grade line to the joist level with masonry 4″ thick. On 12″ foundations, the same procedure may be

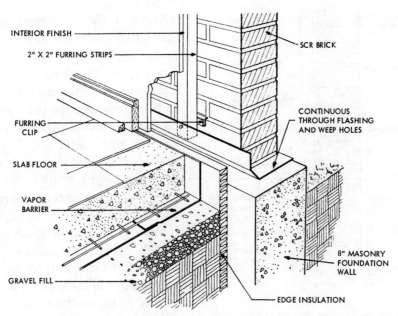

Fig. 5-53. Foundation details with slab-on-ground construction.

Masonry Simplified

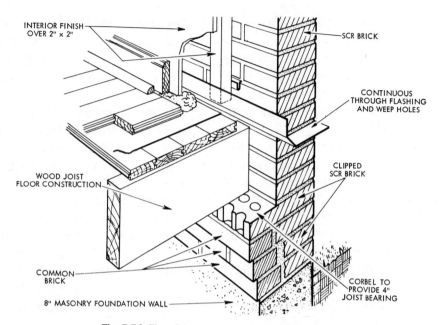

Fig. 5-54. Floor joists supported on corbeled wall.

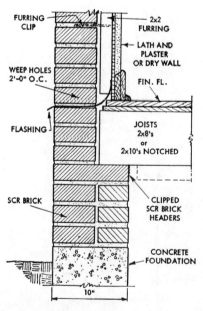

Fig. 5-55. SCR construction on 10" foundation.

followed if the 2" "shelf" on the foundation wall is not objectionable. Otherwise, backup units 6" thick may be used as shown in Fig. 5-56. Whenever constructing walls two units in thickness, the two wythes of masonry should be carefully bonded together with metal ties. It should be borne in mind that the limitations upon the height of 6" walls apply only to that part of the construction that is actually only 6" in thickness.

Head Plate Anchorage. A continuous 2" x 6" head plate is anchored to the SCR brick wall by means of anchor bolts, in the manner shown in Fig. 5-57. When the units are laid with stretcher bond, the core holes

Brick Masonry: Units and Building

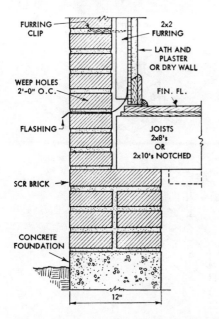

Fig. 5-56. SCR construction on 12" foundation.

line up and no cutting is necessary to locate the anchor bolts. The depth to which the bolts must enter the wall is regulated by local codes but should never be less than three courses of brick. If $3/8''$ bolts are used, they should be spaced no more than four feet apart; $1/2''$ bolts may be eight feet apart. The bolts pass through the core holes and, by staggering their positions, bowing of the head plate will be prevented. The anchor bolt nuts are to be tightened with the fingers and a wrench should never be used on them. Since the furring strips are nailed to the head plate, it should extend about $1/4''$ beyond the inside edge of the wall to properly line up with the strips.

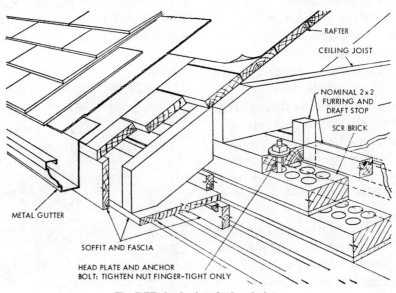

Fig. 5-57. Anchoring the head plate.

115

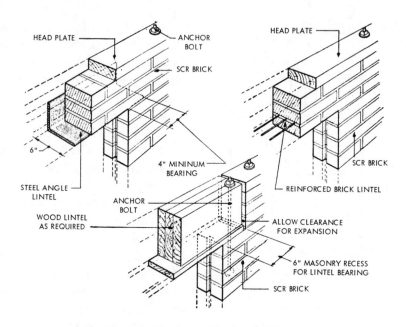

Fig. 5-58. Three methods of lintel construction.

Doors and Windows. Using SCR brick, lintels may be designed in several ways, as shown in Fig. 5-58. When reinforced brick lintels are used, care must be taken that the construction details have been accurately computed to safely bear the load over the opening. Where steel lintels are used, the horizontal width must be 6 inches. In some cases it may be desirable to employ all frame construction above the height of the openings.

Installation of doors and windows in SCR walls raises no unique problems. Stock items can usually be selected that will not necessitate any unusual cutting of brick. The fin of a steel window frame fits into the jamb slot of the brick as shown in Fig. 5-59.

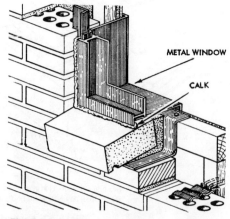

Fig. 5-59. Installation of steel window frame.

Brick Masonry: Units and Building

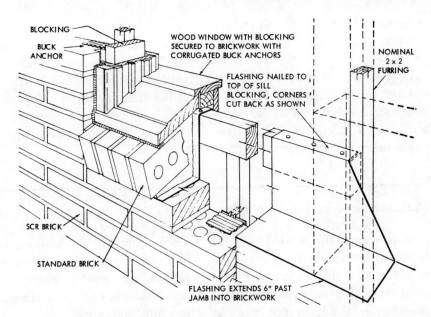

Fig. 5-60. Installation of wood window frame in SCR brick wall.

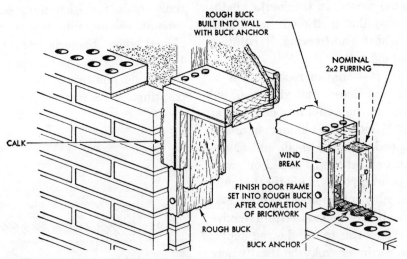

Fig. 5-61. Details of door frame construction in SCR brick wall.

Installation of a wood window frame is illustrated in Fig. 5-60 with the blind stop fitted into the brickwork and the frame anchored in the wall with a buck anchor. Proper calking will make either type of con-

117

Masonry Simplified

struction weathertight. Standard procedures in placing flashing should be followed.

Door construction details are similar and are illustrated in Fig. 5-61. A ¾" strip is fastened to the rough buck and installed in the slot of the brickwork. The addition of calking insures a positive weather stop to prevent the entrance of wind or water. The rough buck is securely held in place by anchor strips in the joints of the masonry. After completion of the brickwork, the finished door frame is set in position in the ordinary way.

Maintenance of Brick Walls

Repointing Old Brickwork. Old brick walls which have been exposed to the weather for many years may need repointing as a means of improving their appearance and to make them more watertight and airtight. This is necessary because the original mortar in the joints weathers away due to the effects of rain, wind, heat, and freezing. Repointing is not difficult.

The first step in repointing is to clean out loose and disintegrated mortar in the joints to a depth of at least ½ inch. The old mortar can be loosened with the small end of the hammer, and then scraped out with the pointing tools made for that purpose. Also, a thin chisel can be used with a hammer to loosen and remove the old mortar to the desired depth. Power grinders may be used where large areas are involved. A stiff brush should be used to remove all dust or remaining loose particles of old mortar. See Fig. 5-62.

Some bricklayers wet the joints before applying the new mortar to improve bond. Using a water jet (sprinkling nozzle) removes dust and loose material thus eliminating the brush-out time.

New mortar is applied to the joints using one of the many special tools made for such purposes. In general, a long, thin trowel is employed whose blade is from ¼" to ½" wide. Before the mortar sets, it should be pointed using a regular pointing tool. Sometimes the mortar

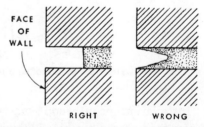

Fig. 5-62. Right and wrong methods of preparing joints for tuckpointing.

is applied with very small trowels so that the new mortar is flush with the bricks and then brushed with a semi-stiff brush just before it sets. The finishing method is largely a matter of taste. The important item is to force the mortar into the joint so that it completely fills the cavity and sticks in place. Repointing an old wall adds surprisingly new beauty to it.

Exposed chimneys are apt to need repointing much more frequently than walls. This is because chimneys are alternately hot and cold and because they are generally more exposed to beating rain, wind, and freezing than most walls. It is wise to inspect chimneys, especially near their tops and at the point where they emerge from roofs, every few years. Chimneys which do not have flue linings should be checked yearly and frequently repointed due to their becoming causes of roof fires when mortar joints are loose.

The process of repointing chimneys is the same as explained for walls.

Mortar For Pointing. Rich mortar mixes should be avoided to eliminate excessive shrinkage after hardening. A high water retention mortar is desired to prevent a rapid loss of water to the brick before tooling and to insure a good bond with the brick and old mortar. For best results the pointing mortar should not be denser than the original mortar. If the density and proportioning of the old mortar is not known, prehydrated mortar is recommended. Prehydrate by mixing the mortar about two hours before using with only a portion of the required water. When ready to use, add the remaining water and rework the mortar. This eliminates much of the initial shrinkage and greatly improves the workability of the mortar. Standard proportions for pointing mortar are one part cement, one part lime, and six parts sand.

Painting Brickwork. Brickwork should generally never be painted for two reasons: beauty and economy of maintenance. Brick are used not only for their structural qualities but also for their naturally pleasing appearance. Also, painting a brick wall creates a seal which prevents the natural escape of moisture, which will cause spalling if the wall is subjected to freezing.

If brick are to be painted, they should be hard-burned and more resistant to freezing in the presence of moisture than ordinary bricks.

Efflorescence. Efflorescence is a light powder or crystallization deposited on the surface of brickwork or concrete and is the result of the evaporation of water carrying water soluble salts to the wall surface. There are two general conditions necessary to produce efflorescence: First, soluble salts present in the wall materials; and second, moisture to carry these salts to the surface of the bricks. In general, good brick contain but little of the soluble salts which produce efflorescence. The mortars and plasters are more frequently the sources of the salts.

Since moisture is necessary to carry soluble salts to the surface of bricks, efflorescence is evidence that there is faulty construction. Wet walls may be due to defective flashings, gutters, and down-spouts, faulty copings, or improperly filled mortar joints. The use of frozen bricks at the time of construction also adds moisture to walls. Any repairable item should be taken care of, including repointing of the joints.

Efflorescence can be prevented by the use of good materials in proper condition and especially by good workmanship in regard to the proper amounts of mortar in all joints and proper flashing. Unless joints are completely filled, moisture is almost certain to enter walls and very possibly cause efflorescence.

Efflorescence can be removed sometimes by scrubbing the wall with water and a stiff brush. If this treatment is unsuccessful, wet the wall thoroughly with water, then scrub with water containing a 10 percent solution of muriatic (hydrochloric) acid (1 part acid to 9-10 parts water by volume). Immediately before and after acid washing, the wall should be thoroughly rinsed with plain water. It is sometimes desirable to follow this rinse with water containing approximately 5 percent of household ammonia.

Cleaning Brickwork. There are many uncertainties about the cleaning of any brickwork which makes the job one for men experienced in the process. Sand blasting frequently is employed as is the steam or steam- and water-jet process. These methods are used mainly on large buildings because of the large amount of equipment required. The use of cleaning compounds may be applied to any size building.

The sand blasting method actually removes a thin layer from the surface of the bricks. This destroys the original texture of the brick and leaves the surface with a coarse texture which may or may not be pleasing. Also, it is necessary frequently to repoint the joints after sand blasting.

The steam or steam-and-hot-water-jet process successfully removes most of the soot and dirt. It is most successful when used for fine-textured, hard-burned bricks.

The cleaning compound method is used to a greater extent than the other methods. On large projects, a cleaning contractor will develop a compound that is best suited to the particular job. Examination and analysis of the material and stains to be removed determines the strength and chemical composition of the solution.

Brick Masonry: Units and Building

Checking On Your Knowledge

The following questions give you the opportunity to check up on yourself. If you have read the chapter carefully, you should be able to answer the questions. If you have any difficulty, read the chapter over once more so that you have the information well in mind before you go on with your reading.

DO YOU KNOW

1. What are the raw materials for making bricks?
2. The three main methods of molding bricks?
3. The difference between actual and nominal dimensions?
4. The difference between grades and types of facing brick?
5. How brick veneer should be secured to the frame wall for which it acts as facing?
6. How parapet wall flashing is attached to the wall?
7. Why headers should be used to a large extent when corbeling?
8. What three types of masonry materials, aside from mortar, are used in fireplace and chimney construction?
9. How bricks are made to fit snugly around steel lintels?
10. How door frames, for example, are secured to brick walls?
11. How big a quarter-closure is when compared to a queen closure?
12. If bed mortar in stretcher courses should be furrowed and if so, how deep the furrows should be?
13. How the facing tier of a solid 12" brick wall laid in stretcher bond is secured to the wall?
14. What kind of bonding is usually used for brick walls?
15. Why bonding is necessary in walls?
16. What principal action can be taken to avoid efflorescence on wall surfaces?
17. When mortar joints should be finished?
18. If header courses are ever used in laying up pilasters?
19. When the backing tier is laid in an 8" brick wall?
20. Why a bed of mortar should be placed on the top of a window frame before the steel lintels are placed?
21. What the nominal size of an SCR brick is?
22. The best widths to plan doors and windows in SCR brick walls?
23. What size furring strips are used on an SCR brick wall?
24. How floor joists can be supported when building with SCR bricks?
25. Whether the limitation on the height of a 6" wall applies to the overall height, regardless of thickness?

Chapter 6
Concrete Block Masonry: Units and Building

The term *concrete masonry* is applied to various sizes and kinds of hollow or solid block, to brick, and to many sizes and kinds of concrete building units, all of which are molded from concrete and laid by masons. The concrete is made by mixing portland cement with water and such materials as sand, gravel, slate, crushed stone, cinders, slag, expanded shale or clay, or other types of aggregate.

Concrete masonry is of great interest to all who are concerned with the planning and erection of masonry projects. This type of masonry is economical and allows easy planning and quick erection. It has excellent sound reduction and insulating properties, great durability to weathering and other destructive agents, and high fire resistance. Concrete masonry is available in a great variety of shapes, textures, and colors, which makes it readily adaptable to all commonly used styles of architecture.

The purpose of this chapter is to describe some of the more commonly used kinds of concrete units; to explain in greater detail the types of construction they can be used for; to set forth helpful and useful facts relative to textures, colors, wall patterns, mortar, and joints; to show and explain typical details of construction; to show how typical erection is carried on; and to present other miscellaneous items of a helpful nature.

Concrete Block Masonry: Units and Building

Materials and Manufacture

The materials used in making concrete building units (block) are portland cement, various kinds of aggregate, and enough water to bind the mixture. The proportions will vary considerably according to the size, density, appearance requirements, etc. The kind of aggregate used in the mixture is the main factor in determining the properties of the finished product. The two kinds of aggregate are normal weight and lightweight.

Normal Weight Aggregate

These are the naturally occurring materials normally used in making all kinds of concrete. They are sand, gravel or crushed stone (usually some form of limestone). The proportion of sand to coarse aggregate will vary according to the strength and texture desired, as will the size of the coarse aggregate. Because of the relatively thin shells of most concrete block units, the size of the coarse aggregate will seldom exceed ⅜". Normal weight aggregate produces a heavier, more dense unit with slightly higher compressive strength than those made with lightweight aggregate. A hollow load-bearing unit of 8" x 8" x 16" nominal size will weigh about 40 to 50 pounds.

Lightweight Aggregate

These aggregates are commercially manufactured products which are sold under several names. They are produced from clays, shales, slag or cinders which are expanded in very high temperature rotary kilns. This produces a porous, rock-like substance which is then ground into the desired sizes. Although the compressive strength of the units made with this type of aggregate is slightly less than normal weight units, they may be, in several instances, interchangeable. Units made with lightweight aggregates usually have better heat and sound insulating qualities. Lightweight units, because their aggregate is a commercially manufactured product, are necessarily more expensive. An 8" x 8" x 16" hollow, load-bearing lightweight unit will weigh from 25 to 35 pounds.

Concrete Block Manufacture

One of the most important aspects of concrete block masonry is its economy. The units cost considerably less than other types because they

123

are made from inexpensive and readily available materials and may be manufactured in huge quantities in a short time. Although manufacturing plants vary in size or capacity, they all employ the same basic steps in manufacture: receiving and storing materials, batching and mixing, molding, curing, packaging and delivery. Modern plants are automated to a greater or less extent. Usually, after the materials are delivered and batched, they are fed into mixers in proportions controlled by electronic meters. After mixing, the materials are fed into molding machines which employ both vibrating and tamping actions. The molding machines are automatically regulated to control the density and texture of the units. There is only enough water in the mixture to make it fluid enough to feed into the machines and so that the units will hold their shape when taken from the molds.

Curing

The method first used for curing concrete blocks was to store the freshly molded units (called *green units*) so that they were exposed to normal atmospheric conditions until thoroughly cured. This usually took about 28 days. The development of modern technological methods, using heated air and steam, revolutionized concrete block manufacture by greatly reducing the time of curing from several days down to a matter of hours.

Unlike clay brick units, concrete blocks must be cured in the presence of moisture. If the green units are subjected to hot dry air, they will lose moisture too rapidly, causing excessive shrinkage and cracking. These units will be weak and brittle and generally unfit for almost any kind of construction. The principal methods are low pressure steam curing and high pressure steam curing.

Low Pressure Steam Curing. This takes place in drying rooms similar to those used in curing clay bricks except that warm moist air is used instead of dry heat. Normal atmospheric pressure is maintained throughout the cycle. After the green units are stacked in the room, they are left to sit for 1 to 3 hours at normal temperatures (70 to 100 degrees) to attain initial hardening. This is called the "holding" period. Then steam is injected into the kiln and the temperature gradually raised to a maximum of 150 to 165 degrees for normal weight units and 170 to 180 degrees for lightweight units. When the maximum temperature is reached, the steam is turned off, allowing the units to "soak in" the residual moisture in the room. The low pressure curing cycle takes about 24 hours in the kiln; the units attain most of their ultimate strength in 2 to 4 days.

High Pressure Steam Curing. This takes place in air-tight steel cylinders called *autoclaves*. Auto-

Concrete Block Masonry: Units and Building

Fig. 6-1. Green units entering an autoclave for curing. (Illinois Brick Co.)

claves are from 6 to 10 feet in diameter and 50 to 100 feet long. See Fig. 6-1. After a holding period of from 2 to 5 hours, steam under pressure is gradually injected into the autoclave until the air pressure is 150 psi and the steam temperature is 366 degrees. This process requires at least 3 hours. The units are then steamed from 5 to 10 hours, depending on the thickness of the units. After steaming, the pressure is quickly released, usually in less than ½ hour. This sudden release of pressure, called "blow-down," creates a vacuum-like atmosphere, causing a rapid loss of moisture from the units without allowing time for shrinkage. The units are ready for use immediately after removal from the autoclave. The entire cycle, from removal from the mold to curing and delivery, usually takes less than 24 hours. High pressure steam cured units are uniformly strong, stable, and highly resistant to shrinkage.

Fig. 6-2 illustrates the manufacture of concrete building units.

Masonry Simplified

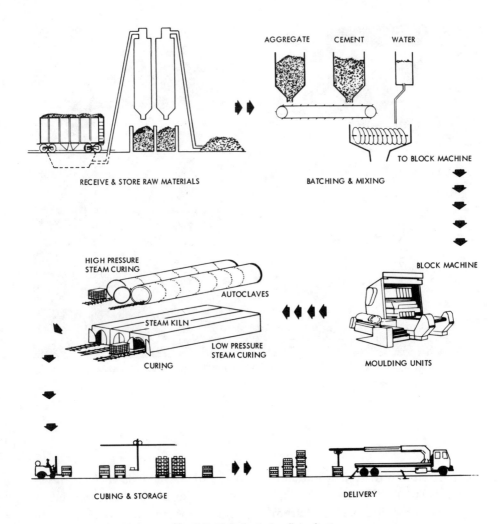

Fig. 6-2. Manufacturing flow chart.

Concrete Block Units

The American Society for Testing and Materials (ASTM) classifies concrete masonry units into two *grades* according to degree of resistance to frost action in different geographical areas, and into two *types* according to the amount of moisture in each individual unit (again in relation to local climatic conditions).

Grade N units are intended for use in areas where they will be subjected to frost action (freezing and thawing), and will be in direct exposure to moisture, such as in above and below grade exterior walls. These units may be used either with or without a protective coating. The minimum allowable compressive strength per unit is 800 psi.

Grade S units are limited to use in above grade exterior walls with weather protective coatings and in walls not exposed to weather. The minimum allowable compressive strength of Grade S units is 600 psi per unit.

All concrete masonry units retain a small amount of moisture after curing. If the structure in which they are laid is subjected to hot, dry conditions, they may shrink slightly as the retained moisture evaporates. This may eventually cause cracking in the mortar joints. For this reason, the ASTM has designated as *Type I* units with specified low maximum moisture content for geographical areas with low relative humidity. The desert regions of the American southwest is an example of where Type I units would be specified.

Type II units have no specified moisture content limitations and are generally used throughout the country where the average relative humidity is moderate to high.

Note: All block manufactured by the autoclave process, as previously described, will be well within the moisture content limitations.

In building specifications, grades and types are usually listed together as Grades N-I, N-II, S-I, and S-II.

Unit Classification

The ASTM also classifies concrete masonry units as building brick, solid load-bearing block, hollow non-load bearing block, and hollow load-bearing block.

Concrete building brick are solid units manufactured in dimensions much the same as the clay masonry units (bricks) described in Chapter 5 and used in the same manner.

Solid load-bearing block have little or no coring and are used mostly where very great compressive strength is required.

Hollow non-load-bearing block are thin shelled, lightweight units intended primarily for use in non-load-bearing partitions, but in some cases may be used in above grade, non-load-bearing exterior walls if protected from the weather.

Hollow load-bearing block have by far the greatest range of uses of all the types of units. They combine the qualities of compressive strength with light weight and flexibility of design, size, and shape. Throughout this chapter, wherever the term "block" is used alone, it will refer to hollow load-bearing units. Most of the concrete block manufactured is of this type.

Masonry Simplified

Shapes and Sizes of Concrete Masonry Units

Typical shapes and sizes of concrete masonry units are illustrated and named in Fig. 6-3. It should be understood that both heavy and lightweight units can be obtained in these shapes and sizes, and that the typical units illustrated in Fig. 6-3 constitute only those most commonly used. Many other shapes and sizes are available or can be made to order. Most masons refer to concrete masonry units as *concrete blocks* or simply as *block*. All of the three core block illustrated in Fig. 6-3 are obtainable as two core block, the difference in structural properties between the two types being only slight. The dimensions shown are true whether the two or three core variation is used. The following paragraphs describe briefly the qualities and uses of the various concrete masonry units shown in (A) through (BB) of Fig. 6-3.

Stretcher Block. The block shown at (A) is perhaps the most used of all concrete block in the construction of farm buildings, garages, etc., and for all types of buildings where stucco, or other surfacing materials such as brick veneer, is to be used as exterior surfacing.

Corner Block. (B) is used for corners and for simple window and door openings, numerous illustrations of which are shown in this chapter.

Double Corner or Pier Block. (C) is designed for use in laying piers or pilasters, or for any other purpose where both ends of the block would be visible.

Bullnose Block. (D) serves the same purpose as corner block, but it is used where rounded (bullnose) corners are desired.

Wood Sash Jamb Block. (E) is used with stretcher and corner block around the more elaborate window openings. The recess in the block allows room for the various casing members as, for example, in a double-hung window.

Header Block. (F) and (G) are the same as stretcher block, except that a shelf has been provided to facilitate bonding them with brick masonry. (Fig. 6-41, page 149 illustrates the use of a header block.)

Solid Top Block. (H) is the same as stretcher block except that the upper four inches of the block are solid. It is used as the bearing surface on which floor and ceiling joists rest. It may also provide the bearing surface for light girders, such as residential center beams, although gir-

Concrete Block Masonry: Units and Building

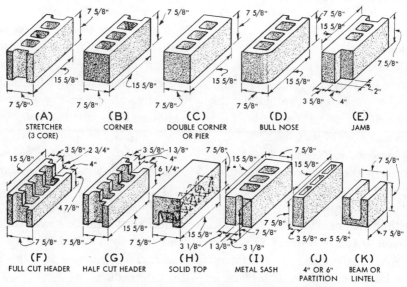

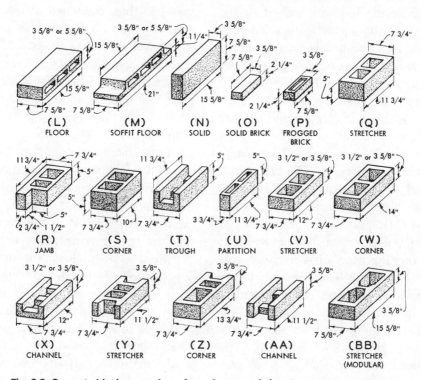

Fig. 6-3. Concrete blocks come in various shapes and sizes to serve many purposes.

ders which are intended to carry heavy loads are usually supported by pilasters.

Metal Sash Block. (I) is used for window openings in which metal sash is to be employed. The slot in the block allows room to anchor the jamb members of such sash.

4 in. or 6 in. Partition Block. (J) is used in constructing non-bearing partition walls. It comes in 3 5/8" and 5 5/8" widths, allowing construction of either a 4" or 6" wall.

Beam or Lintel Block. (K) is used for the construction of reinforced block beams or lintels. The block is laid end-to-end forming a channel. Reinforcing bars are placed in the channel, after which it is filled with concrete. The beams or lintels may be cast in place on shoring or precast and installed later. See Fig. 6-46.

Special Concrete Blocks. The blocks shown in (L) through (BB) of Fig. 6-3, as well as blocks of other shapes and sizes, are used for a large variety of purposes. The filler block used in the construction of concrete filler block floors are shown at (L) and (M). The solid block, shown at (N), is used primarily as a facer unit in framing floor joists in a concrete masonry wall. The concrete bricks shown at (O) and (P) are used in laying concrete brick walls. The remaining units are variations of the blocks previously discussed, and provide some of the range of sizes necessary to meet various construc-tion problems. The variation in sizes also allows the erection of patterned walls, of which more is said in the section dealing with wall patterns.

Actual and Nominal Block Sizes. The sizes shown in Fig. 6-3 for the various blocks are the *actual* sizes. For example, the block shown at (A) has actual dimensions of 7 5/8" x 7 5/8" x 15 5/8". However, it is common practice to speak of or designate such block in terms of *nominal* dimensions. Using this system, these units are then called 8" x 8" x 16" block. This practice is followed in the succeeding pages.

Actual dimensions are made fractional so that, combined with a 3/8" mortar joint, all dimensions come out to even inch sizes. Thus, a 15 5/8" stretcher block with its 3/8" mortar joint, equals 16 inches. The same explanation holds true for heights and widths. The 3/8" mortar joint has been adopted as the standard-sized joint for concrete masonry construction. This does not mean that mortar joints cannot vary in size. In general, the thinner the mortar joint, the stronger the wall. It is not recommended that mortar joints exceed 1/2" in thickness. Since the *actual* dimensions of concrete masonry units make allowance for a 3/8" mortar joint, it is easier in calculating wall sizes to adhere to the standard mortar joint thickness.

Although the majority of manufacturers observe the standard sizes

Concrete Block Masonry: Units and Building

for concrete masonry units shown in Fig. 6-3 there remain a few whose block may vary slightly in size. It is always wise to ascertain the exact sizes of block available before planning wall dimensions.

Special Concrete Blocks

Besides the standard building units described above, there are other types of units available for special decorative purposes. They may be the regular building units with special surfaces, or specially molded shapes. Some of the more common types are described in the following paragraphs.

Scored Block. Regular stretcher and corner units may be ordered with scoring on the outside face. This creates an artificial joint which gives the appearance of different sized units in the wall. See Fig. 6-4.

Textured Block. Regular sized building units may be specially molded with a raised pattern on the face surface. This allows for a great variety of geometric designs and creates interesting light and shadow patterns. See Fig. 6-5.

Face or Glazed Block. These blocks are also available in most regular building sizes. They have one or more surfaces with a glaze or facing of hard, smooth material such as ceramic or stone. Because of the smooth surface, they may be washed easily and are often used in walls where sanitation is important, such

Fig. 6-4. Scored block.

as in hospitals, kitchens, toilet facilities, etc. See Fig. 6-6. They are also available in a variety of patterns.

Slump Block. A mixture of special consistency is used in the manufacture of this block so that when the units are released from their molds before complete setting they will sag or *slump*. This method of manufacture produces irregularities in the blocks which provide an interesting medium artistically. This type of block may be laid in coursed or ashlar patterns in random lengths of 12" to 24". The blocks vary from 1⅝" to 3⅝" in height. Many of the slump block are integrally colored.

Slump block may be laid in coursed or ashlar patterns, as shown in Fig. 6-7. Walls patterned with slump block are especially popular for interiors and exteriors of homes and commercial buildings, as well as for fireplaces, planter boxes, chimneys, and other decorative details.

Split Block. Rough hewn textures are quite often desirable in home

Masonry Simplified

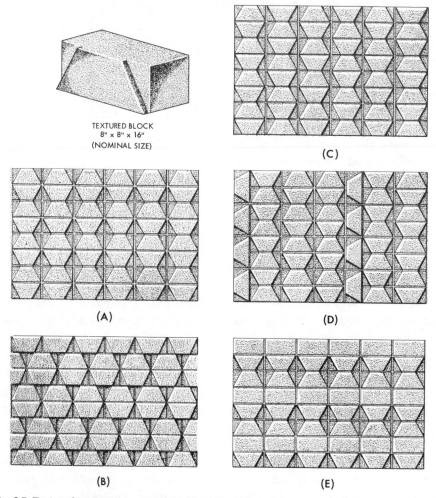

Fig. 6-5. Textured or shadow masonry block add beauty and variety to interior and exterior walls.

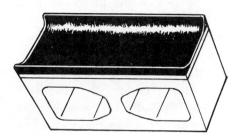

Fig. 6-6. Faced or glazed block.

Concrete Block Masonry: Units and Building

Fig. 6-7. Slump block.

 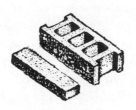 

Fig. 6-8. Split block.

construction, or in churches and other public buildings. Split block is produced by splitting or fracturing concrete masonry units lengthwise. The colors of exposed aggregates and the natural color of the cement binder produce a wide range of color choices. A great number of these units have a coloring pigment intermixed. The use of such units presents rough-hewn walls for both exterior and interior walls. See Fig. 6-8.

Color. Many manufacturers make blocks in a variety of colors. The colors are achieved by the use of nonfading mineral pigments or by the use of special cement paints. Colored blocks produce pleasing walls, especially for residences. Almost any color may be reproduced.

Wall Patterns

The following pages illustrate a few typical wall patterns built with concrete masonry units. Many of the units which appear here are relatively new, but they are receiving wider recognition continually and are generally available. The variations achieved in these walls are the result of a number of factors. Changing the course heights and bonding patterns are the most obvious. Variation in the treatment of mortar

Masonry Simplified

joints can also produce interesting effects, and some examples of this are shown. Finally, changing of block styles affords still another method of altering the appearance of a wall.

Fig. 6-9 shows four examples of wall pattern variations which may be made using the standard 8" x 8" x 16" block. Both examples in the top row of Fig. 6-9 have tooled horizontal joints, but the one to the right has flush rubbed vertical joints, while the one to the left has tooled vertical joints. In this manner, a wall laid in the typical running bond (left) may be given a horizontal accent (right). The lower left example in Fig. 6-9 shows a basket weave pattern, and the lower right the vertically stacked pattern.

A variation on the basket weave pattern shown in Fig. 6-9 is made in Fig. 6-10 using 4" x 4" x 16" units.

Split block adapts itself widely in wall patterning. The illustration to the left in Fig. 6-11 shows split block in quarter-bond pattern, while the illustration to the right shows split block in several lengths and heights laid in random ashlar pattern. A third variation in wall patterning using split block is shown in Fig. 6-12. Here, the horizontal joints were raked, certain units were separated, and the mortar recessed to produce the diamond pattern shown.

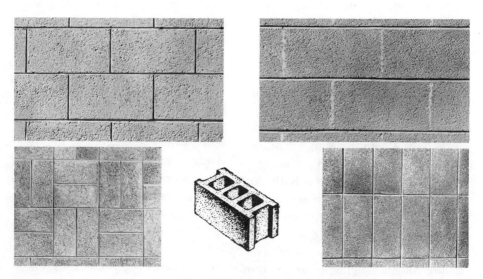

Fig. 6-9. Standard block.

Concrete Block Masonry: Units and Building

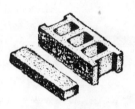

Fig. 6-10. 4" x 4" x 16" units.

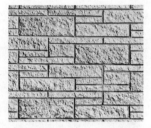

Fig. 6-11. Split block.

Fig. 6-12. Patterns may be achieved by light and shadow effects.

The striking light and shadow effect in Fig. 6-13 is created with units 4" high. The mortar joints have been raked deeply to create the desired contrast. A coat of paint may be added, as has been done here, to emphasize the highlights. The clapboard effect in Fig. 6-14 is achieved

Masonry Simplified

Fig. 6-13. 4" high units.

Fig. 6-14. Shadow block.

through the use of shadow block. The lower edge of each shadow block unit projects ⅝" from the face of the block below. This type of wall patterning harmonizes well with low ranch and Colonial style homes.

The wall patterns in Fig. 6-15 are made with similar units, but the treatment in each is totally different. To the right, polished units with a face size of 2" x 16" are employed. To the left, units with a face size of 2" x 16" (called concrete roman brick) are again used, but no special treatment was given the units themselves, and the bonding pattern is entirely different.

The effect in Fig. 6-16 results from a treatment known as "extruded mortar". It is produced by allowing the excess mortar to set as it extrudes without being trimmed off. The open lattice pattern in Fig. 6-17 is produced with overlapping 8" x 8" x 16" units. In making the interesting wall section in Fig. 6-18 the block sizes used were: 8" x 8" x 16", 8" x 8" x 8", and 12" x 8" x 8".

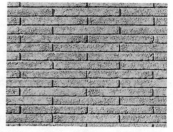

Fig. 6-15. Concrete Roman brick.

Concrete Block Masonry: Units and Building

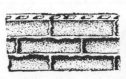

Fig. 6-16. Extruded mortar.

Fig. 6-17. 8" x 8" x 16" block.

Fig. 6-18. Patterned wall.

Still another interesting way to vary wall patterns is to use different types of units in conjunction. An example of this is shown in Fig. 6-19, where ordinary stretcher units are combined with core blocks to produce an attractive interior partition.

The preceding illustrations show only a few of the many possibilities for wall patterns in concrete masonry. After the size and strength requirements are determined, the only limitation is the imagination of the builder.

Fig. 6-19. Patterned partition wall using core blocks.

Masonry Simplified

Concrete Block Construction

Units

As noted earlier, the most common concrete building unit is the *hollow load-bearing block*, measuring nominally 8″ x 8″ x 16″. The nominal dimensions include the actual manufactured size of the block *plus* the thickness of the mortar joint, ⅜″. (Note: The ⅜″ mortar joint is standard throughout the building industry and is always used unless another thickness is definitely specified.) Thus, the actual size of the block is 7⅝″ x 7⅝″ x 15⅝″. Fig. 6-20 shows a typical block of this type and the nomenclature of the parts that will be used in the remainder of this chapter.

Note in Fig. 6-20 that the cores taper in toward the top of the block affording a wider face shell. The units are always laid with the wide face shell up to provide a greater area for bed joint mortar. The unit shown in Fig. 6-20 is made with either three cores or two cores. Two-cored units (Fig. 6-20, right) have slightly different physical properties from the three-cored units. Two-cored units have larger holes and thus have less concrete mass and more core space; they weigh less and cost less. Also, as air does not readily conduct heat, two-cored blocks have slightly better insulating quality.

Note in Fig. 6-20, right, that the inside of the side face shells in the two-core units taper in toward the center web. This provides greater lateral strength than the three-cored units.

In most concrete structures, the main stress is from the top (compression). As three-cored blocks have greater concrete mass and

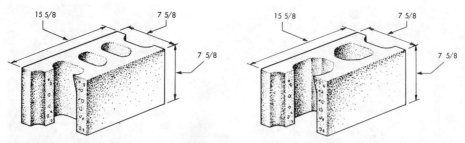

Fig. 6-20. Standard nominal 8″ x 8″ x 16″ hollow load bearing block. Left: Three cored unit. Right: Two cored unit.

Concrete Block Masonry: Units and Building

larger bearing surface area, they have greater compressive strength and are specified for most jobs.

Preparation of Materials

As previously noted, the presence of excess moisture in concrete blocks will cause shrinkage upon drying. For this reason, blocks should never be wetted before laying. On the job, they should be stockpiled on raised platforms in order to prevent absorption of ground moisture. At the end of the work day, the stockpiled blocks should be covered with a watertight tarpaulin to protect them from rain. If the blocks are accidentally wetted, the work must be postponed until dry units are available. If practical, they should be stored indoors.

Mortar for concrete block masonry should be mixed according to the specifications for brick masonry as outlined in Chapter 4. For example, Type S mortar would be used for most load-bearing walls, particularly those exposed to severe weathering, and below grade structures exposed to earth. Type M mortar would be used in such structures as pilasters, columns, and piers which support heavier than normal loads. Types N and O might be specified for interior non load-bearing partitions.

Laying a Concrete Block Wall

Having learned what the common types of concrete masonry units are, and how to mix mortar properly, the next step is to learn how to correctly handle these materials in actual construction. The remainder of this section will describe the practical application of basic concrete masonry skills by showing how a simple concrete block wall is planned and erected.

Block Planning. The most economical concrete masonry walls are constructed using two standard block, namely stretcher and corner blocks, shown at (A) and (B) of Fig. 6-3. Good construction demands either that joints be staggered or that adequate reinforcement be used. Good appearance demands that the block be used uniformly in all courses. To comply with these demands, the designer of concrete masonry walls must carefully consider the lengths of the blocks plus mortar joints in deciding on the lengths of walls and the location and size of window and door openings. Heights of walls and heights of window and door openings also must be considered in connection with the height, plus mortar joint, of the block.

Concrete block are made with all dimensions $3/8''$ less than their nominal size. A block whose actual measurements are $7 5/8'' \times 7 5/8'' \times 15 5/8''$ may be considered for planning purposes as $8'' \times 8'' \times 16''$. This means that in planning concrete masonry walls, the mortar joints *and* the masonry units themselves may be figured as a single dimensional size.

In planning a building, architects and designers select some basic grid dimension as a basic planning size. This grid may be of any size and value, but it is generally some multiple of 4 inches. By using such a system, called modular coordination, the architect can design his building to afford a minimum of waste material through a minimum amount of cutting. Planning in this way contributes considerably to the reduction of construction costs.

Concrete block structures should generally be laid out on an 8″ grid or some multiple of 8 inches. In planning this way, wall lengths, window and door openings, and wall heights will automatically conform to the standard sizes of concrete masonry units.

It is always wise for the mason to check dimensions given for a building before construction has actually begun. Suppose, for example, that in planning a small building whose walls are to be constructed of 8″ x 8″ x 16″ concrete block, it is desirable from the architectural standpoint to have one of the walls exactly 12′0″ long. Before this length can be definitely decided upon, the lengths of the block plus mortar joints must be considered.

Each ordinary stretcher block is *actually* 15⅝″ long. Adding a ⅜″ mortar joint to a stretcher makes it exactly 16″ long. Corner block are 7⅝″ wide.

One course of the wall in question is shown in Fig. 6-21, top. It can be seen that two corner blocks and eight stretchers fit nicely into the 12′-0″ wall length. This was determined in advance as follows:

The 12′-0″ dimension equals 144 inches. There must be a corner block at each corner. The two corner blocks, with their mortar joints, are 8″ wide and make a total of 16″ which subtracted from 144″ leaves a remainder of 128 inches. Dividing 128″ by 16″ gives a quotient of exactly 8. In other words, 8 stretcher blocks, allowing ⅜″ joints, fit between the two corner blocks, as shown in Fig. 6-21, top. However, these calculations count the mortar joint at *A* twice, which means that the wall will actually be ⅜″ short of the 12′-0″ dimension. This is not serious, for each of the other joints can be made a trifle larger to take up the slack.

To check the block layout shown in Fig. 6-21, top, note the smaller dimensions shown. Adding these will equal 11′-11⅝″ or 143⅝ inches. There are eight stretchers which, with their ⅜″ mortar joints, equal 16″ each. Eight times 16″ is 128 inches. The corner block at *C* with its joint equals 8″ which, added to the 7⅝″ corner block at *B*, gives 15⅝ inches. Adding 15⅝″ and 128″ gives a total of 143⅝″, which is just

Concrete Block Masonry: Units and Building

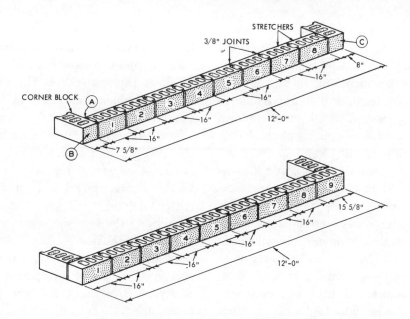

Fig. 6-21. First and alternate courses of a well-planned block wall.

⅜" less than the 12'-0" dimension of the wall. The missing ⅜" is absorbed by the other nine joints, as was pointed out in the last paragraph.

The next regular course above the course shown in Fig. 6-21, top, would have the corner blocks laid in the opposite direction, as indicated in Fig. 6-21, bottom. In this course, there are 9 stretchers (including the corner blocks, since they are in line with the regular stretchers) and 8 joints. There is one less joint than in the under course, top. Therefore, one stretcher (No. 9) takes up only 15⅝" in the wall. The length can be checked by multiplying 16" by 8 and adding 15⅝", which gives a total of 143⅝ inches. In this course again, each of the 8 joints must have a little added to it to fill out the required length of 12'0".

Note: Instead of dividing reductions or increases among all joints, masons generally reduce or increase one or two joints. This is much easier and quicker, and where only small reductions or increases are involved, is hardly noticeable.

Following this procedure, the length dimensions for the walls of each job should be carefully considered, no matter what shapes and sizes of block are to be used. The heights of block walls should be checked with equal exactness. If wall dimensions do not conform to the modular sizes of the block, an attempt should be made to change the wall dimensions.

In some cases it is impossible to change wall dimensions to conform with modular planning of the block. In this case, cut block must be used. Modern masonry saws ease the operation of cutting concrete masonry units, reducing the time necessary to cut block while increasing accuracy. One method of amending walls made unsightly by cut blocks would be to tool the joints flush and then paint.

Layout. Before any concrete block is actually laid in a wall, it is always wise to check the designer's dimensions and to determine an accurate block layout. To achieve this purpose, one course of blocks should be set around the footing without the use of mortar (Fig. 6-22). This allows you to determine if any cutting of blocks will be necessary. It also allows you to find if adequate allowance has been made for mortar joints. A chalked snap-line is sometimes used to mark the position of blocks in the first course. This helps in aligning blocks correctly as they are laid.

Once the block layout has been checked, the first course of the wall is laid. As shown in Fig. 6-23, a full mortar bed is spread and furrowed with a trowel to insure complete bedding of the face shells *and* webs of blocks in the first course. The corner block is laid first, and great care should be taken in positioning it correctly, as this block will act as a guide for the entire corner (Fig. 6-24).

All block should be laid with the thicker end of the face shell up, as this provides a larger mortar bedding area. Only the ends of the face shells are buttered for vertical joints, as shown in Fig. 6-25. To speed laying of the block, you can apply mortar to the vertical face shells of three

Fig. 6-22. Checking the layout.

Fig. 6-23. A full mortar bed.

Concrete Block Masonry: Units and Building

Fig. 6-24. One block-guide to entire corner.

Fig. 6-26. Aligning block in the first course.

Fig. 6-25. Buttering face shell for vertical joint.

Fig. 6-27. Bringing block to grade.

or four blocks in one operation. Each block is then lowered into its final position and pushed downward into the mortar bed, producing well-filled vertical joints.

Another method is to first lay the block in the wall. Then apply the mortar to the face shell. Some masons prefer this method.

After the first few block have been laid, the mason's level is used as a straightedge to assure correct alignment of the block, as shown in Fig. 6-26. The block are brought to proper grade by use of the level, as shown in Fig. 6-27. Position block by light taps with the trowel. Some masons use the edge of the trowel for tapping so as to reduce the amount of mortar scattered around, while others prefer to use the trowel handle. Blocks are plumbed as shown in Fig. 6-28. Take great care in aligning, leveling, and plumbing the first course as it is essential in building a straight, true, wall in a minimum time.

Succeeding courses in the wall are usually laid with what is called "face-shell bedding". This means

Masonry Simplified

Fig. 6-28. Plumbing block.

five or six courses higher than the center of the wall to act as guides in laying the rest of the block. As each course is laid at the corner, it is carefully checked with a level to make sure it is level, plumb, and in alignment. Each corner block is likewise checked with a straightedge to make certain that the faces of the block are in the same plane, as shown in Fig. 6-30. This will further insure a straight, true wall.

that mortar is applied only to the horizontal face shells of the block, as shown in Fig. 6-29. Mortar for vertical joints can be placed on the vertical face shells of either the block to be placed, or the block previously laid. Some masons butter the vertical face shells of both the previously laid block and the block to be laid. Well-filled vertical and end joints are essential as moisture has a tendency to seep through. After the first course is laid the corners are built up

Fig. 6-30. Aligning block faces.

Fig. 6-29. Face shell bedding.

A story- or course-pole, which is a board (usually 1" x 2") with markings 8" apart, is used to find the top of the masonry for each course, Fig. 6-31. In building corners, each course is stepped back half a block. This is sometimes called *racking the*

Concrete Block Masonry: Units and Building

Fig. 6-31. Use of the story pole.

lead. Horizontal spacing of the block may be checked by placing a level diagonally across the corners of the block, as shown in Fig. 6-32.

Fig. 6-32. Checking horizontal spacing.

This is called "straightedging the rack". Each block must touch, and be completely flush with the level if the wall is to be plumb and true and in line with the corners. This step must be repeated often to prevent bulges or depressions in the wall and to keep the courses in line with the corners.

In laying block for the wall between corners, a mason's line is stretched from corner to corner for each course. The outside top of each block is then laid to this line. The manner in which you handle the block is important. Only practice can determine the most practical way for each individual to handle block. If you tip the block you are laying slightly toward yourself, you can see the upper edge of the course below. This will allow you to place the lower edge of the block you are handling directly over the course below, as shown in Fig. 6-33.

All final adjustments to a block must be made while the mortar is soft and plastic. Once the mortar has stiffened, shifting the block will break the mortar bond and cause cracks between the block and the mortar. If the block is disturbed after the mortar has begun to set, the block must be removed, stripped of all the old mortar, and relaid in fresh mortar.

As shown in Fig. 6-34, each block is leveled and aligned by tapping lightly with the trowel handle. In laying block between corners, the

Masonry Simplified

Fig. 6-33. Handling the block.

Fig. 6-34. Aligning the block by tapping.

level is used only to check the face of each block to keep it lined up with the face of the wall.

Mortar should not be spread too far ahead of the actual laying, as it will stiffen and lose its plasticity. This will decrease its bonding power and weaken the structure of the entire wall. Excess mortar squeezed from between the block as it is laid is cut off with a trowel (Fig. 6-35). This excess mortar is usually thrown back on the mortar board to be reworked into fresh mortar. Some masons apply the excess mortar to the vertical face shells of the block just laid. Any mortar which begins to stiffen after it has been spread should be removed and reworked on the mortar board. Mortar applied to the vertical joints of the block just laid and to the block being set insures well-filled joints (Fig. 6-36). Dead mortar picked up from the scaffold or floor should never be used. Full mortar bedding may be required in some localities, and should always be used where the greatest possible structural strength is necessary, as in pilasters, piers, and columns. This requires mortar on the cross webs of the block as well as on the face shells.

The installation of the closure block requires great care. The closure block should be laid in the opening without mortar to determine if there is adequate allowance for the mortar joints. If the block must be cut, care should be taken that the cut is accurate, since oversized joints are just as objectionable as joints which are too thin. All four vertical edges of the closure block and all edges of the opening should be buttered with mortar. The closure block should be lowered carefully into place (Fig. 6-37). If any of the mor-

Concrete Block Masonry: Units and Building

Fig. 6-35. Cutting off excess mortar.

Fig. 6-37. Installing the closure block.

Fig. 6-36. Well-filled vertical joints.

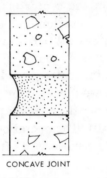 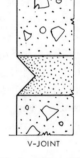

Fig. 6-38. Concave and V-shaped mortar joints.

tar falls out, leaving an open joint, the closure block should be removed, fresh mortar applied, and the operation repeated.

Well tooled joints will produce weathertight walls of neat appearance. Joints should be tooled after the mortar has stiffened, but before it has hardened. The length of time which it takes to set will be less in warm weather than in cold. Learn to check the mortar with your finger tips to determine when it is ready to strike. Tooling of the joints compacts the mortar and forces it tightly against the masonry on each side of the joint. Unless otherwise specified, all joints in concrete masonry walls should be tooled either concave or V-shaped. See Fig. 6-38.

The jointer, sometimes called a sled-runner, for tooling horizontal joints should be at least 22" long, preferably longer, and upturned at one end to prevent gouging of the mortar. A handle should be located

147

Masonry Simplified

Fig. 6-39. Tool used for horizontal joints.

Fig. 6-40. Tool used for vertical joints.

approximately in the center of the tool for ease in handling. A tool made from a ⅝" round bar is satisfactory for concave joints (Fig. (6-39). Horizontal joints should be tooled first, followed by striking the vertical joints with an S-shaped jointer (Fig. 6-40). The operation is then repeated to remove burrs from the joint. In some regions, the joints are *trowel struck*. In this case, vertical are struck first and each joint is struck only once.

Any mortar burrs on the wall should be trimmed off after the joints have been tooled. This may be done with a trowel or by rubbing with a burlap bag.

Construction Details

After learning and practicing the basic skills necessary for laying a concrete block wall that is plumb and level with tight solid mortar joints, the next step is to learn the more complex construction details the mason will commonly encounter. The following section of this chapter will show in detail such structures as combination walls, various supporting structures, door and window openings, etc. A thorough understanding of these details is essential to good masonry.

Wall Details

The basic wall construction previously described will many times have modifications or additions in order to accommodate heavier loads or additional structures. The following paragraphs show details of these constructions.

Brick and Block Walls. Fig. 6-41 shows the use of concrete blocks as backing for a brick wall. Note that at intervals, a header course of brick is employed to tie the tiers of brick and block together. The block used

Concrete Block Masonry: Units and Building

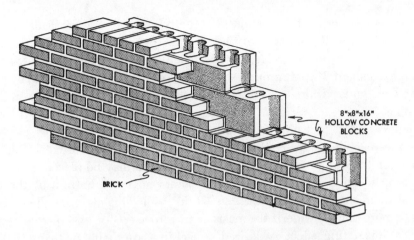

Fig. 6-41. Masonry wall having concrete block backing.

in these courses are specially molded units with a notch provided to receive the brick headers. These units are called "full cut header block".

Cavity Walls. A cavity wall is a form of masonry wall consisting of two parallel wythes of masonry separated by a continuous air space, usually about 2" wide. One of the advantages of cavity wall construction is that the air space acts to prevent rain or moisture that has seeped through the outer wythe from penetrating the inner wythe. If moisture tends to collect in the cavity between the walls, flashing and weep holes are usually located at the base of the outer wall. The cavity also interrupts the continuity of the masonry and provides the additional insulation effect of an air space. An improvement of over 25 percent in insulating value was found for unventilated cavity walls compared with solid walls of the same material.

Cavity walls when properly built will withstand reasonable impact and the usual floor and roof loading of a two-story building. The fire resistance of cavity walls is not much different from walls with the same quantity of solid materials except for the ability to withstand loads. Cavity walls which are loaded toward the inner wythe have the greatest stability during exposure to fire. The advantages which cavity walls display may be lost through improper design, particularly with respect to flashing, openings, ties, and wall intersections.

The two wythes of a cavity wall should be securely tied together with non-corroding ties. For each 3 sq. ft. of wall surface, a rectangular tie of No. 6 gage wire should be used. The ties are embedded in the horizontal joints of both walls. Additional ties are required at all openings, with ties spaced about 3' apart around the perimeter and within 12" of the opening.

Generally, 10" cavity walls should not exceed 25' in height. Neither the inner nor the outer walls should be less than 4" thick, and the space between them should not be less than 2" nor more than 3" wide, if the most efficient insulating effect is desired. Usually the outer wall is 4" thick and the remaining wall thickness is made up by the air space and the inner wall.

If concrete masonry cavity walls are properly designed and built, with well tooled mortar joints and a covering of paint or stucco, the walls should be weathertight. This means that in general there should be no need to include special flashing and weep holes. However, in limited areas subjected to driving rains, or where experience has shown that water collects in the walls, extra flashing and weep holes will be necessary.

Under severe moisture conditions, the heads of windows, doors, and other wall openings, and the bottom course of masonry, are flashed so that moisture entering the wall cavity will be directed toward the outside wall. Only rust-resisting metal or other tested and approved materials should be used for flashing. Weep holes should be placed 2 or 3 units apart in the vertical joints of the bottom course. Weep holes should never be placed below grade.

Weep holes can be formed by placing well-oiled rods in the mortar joint and extracting it after the mortar has hardened; or short pieces of hemp rope, sash cord or fiberglass insulation may be placed in the end joints near the bottom of the outside wall. This material allows moisture to seep out and insects cannot get in. Care should be taken that the inside openings of weep holes are not clogged by mortar droppings. Place a 1" x 2" board across a level of wall ties to catch the droppings. As the masonry reaches the next level for placing ties, the board is raised, cleaned, and laid on the ties placed at this level.

Fig. 6-42 shows the basic arrangement of a 10" cavity wall. As shown in the illustration, the outer wythe is constructed of partition block which is nominally 4" thick. The inner wythe is constructed of the same size block. To increase the size of a cavity wall, the inner wythe is generally thickened. Thus, for a 12" cavity wall, the outer wythe would still be made of 4" nominal block, but the inner wythe would employ 6" block. One type of wall tie is shown in the illustration, and the spacing recommended is indicated.

Horizontal Reinforcement. Reinforcement added to the horizontal mortar joints of concrete masonry

Concrete Block Masonry: Units and Building

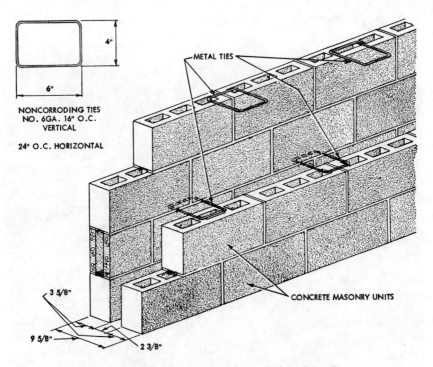

Fig. 6-42. Basic arrangement of a ten inch cavity wall.

walls develops appreciably higher bond stress in the wall and aids in reducing cracks in the masonry units due to shrinkage. Concrete masonry units tend to expand when dampened and to contract when dried. While this does not affect independent units which are not part of a wall but are laid into a wall, such shrinkage or expansion tends to create cracks in the units. This shrinkage may be reduced either by the use of control joints or horizontal reinforcement. One method of horizontal reinforcement is to use two $\frac{1}{4}''$ round steel reinforcing bars, one placed in each face shell joint. These bars should be placed in alternate courses above and below window openings. Bars should also be carried over door openings and should be bent around corners. Bars should be lapped a minimum of 10 inches. See Fig. 6-43.

Horizontal reinforcement may also be achieved by the use of prefabricated welded cross wires. Such reinforcement can be used in either single wythe walls or to tie brick faced or cavity walls together. The cross wires in this type of reinforcement act to replace the more usual

Masonry Simplified

Fig. 6-43. Steel bars used as horizontal reinforcement.

wall ties necessary for brick faced or cavity walls. Prefabricated reinforcement may be purchased with cross wires which are crimped into a drip. This drip provides a barrier to any moisture which collects on the cross wire, preventing it from entering the inner wythe of a cavity wall. Horizontal reinforcement increases the ability of a concrete masonry wall to withstand side pressure caused by water, soil, or wind pressure considerably.

The degree to which horizontal reinforcement is used is determined by the qualities desired of a concrete masonry wall. Used in every third course and above all window and door openings, horizontal reinforcement aids materially in developing lateral strength in a wall, while largely reducing cracking due to shrinkage. Horizontal reinforcement used in every course of a concrete masonry wall reduces shrinkage cracking to a minimum while developing walls with great lateral strength. Fig. 6-44 shows one type of modern, prefabricated horizontal tie.

Concrete Block Masonry: Units and Building

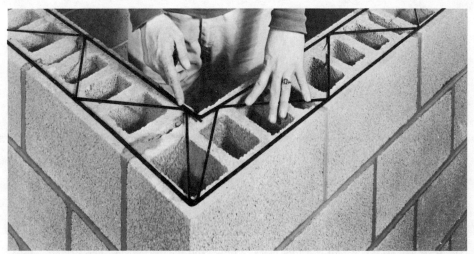

Fig. 6-44. Horizontal joint reinforcement provides added lateral strength for a concrete block wall. (Dur-O-Wal Products, Inc.)

Steel Bar Reinforcing. In areas where concrete block walls are subjected to greater than normal lateral stresses such as earthquakes or high winds, the walls may be reinforced with deformed steel bars. The bars are placed in the footing or foundation at regular intervals so as to run vertically through the head joints and block coring in each course. The cores or joints are filled with mortar or grout as each course is laid up. This is called "rodding the core". This type of reinforcing is also employed at corners and at door and window openings. See Figs. 6-45 and 6-47.

Steel bars are used in the same manner and for the same reason in such horizontal structures as bond

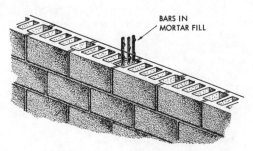

Fig. 6-45. Steel bars used for vertical reinforcing.

153

Masonry Simplified

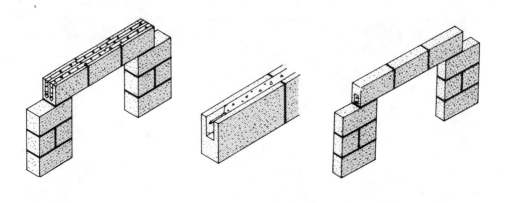

LINTEL USING BOND LINTEL BLOCK--MAY BE ASSEMBLED IN PLACE ON SHORING

ASSEMBLY OF LINTEL USING ANY CUT STRETCHER UNIT-- NOT ASSEMBLED IN PLACE

PLACEMENT OF THE ASSEMBLED LINTEL AFTER INVERTING FROM POSITION SHOWN AT LEFT

Fig. 6-46. Lintels made with concrete block, mortar, and reinforcing steel bars. (Illinois Brick Co.)

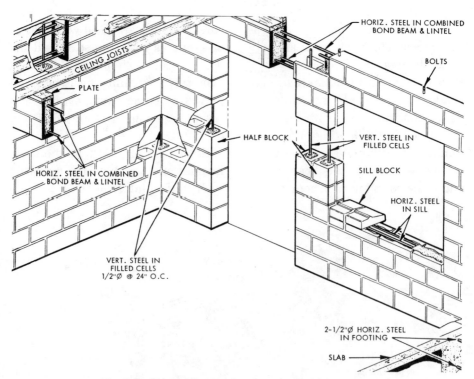

Fig. 6-47. Steel reinforcing may be used in several ways.

154

Concrete Block Masonry: Units and Building

beams and lintels. See Figs. 6-46 and 6-47. Note: A bond beam is usually the top course of each story height.

Fig. 6-47 illustrates many ways in which steel rods are used as reinforcement in concrete block construction.

Note: In Fig. 6-47 that the diameter of the steel reinforcing bars is given as the actual diameter in inches and fractions. This is only to give examples of typical sizes of bars used for specific purposes. On regular working drawings, specifications, and shipping tags, the size is given as a number corresponding to the diameter of the bar in eighths of an inch. For example, a one-half inch bar would be listed as #4, a one-inch bar as #8, etc.

Fig. 6-48 shows the use of a precast solid concrete lintel in combination with a steel angle in a brick-faced wall with concrete block backing.

Joist Support. Joists usually support a rather heavy floor load, and careful provisions must be made to support their ends in the concrete block walls. There are several ways in which such support may be secured. Fig. 6-49 shows several typical methods, all of which are acceptable.

Fig. 6-49(A) illustrates one of the simplest methods. Here the ends of regular stretcher blocks are cut to allow room for the joist ends. All joists must be beveled as shown at (E). In the event of a fire or accident which results in a collapse of the

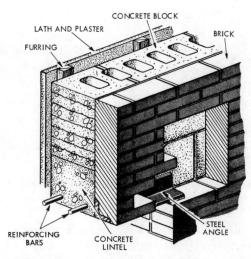

Fig. 6-48. Solid reinforced concrete lintel in a brick and block wall.

155

Masonry Simplified

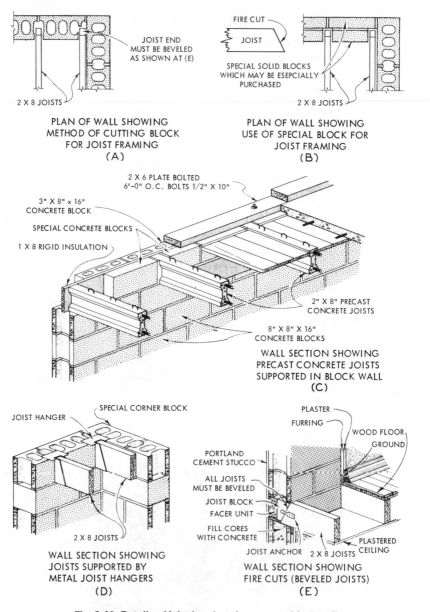

Fig. 6-49. Details of joist bearings in concrete block walls.

floor, beveled joists will fall free from their positions in the wall. Joists with square ends will act as levers, prying their way free and destroying the wall. The course of blocks directly under the joists

Concrete Block Masonry: Units and Building

should be solid or have all cores filled with concrete. The bearing of the joists should be a minimum of 3 inches.

In Fig. 6-49(B), the use of special blocks for joist framing is shown. All other conditions explained for (A) apply here.

At Fig. 6-49(C), the method of supporting precast concrete joists in a block wall is illustrated. Note that special size blocks are required.

In Fig. 6-49(D), the use of metal joist hangers is illustrated. Here the joists need not be beveled. The hangers, of which several kinds are available, are set into the blocks at their vertical joints.

Fig. 6-49(E) is a more complicated method of support. This method is of particular use for residential construction.

It will be noted that all joists shown in Fig. 6-49 are 2" x 8" in size. This was done in order to simplify the presentation of construction details in floor framing. Normally, 2" x 10" joists are necessary, and when framing around such joists, special blocks, as shown at Fig. 6-49(B), are required.

Pilasters and Columns. Sometimes the loads on beams are so great that a large bearing surface and pilaster or pier are required. In such cases, pilasters or piers as shown in Fig. 6-50 are necessary. Pilasters can be made of standard stretchers and corner blocks as part of regular walls, or specially sized units may be used. Piers are constructed with double corner or pier blocks. All cores in the pilaster or pier blocks should be filled with concrete for maximum strength under heavy loads.

Columns built of concrete block can be made using two or three blocks in each course, as shown in the pier and pilaster construction in Fig. 6-50. In some parts of the country, circular blocks having a 12" diameter can be purchased to use in column building.

Fig. 6-51 shows details of a corner pilaster and a typical mid-wall pilaster.

Columns are free-standing units constructed in the same manner as pilasters and piers. They are used mainly as center supports for joists, beams, etc.

Beam Supports. In most residences and other larger buildings, there are one or more beams which have to be supported at one end by the walls or foundations. Such beams generally carry heavy loads and therefore must be carefully supported.

In Fig. 6-52, left, a typical case where a block wall supports one end of a beam is shown. The beam should have at least 3" of bearing. This is important where heavy loads are involved in order to avoid the shearing off of the edges of blocks. All blocks under the beam (see blocks

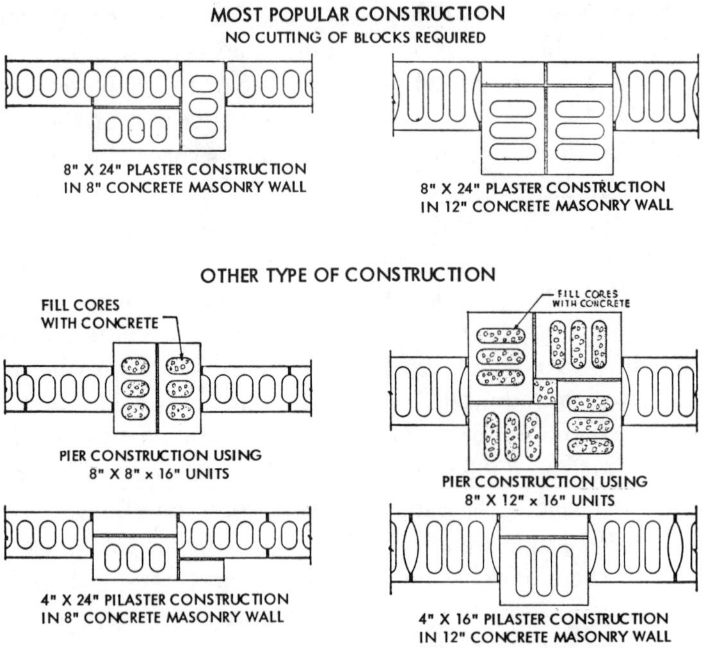

MOST POPULAR CONSTRUCTION
NO CUTTING OF BLOCKS REQUIRED
8" X 24" PLASTER CONSTRUCTION IN 8" CONCRETE MASONRY WALL
8" X 24" PLASTER CONSTRUCTION IN 12" CONCRETE MASONRY WALL
OTHER TYPE OF CONSTRUCTION
FILL CORES WITH CONCRETE
FILL CORES WITH CONCRETE
PIER CONSTRUCTION USING 8" X 8" x 16" UNITS
PIER CONSTRUCTION USING 8" X 12" x 16" UNITS
4" X 24" PILASTER CONSTRUCTION IN 8" CONCRETE MASONRY WALL
4" X 16" PILASTER CONSTRUCTION IN 12" CONCRETE MASONRY WALL

Fig. 6-50. Pilasters and piers used for added strength. (National Concrete Masonry Association.)

FILL ALL CORES OF PILASTER WITH CONCRETE
CORNER
STRETCHER CONCRETE BLOCK
DOUBLE CORNER (OR PIER) BLOCK
CORNER BLOCKS
1/2 STRETCHER
CONCRETE FLOOR
FILL
FOOTING

Fig. 6-51. Corner and mid-wall pilasters.

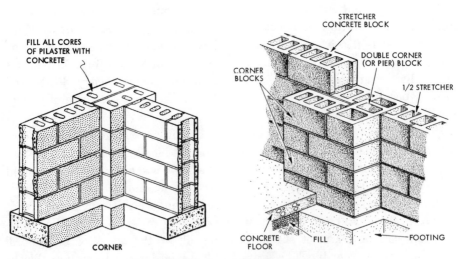

Concrete Block Masonry: Units and Building

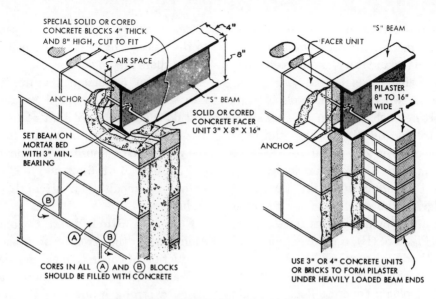

Fig. 6-52. Typical method of supporting beams in concrete block walls.

A and *B* in the elevation view) should have their cores filled with concrete. This practice strengthens each block and makes a practically solid concrete column capable of supporting great loads. Anchor pins should be placed at least 4″ from face of wall.

The top view of the section at Fig. 6-52, left, shows how facer units and special solid or cored blocks are filled in around the beam to make the wall solid. The special blocks should be placed snugly against the beam web to hold the beam in place, although it is desirable to leave a small amount of play to allow for expansion of the beam due to temperature changes.

Sometimes beams carry such heavy loads that the 8″ block wall is not sufficiently strong by itself to safely support the beam ends. In such cases, larger blocks can be used. Should this be inadequate, the wall or foundation can be given considerable added strength by pilasters, as shown in Fig. 6-52, right. Such pilasters usually are constructed of concrete blocks, but bricks or small concrete units also may be used. The pilasters need not be more than 12″ to 24″ wide and should extend from the beam down to the footing.

159

Concrete Floor Support. In buildings where fireproof construction is mandatory, reinforced concrete floors are necessary. Fig. 6-53 shows two typical methods of supporting such floors in block walls.

In Fig. 6-53(A), where the floor thickness is not equal to 8", facer units and concrete bricks can be used together with the thickened edge of the concrete floor to build up and provide support for the wall. In Fig. 6-53(B), a more complicated method is required because the thickness of the precast concrete joists plus a concrete floor amounts to more than 8 inches.

Note in both (A) and (B) that an air space is provided to allow for expansion or contraction of the floor due to changes in temperature. This space can be filled with insulation to help make the wall more resistant to the passage of heat at the points where there are no cores.

The most important features to keep in mind with regard to joist and floor supports are that at least 3" of bearing are required, and that the walls must be built up, as shown in Figs. 6-49, 6-51, 6-52, and 6-53 so as not to impair their strength or stability.

Concrete Block Joist Floors. Along with the greatly increased production of concrete block for use in walls and partitions, there has arisen a strong interest in the use of hollow concrete units as fillers in the construction of one-way ribbed slab floors. Concrete filler blocks are made of lightweight aggregate and have hollow spaces which total 40 to 60 percent of the gross volume of

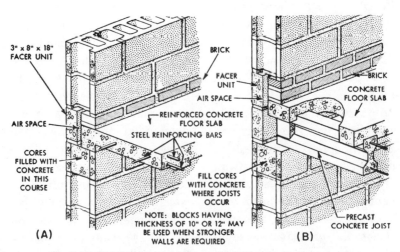

Fig. 6-53. Details of support for concrete floor in a masonry wall.

the unit itself. This effects a marked reduction in the dead load of floors constructed of concrete filler block, as compared with a solid slab of equal load carrying capacity.

This type of construction can be used in both floor and roof finishing, and combines precast concrete block with the cast-in-place concrete slab and joists. The concrete blocks provide both the forms for the joists of the floors and a flat ceiling which can be painted or plastered direct. In the example shown in Fig. 6-54, where filler block units $15\frac{5}{8}''$ long are used, the design for this type of floor system requires a uniform spacing of joists 21" on centers. The joists, then, are actually $5\frac{3}{8}''$ wide. The joist depth is the sum of the thicknesses of the block filler and the concrete slab. While the dead load of such a floor is less than a slab floor of equal load carrying capacity, the total depth of the floor is increased.

This method of floor construction is simple to build and requires no

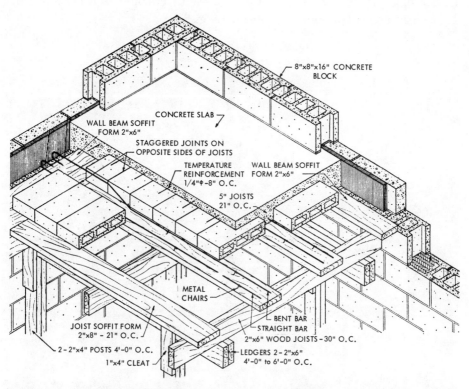

Fig. 6-54. Basic method used in laying a concrete filler block floor. (Portland Cement Association.)

Masonry Simplified

special construction methods. In addition to cutting down dead load and providing ceiling surfaces which may be plastered directly, there is a marked improvement in the heat insulating qualities of the slab. A minimum of form lumber is required in construction, and the finished floor has excellent fire resistance.

Fig. 6-54 illustrates a basic method used in laying concrete filler block floors. The filler block may be of the size and shape shown in (L) of Fig. 6-3, or they may be the soffit block type shown at (M) of Fig. 6-3. Either block is acceptable but the soffit block presents a more pleasing appearance on the finished ceiling surface. Fig. 6-54 is not intended as a final working drawing, and it is recommended that each job be designed and built under the supervision of a competent architect or structural engineer.

Control Joints. Cracks may occur in masonry walls due to unusual stresses, and increasing use is being made of expansion or control joints to control this problem. The joints are built into the walls at the most favorable location in such a manner as to permit slight wall movement without cracking the masonry. See Fig. 6-55.

Fig. 6-55. Control joints are vertically continuous.

Concrete Block Masonry: Units and Building

The spacing and location of these joints depends upon several factors, among them: the length of the wall, architectural details, and especially on the experience records as to the need for control joints in the locality in which the wall is to be erected. Control joints may be placed at junctions of bearing and nonbearing walls, at junctions of walls and columns or pilasters, and in walls weakened by chases and openings. Joints are ordinarily spaced at 20- to 25-ft. intervals in long walls, depending again on local experience and judgment. Control joints should also be used in junctions of walls in L, T, and U-shaped buildings.

Control joints must have a certain amount of "give" in order to relieve the horizontal stresses resulting from moisture and temperature movement. For this reason mortar is not used in the joints. In order to maintain watertightness, the open joints are partly filled from the outside with an elastic calking compound which is usually applied with a calking gun. The two main types of control joints are built-in and manufactured.

Built-in control joints may be made using either some of the common regular types of units or especially designed units made specifically for control joints. Fig. 6-56 shows a simple control joint using regular stretcher units of full and half lengths. Fig. 6-57 shows a similar joint employing jamb blocks.

While the purpose of control joints is to provide flexibility for horizontal stress, lateral stability must be maintained. In such joints as shown in Fig. 6-56 and 6-57, this is provided by the use of Z shaped metal tiebars.

Fig. 6-58 shows the use of specially designed control joint blocks. Lateral strength in this joint is provided by the tongue and groove design.

Fig. 6-56. Simple control joint using "Z" tiebars.

Masonry Simplified

Fig. 6-57. Control joint using jamb blocks.

Fig. 6-59. Control joint using building paper to break bond.

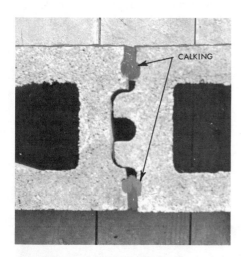

Fig. 6-58. Special control joint blocks.

calked. Roofing felt may be used in place of paper for this type of joint.

Fig. 6-60 shows a modern manufactured control joint. The unit shown in the illustration is standardized and readily available under several brand names. It is made of rubber or other synthetic resilient materials. Its dimensions are such as to fit into standard block units pre-notched for metal sash windows.

Fig. 6-59 shows a control joint constructed with building paper inserted in the end core of a regular stretcher block. The core is filled with mortar for lateral strength. Again, the outsides of the joint are

Fig. 6-60. Manufactured control joint.

Concrete Block Masonry: Units and Building

The advantages of this type of control joint are that it is quickly and easily installed and that the unit itself provides both horizontal flexibility and lateral stability.

Concrete Block Wall Intersections. One method of building the intersection when two concrete bearing walls meet at right angles is shown in Fig. 6-61(A). This method

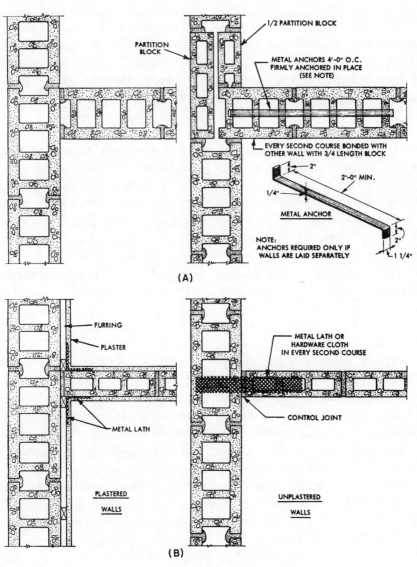

Fig. 6-61. Concrete block wall intersections.

165

makes no alteration in the exterior appearance of the wall because the partition block used in the alternate courses has the same face dimensions as standard stretcher blocks. The alternate courses shown indicate the method of obtaining a good bond at the intersection. Anchors are required only if the walls are laid up separately.

Proper bonding at the intersection of a bearing and nonbearing wall is much more easily accomplished. As may be seen in Fig. 6-61(B), plastered walls of this type need only metal lath fastened to both wall surfaces. In the case of unplastered walls, metal lath or hardware cloth is inserted in every second course. A control joint is then made at the point of intersection.

Note: Intersecting load-bearing walls should not be tied together in masonry bond except at corners. One wall should terminate at the face of the other wall with a control joint at that point. For lateral support, bearing walls are tied together mechanically as shown in Fig. 6-61(A).

Window Opening Details. The location and size of windows in concrete block walls require careful consideration in order to avoid the cutting of block. Note the window opening shown in Fig. 6-62. It can be seen that the opening width C, the height B, and the height of the bottom of the opening above the foundation, A, are all in terms of an exact number of blocks horizontally and vertically. This is the ideal situation and should be carefully planned for any window opening in a block wall.

To accomplish ideal window planning in block walls, the size of windows must be selected to accommodate the blocks both horizontally and vertically with accommodations made for masonry sills. Any deviation from this rule results in unsightly walls, occasionally poor bond between various blocks, and always increased labor costs. The location and size of doors in concrete block walls require the same considerations necessary for windows.

The masonry work on concrete block walls should never be started until the mason doing the work has carefully checked all of these items. The time thus spent will be of much value and will save a great deal of extra labor, expense, and disappointment.

There are many varieties of windows from the standpoint of the size, shape, and assembly of their various parts, such as casings, stools, aprons, stops, etc. However, all varieties are nearly enough alike that a few typical examples will amply illustrate the general details.

As far as the mason is concerned, when laying a concrete block wall, a window has three principal parts, namely, the *head, jambs,* and *sill.* The head, as the name implies, is the

Concrete Block Masonry: Units and Building

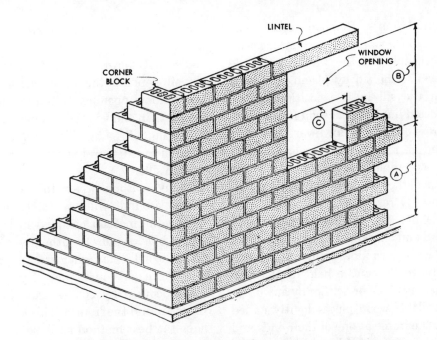

Fig. 6-62. Locations of window openings must be planned.

horizontal top of the window, the jambs are the vertical sides, and the sill is the horizontal bottom.

Although carpenters are usually responsible for installing window frames in masonry walls, it is necessary for the mason to know the basic installation procedures in order to provide properly located and sized openings. The following is the procedure for setting the frame in place and laying blocks around it for a simple window in a regular 8 inch concrete block wall.

You must first determine from the plans commonly provided for construction work, the position of the window in terms of the number of courses under the sill and the number of block on either side of it. Then the wall should be laid up to and including the course under the sill. Thus wall AB would be laid up to and including the course marked A in Fig. 6-63.

Next apply mortar to the blocks which will be directly under the sill. In Fig. 6-63, these block are shown as C and D and the method of mortar application is shown in the mortar detail for the sill. The width and depth of the mortar should be the same as previously described. Then set the sill gently in place and press it down until the mortar joint is the same as for the other hori-

167

Masonry Simplified

zontal joints. Remove excess mortar and smooth the joint surfaces on both sides of the wall. Check the level of the sill by placing the level on the sill in the position of the dotted line, X. If the sill is not level, press or force it downward at either end as required. If leveling cannot be accomplished in this manner, remove the sill, apply more mortar, and try resetting.

When the sill has been properly laid and checked, the next step is to place the window frame on the sill. The frame can be held in place by pieces of 2 x 4 or 1 x 4 or by any other available wood pieces lightly nailed to the frame at one of their ends and supported or held in place at their other ends by flooring, stakes, etc. Carpenters generally put such frames in place for the masons. In any event, the frames must stand vertically plumb and square with the wall.

The sill shown in Fig. 6-63 is a slip sill which is the same length as the window opening. Sills that are longer than the window opening are called lug sills. Lug sills are laid in a full mortar bed in the same way as for slip sills.

Many masons prefer to lay the blocks around window frames following the method outlined in laying a block wall. In other words, after the sill is laid, they stretch the line for each course and lay the blocks course by course, including those around the frames. This is perhaps the best method because it assures all courses being laid in proper alignment. Regardless of the method used special care must be taken to keep all blocks level, plumb, and in alignment. Unless you keep all the

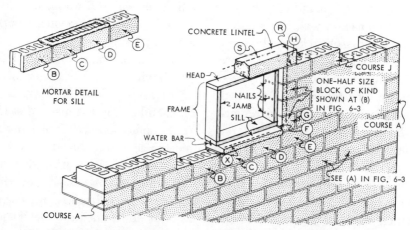

Fig. 6-63. Part of a concrete block wall window frame, sill, and lintel.

Concrete Block Masonry: Units and Building

blocks exactly one above the other you will not have a vertical wall.

Note the joint F between the sill and the block G in Fig. 6-63. This joint must be well filled in order to avoid a moisture leak.

Before block G is laid, the end of the sill should have mortar applied to it so that joint F contains mortar and is the same thickness as other vertical points.

When the blocks around the window frame have been laid, including the course at J on both sides of the frame, the concrete lintel can be placed. If the window opening is wide and the lintel, therefore, long and heavy, some time should be allowed for the mortar in the courses below to harden before the lintel is set.

To set the lintel, apply mortar to the blocks (see H) on either side of the frame. Then set the lintel gently down into proper position and adjust until the mortar joint is the same thickness as other horizontal joints. Special care should be taken to see that the lintel is perfectly level. This can be accomplished by setting the level in the positions of dotted lines R and S. Above the lintel, the blocks are laid as in the balance of the wall. The lintel should have at least an 8" bearing surface on each end as shown at H in Fig. 6-63.

Fig. 6-64 shows details of a finished window of the type that would be used in preceding construction.

Fig. 6-65 shows section views of the head, jamb, and sill for a common kind of a metal sash window. The head sections consist of steel (angle) lintels used in conjunction with precast concrete units plus nailing strips, the metal sash fittings, etc. The jamb section consists of a metal sash block plus the metal sash fittings, nailing strip, and plaster. The *slip sill* shown in the sill section is not *let in* to the wall like the lug sill. The metal fittings are anchored to the concrete sill by bolts set into the sill.

Openings for more complex windows, such as double hung, casement, sliding, etc., are planned and executed in the same manner as for the simple window. When the level is reached, the window unit is mounted, leveled and braced in place. In most modern construction, the window units are factory assembled to the designer's specified dimensions. Some windows have built-in sills and are mounted either directly on the top of the concrete block units or a wood nailing block may be placed on the block. Fig. 6-66 shows a typical double hung window in a concrete masonry wall.

Door Details. Fig. 6-67 shows part of a wall for which a door is indicated. The details for this door are quite simple. Fig. 6-67 shows the frame for that door in the partly laid block wall. The purpose of this illustration is to show the positions of

Masonry Simplified

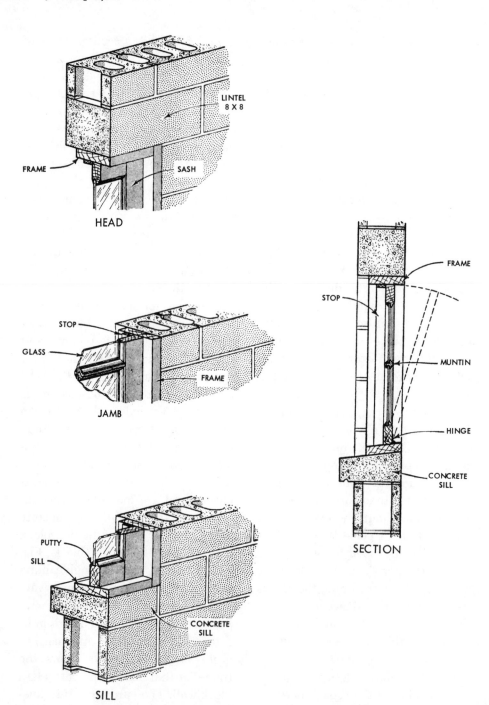

Fig. 6-64. Typical details for a simple window in a concrete block wall.

Concrete Block Masonry: Units and Building

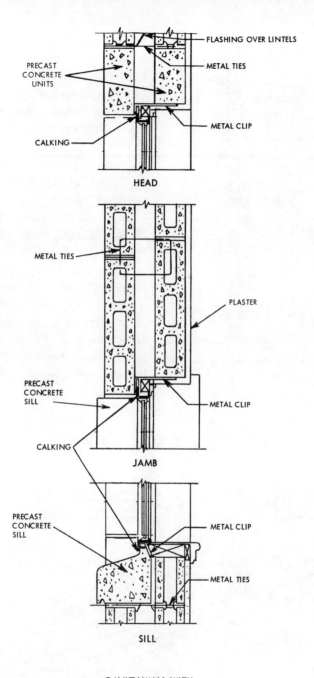

Fig. 6-65. Metal sash casement window in a concrete block wall.

Masonry Simplified

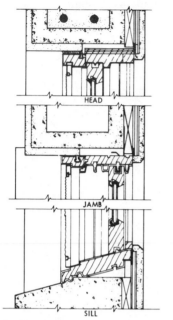

Fig. 6-66. Details of a double-hung window in a concrete block wall.

the frame for the jamb and head in a wall. The procedure for laying blocks around door frames is the same as that given for windows.

For the most part, door details are greatly similar to window details as far as concrete block walls are concerned. In most cases, the mason is guided by the doors and frames purchased for each job, in that the type of frame determines the type of lintels and jamb blocks to be used.

There are various types of door frames available, just as there are a variety of window frames. For residential or store and office construction, doors having details similar to those shown in Fig. 6-68 can be used to advantage. In the head

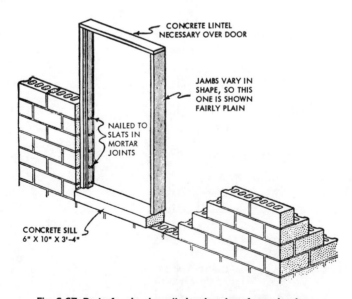

Fig. 6-67. Part of a simple wall showing door frame in place.

172

Concrete Block Masonry: Units and Building

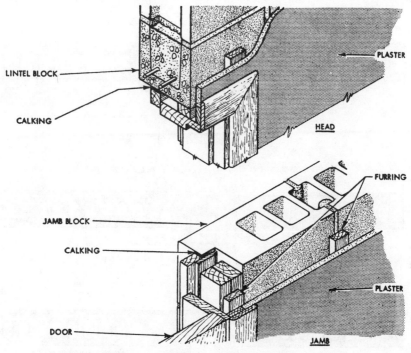

Fig. 6-68. Details for a finished door in residential construction.

section of this particular door it should be noted that lintel block has been employed. This type of lintel is constructed on-the-job by forming a channel with the lintel block. Reinforcing bars are then placed in the channel, and the channel is filled with concrete. Such lintels provide a surface texture which matches the rest of the concrete masonry wall. If the lintel is built in place, it must be firmly supported during construction. In other respects, construction details for the door in Fig. 6-68 are very similar to the details for windows. Jamb blocks are used for the vertical sides of the opening and a precast concrete sill is employed.

Laying Concrete Block Chimneys. When chimney blocks of the general kind shown at Fig. 6-69(A) are laid, the procedure is as follows:

First mark with chalk the exact position for the first course on the footing. The position can be ascertained from the blueprints of the building the chimney is to be in.

Apply mortar to the footing so that it will be under the areas marked X in Fig. 6-69 wide enough to cover the areas and about 1″

173

Masonry Simplified

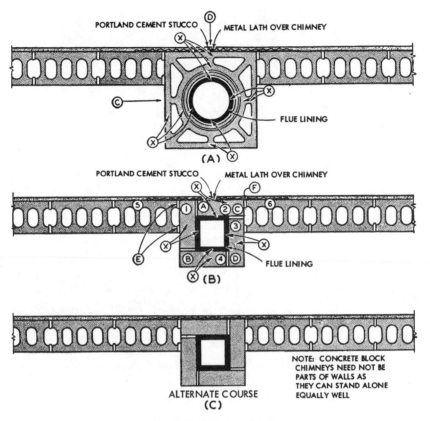

Fig. 6-69. Chimney details.

deep. Note that this includes the flue lining which is part of each block. Care must be taken to have mortar all the way around under the areas marked X. Place the first block and press it gently down into the mortar to make a joint of the same thickness as used for walls. Then remove excess mortar. Check the level of this block by placing the plumb rule across the block in the direction of the arrows at C and D. Make any adjustments necessary relative to leveling. This block must be *absolutely* level.

Apply mortar to the first block at all areas marked X all the way around the block. Then place the second block gently in position. Press it down to form proper thickness of joint, remove excess mortar and smooth the joint, and check its level, as for the first block. Also, place the plumb rule in a vertical position, up against the four sides of the chimney to make sure the sides

Concrete Block Masonry: Units and Building

are perfectly in line and vertical. The mortar joint for the flue lining must be smooth and should not protrude between the surface of the lining. Each such joint must be smoothed as each block is laid.

Place succeeding blocks in like manner, being sure to watch every detail explained for the first two blocks.

When chimney block of the kind shown at Fig. 6-69(B) are being laid, the procedure is as follows:

The exact position of the chimney should be marked on the footing. Then place mortar so that it is completely under all areas marked X. In other words, each block must have a full mortar bedding. Also place mortar at the ends of block 5 at the points marked E. Chimney block 1 should then be pressed into position, making sure the joints at E and the joint between the block and the footing are of the proper thickness. Make sure that the end of the block lines up with the face of block 5. Check the level of block 1. Next apply mortar to block 2 at the end where it touches block 1 (joint A) and on block 6 at the point F. Press this block into position as for block 1, making sure it lines up with the face of block 6. Place blocks 3 and 4 in the same manner, taking care to make joints C, D, and B carefully.

Most masons prefer to set the first length of flue lining in place before more courses of blocks are laid.

Blocks 1, 2, 3, and 4 must fit tightly against the flue lining as shown at Fig. 6-69(B).

For the second course of blocks, apply mortar all around, completely covering the first course. For this course, place the blocks as shown at (C). This is necessary to create good bond.

Place succeeding courses alternating them as explained. As each length of flue lining is added, a careful mortar joint must be made between lengths. Special care must be taken that all flue lining joints are smooth on the inside of the flues.

As the chimney construction progresses, frequent checks by the use of the plumb rule should be made to see that the chimney is perfectly vertical. All excess mortar should be removed and the joints smoothed.

Note in Fig. 6-69 that metal lath should be applied to the exterior wall surface at and on either side of the chimney if and before stucco is applied. The lath can be secured to the wall by the use of nails driven into the blocks if they are the lightweight kind or into the joints if heavyweight blocks are used. The metal lath tends to avoid wall cracks near the chimney at times when the chimney sways slightly due to wind, etc.

Cornice Details. Fig. 6-70 illustrates cornice details for two types of roof. Note that in both cases the top two blocks have their cores filled

Masonry Simplified

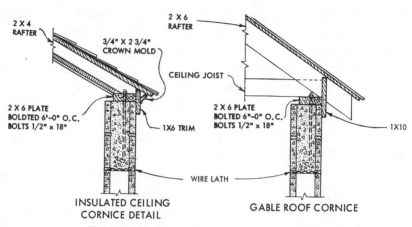

Fig. 6-70. Cornice details in concrete block walls.

with concrete. This practice tends to better distribute the weight coming from rafters and ceiling joists. Some masons make it a practice to fill only those cores where the anchor bolts occur. Either method is acceptable.

The cores are filled by laying wire or metal lath under the top two courses of blocks, as indicated. This serves to keep the concrete in place while it is drying. If metal lath is not available, wads of paper can be shoved into the cores below the top

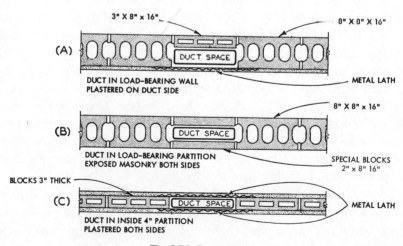

Fig. 6-71. Duct details.

Concrete Block Masonry: Units and Building

two courses to keep the wet concrete in place.

Duct Details. The ducts used in connection with heating and air conditioning systems for residences, stores, etc., are generally from $3\frac{1}{2}''$ to $3\frac{5}{8}''$ thick. They can be built into block walls easily, as indicated in Fig. 6-71.

Waterproofing Concrete Block Foundations

There are many methods of waterproofing which may be used successfully. Those discussed here are typical and are recommended by average building codes, the Portland Cement Association, and by the National Concrete Masonry Association.

Perhaps the simplest form of waterproofing for foundations built in soil which is just damp, is to apply hot tar or asphaltum (asphalt) to the outside surfaces. These materials are moistureproof and constitute satisfactory waterproofing. This is shown in (A) of Fig. 6-72.

Where excessive dampness or severe conditions of water in the soil occur, the exterior surfaces of the foundation can be satisfactorily waterproofed as shown in (B) of Fig. 6-72. Two or more plies or lay-

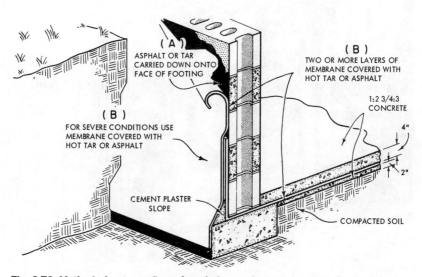

Fig. 6-72. Method of waterprofing a foundation against severe groundwater pressure.

Masonry Simplified

ers of membrane (felt, for example) coated with tar or asphaltum are used. Note in Fig. 6-72 that for extraordinary conditions the waterproofing is applied at the joint between the footing and the foundation and that the floor is built in two layers with the waterproofing placed between the layers. In addition, note that a mix of 1:2¾:3 concrete can be used as a means of further retarding the flow of water through the foundation. Such concrete becomes dense in setting and, because of this characteristic, tends to prevent the flow or seepage of water through it. The slope at the point where the foundation meets the footing also helps to make that joint waterproof.

Fig. 6-73 shows a slightly different treatment of the membrane and tar or asphaltum treatment.

Another waterproofing method frequently used where there is excessive water in the soil is shown in Fig. 6-74. The clay or concrete tile is laid around all sides of the footing with a gravel or cinder fill covering it to the depth shown. The fill material allows the water to flow directly to the tile where it collects and drains off to some point away from the basement where it cannot

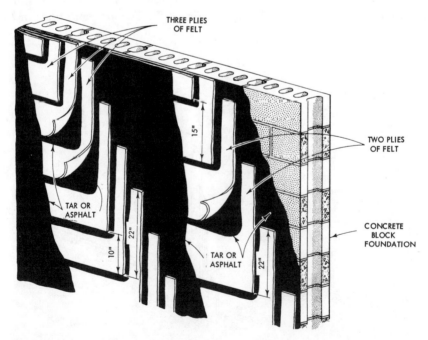

Fig. 6-73. Method of lapping three and two layers of felt waterproofing for a foundation.

Concrete Block Masonry: Units and Building

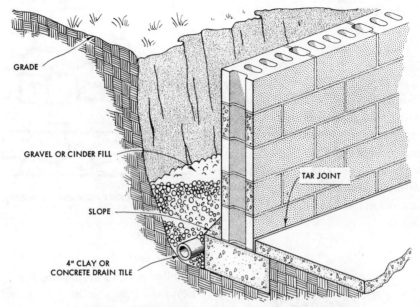

Fig. 6-74. Footing drains for use with all kinds of foundations.

do any harm. Occasionally, the water drains into a sump from which it is pumped to the regular sewer. When soil water conditions are really severe, drain tile are used in conjunction with the waterproofing methods explained in Fig. 6-73.

Note the tar joint between the floor and foundation in Fig. 6-74. This method of calking is effective and helps materially to keep moisture out of the basement in cases where no waterproofing materials were built into the floor.

Stone Masonry

Natural stone, such as limestone, granite, or marble, has been used in masonry for centuries. Stone is available in wide varieties of size, shape, color, and texture, allowing for any number of wall patterns. Due to the different sizes and uneven surfaces of individual units, stone masonry is slower and more difficult than brick or block masonry and, therefore, more expensive. The use of stone at the present time is limited mostly to veneers on building walls and decorative facings on such structures as barbecues, fireplaces, and garden walls. Garden walls may also

179

Masonry Simplified

be constructed of solid stone. The following figures illustrate some of the more common wall bond patterns.

The type of wall shown in Fig. 6-75 is called a random rubble wall. This type of wall employs stones of many sizes and shapes, such as are found in fields and stream beds. Although there is little or no coursing in random rubble, note that the bed joints are made as horizontal as possible.

Fig. 6-76. Coursed rubble stone masonry.

Fig. 6-75. Random rubble stone masonry.

Fig. 6-76 shows another kind of rubble masonry employing stones that are roughly squared. These stones are laid in such a way as to produce approximately continuous horizontal bed joints. This is called coursed rubble masonry.

A solid, freestanding rubble stone wall must be *bonded* for strength and stability just as a brick or block wall. That is, at intervals, there must be a unit which passes all the way through the wall. This is called a bond stone. See Fig. 6-77. Bond stones should be placed as frequently as possible; at least in every 6 to 10 square feet of wall.

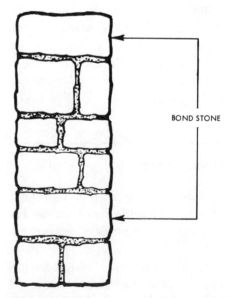

Fig. 6-77. Rubble stone masonry showing bond stone.

Concrete Block Masonry: Units and Building

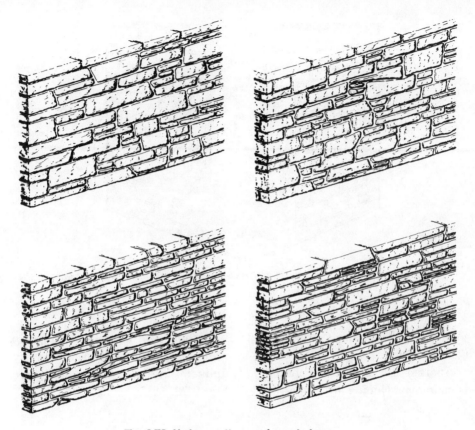

Fig. 6-78. Various patterns using cut stone.

Fig. 6-78 shows a few of the patterns possible using cut stone. The roughly rectangular shapes are laid in the same manner as brick and are commonly used as veneer.

As in all masonry, each joint should be completely filled with mortar. Mortar for stone masonry is usually specified to be 1 part cement, 1 part lime, and 6 parts sand. This formula has been found to be quite strong while still supplying the workability necessary for laying random size stone.

Again, as in all masonry, stonework construction is begun at the ends or corners of the wall. If the wall is to be coursed, the mason's level is used to level the top of each course.

In order to accent the natural rugged beauty of stone, the mortar joints are raked out about $\frac{1}{4}$ to $\frac{1}{2}$ inch. This may be done with regular

181

Masonry Simplified

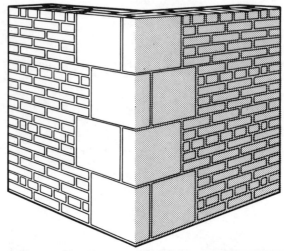

Fig. 6-79. Quoins may be used in corners and around openings in masonry walls.

jointing tools, pointing trowels, or calking trowels.

Another use of stone in masonry is as *quoins*. Quoins are stones used in corners and around openings in walls for contrast with the masonry units in the rest of the wall. See Fig. 6-79.

Maintenance

Concrete block construction, when properly built, requires little or no maintenance. As mentioned, if mortar is accidentally dropped on the face of a block, the burrs may be removed after they are dry by using a trowel or wire brush. If joints must be patched, use good fresh mortar.

Painting Concrete Block

Mixtures of portland cement with water as a vehicle to which waterproofing agents, colored mineral oxides, and other ingredients have been added can be used readily to waterproof concrete block walls as well as to add to their beauty. It is recommended that portland cement paint ready to use be purchased rather than to attempt to make it on the job. Practically all building material dealers sell such products.

If block walls are to be painted for waterproofing, or weather resistance, two coats, called the *seal* and *finish* coats, should be used. Wall surfaces should be dampened just prior to the application of the seal coat. The seal coat should be *scrubbed* into the wall surface as this practice avoids pin holes in the paint. Ordinary scrub brushes with stiff fiber bristles have been found to be the most satisfactory because they force the paint into the pores of the blocks. Use ample amounts of the paint in the brush and completely cover the wall surfaces, including the joints.

The finish coat can be applied within 48 hours after the seal coat has been put on. Enough paint should be applied over the seal coat to cover it completely. The finish coat should be sprayed on, providing ordinary paint spraying equipment is available. When spraying, the nozzle of the spray should be manipulated so that the spray hits every point of the wall surface from four or five angles. If the work must be done by hand, a six-inch brush will be found to be most practical.

Seal and finish coats should be kept in a moist condition for a period of at least 48 hours following application. This can be accomplished by spraying with water at intervals after the paint has set.

If the wall is constructed of dense block and good joints and is to be painted for color only, one coat is sufficient. Interior walls should receive only one coat to preserve sound absorption qualities.

Note: Care should be taken to protect hands in using portland cement paint as prolonged contact may irritate tender or sensitive skin.

Besides portland cement paint, there are several other types of paint suitable for painting concrete block. The main types are oil-base, varnish base, lacquer base, and water-thinned or latex.

Although all types have been found suitable for exterior, above grade use, oil base paints provide a more effective barrier against penetration of rainwater. Varnish base paints are particularly suitable for interior walls. They provide a smoother surface and dry more rapidly than oil base paints. They are available in gloss ranges from flat to full gloss. Lacquer base paints are highly resistant to alkalis and can be applied on damp surfaces. Water-thinned, or *latex*, paints have the widest range of use. They have been found to be suitable for above grade exteriors and interiors and

Masonry Simplified

are particularly suitable for interior basement walls because of their alkali resistance and ability to "breathe"; that is, to contain moisture behind the paint film without blistering.

All of these paints may be applied without any special coat although two or three coats are generally required for the desired results. The surface should be thoroughly cleaned before applying. These paints may be applied successfully by brush, roller, or spray, providing good workmanship is employed. As with any commercial product, the manufacturer's directions on the container must be carefully read and followed.

Checking On Your Knowledge

The following questions give you the opportunity to check up on yourself. If you have read the chapter carefully, you should be able to answer the questions. If you have any difficulty, read the chapter over once more so that you have the information well in mind before you go on with your reading.

DO YOU KNOW

1. What type of concrete blocks are generally used in buildings in which the exterior is surfaced with stucco?
2. What type of 8" x 8" x 16" concrete blocks are generally used around window openings?
3. What type of concrete masonry sill actually extends into the walls?
4. What maximum mortar thickness should be used in building a concrete masonry wall?
5. What is meant by the term "modular coordination"?
6. What special processing must be given to the blocks in a wall directly under the point where a beam end is supported by the wall?
7. What can be done to strengthen a block wall under a beam bearing area?
8. What size joists can be used to the best advantage in connection with block walls?
9. What minimum amount of bearing should be allowed for beams in block walls?
10. What the preferred methods for fastening wooden members to concrete masonry walls are?
11. What the function of the cavity in a cavity wall is?
12. What the two methods of reducing shrinkage and expansion of concrete masonry units that are laid into a wall are?
13. At least two methods of waterproofing a concrete block foundation wall?
14. How anchor bolts are secured in concrete block walls?
15. What the purpose of allowing a certain amount of air space between the floor joists and a concrete masonry wall is?
16. What tool should be used for tooling horizontal joints?
17. The different types of paint for concrete block?

184

Concrete: Basic Materials and Mixing

Chapter 7

Concrete is one of the most interesting building materials in use today. Because of its capability of being molded or formed to almost any size or shape and its outstanding strength and durability characteristics, it is probably the most versatile building material of all.

If the cement mason is to go about his work properly and efficiently, it is essential that he thoroughly understand the nature of his materials. This chapter will cover the components of concrete, the characteristics of the different types of concrete, and the mixing of concrete.

The materials used to make concrete are portland cement, water, and aggregate. Water added to cement reacts with the cement to form what is called a *cement paste*. As it is mixed, either by hand or machine, the cement paste forms a coating on all particles and pieces of the aggregate. When the mixture has been placed as, for example, for a sidewalk or structural part, a chemical reaction takes place in the cement paste which causes it to harden. This hardening process binds all of the aggregate together, forming a permanent and dense mass which is known as concrete.

Fig. 7-1 shows a piece of concrete which has been sawed in half. As is characteristic of good concrete, the particles and pieces of aggregate are held together by the hardened cement paste.

Masonry Simplified

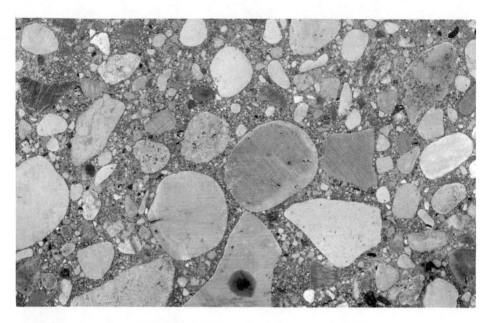

Fig. 7-1. A piece of concrete which has been sawed in two to show how the fine and coarse aggregates combine to form a solid mass.

Aggregates

The sand and crushed stone or gravel used in concrete are known as aggregates; sand is further classified as *fine aggregate* and the crushed stone or gravel as *coarse aggregate*. The fine aggregate is of varying sizes so that the smaller particles tend to fill the spaces (voids) between the larger particles. For the same reason, the coarse aggregate also is composed of varying sizes. When the fine and coarse aggregates are mixed together, the fine aggregate tends to fill the small voids between the smaller pieces of the coarse aggregate. This results in a dense, solid mass.

Fine aggregates consist of natural or manufactured sand with particle sizes up to about ¼ inch. (Note that this is about the same as mortar sand described in Chapter 4.)

Coarse aggregates are those with particle sizes greater than ¼ inch. The maximum size specifications for coarse aggregates will vary according to the purpose for which a given batch of concrete is intended. For instance, a floor requiring a smooth finished surface and bearing

Concrete: Basic Materials and Mixing

only light loads may require as little as ⅜ inch maximum size particles, whereas a heavy duty foundation may call for up to 2½ inch particles. Even larger particles may be specified for such massive structures as dams or bridge abutments.

The aggregates for concrete must also be clean. Aggregates obtained from a reliable dealer will almost always be clean and graded. However, if the contractor must dig his own from natural sand or gravel beds, he should test samples for impurities. The two main impurities that might be present are silt and organic matter. The tests for the presence of these impurities as described for mortar sand in Chapter 4 may be applied to concrete aggregates as well.

Fine Aggregate (Sand)

The term fine aggregate, or sand, applies to any finely divided material of rock or mineral origin, the particles of which have a diameter ranging from $\frac{1}{20}$ to 2 mm., which will not injuriously affect the cement, and which is not subject to disintegration or decay. Sand is almost the only material which is sufficiently cheap and which will fulfill these requirements, although stone screenings (the granulated or pulverized material resulting from stone crushing) and powdered slag have been used as substitutes.

Quartz sand is the most durable and unchangeable. Sands which consist largely of grains of feldspar, mica, etc., which will decompose upon prolonged exposure, are less desirable than quartz.

Grading. Grading means the particle distribution in a batch of aggregate. The most satisfactory sand is a mixture of coarse and fine grains, with coarse grains predominating. It makes a denser, stronger concrete than does fine-grained sand when both sands are mixed with the same quantity of cement. In other words, very fine sand may be used alone but it makes a weaker concrete. In a given quantity of very fine sand, there are more grains or particles than in the same quantity of coarse sand. More water is required to mix mortar or concrete using very fine sand. The water forms a film and separates the fine grains, thus producing a larger volume of concrete but with less density and strength.

A well-graded sand has particles ranging in size from very fine up to those which will pass through a screen having meshes ¼″ square. As previously explained, the larger particles should predominate in quantity. Fig. 7-2 shows a good sand for making concrete. Note that the particles are of various sizes and that the larger sizes predominate. Well-graded sand not only makes stronger concrete but allows a more economical use of cement paste in

Masonry Simplified

Fig. 7-2. Fine aggregate for concrete. Top: Good concrete sand. Bottom: The same sand which has been screened to show the different particle sizes.

filling the voids and binding the aggregate together.

The particles of sand may be either round or angular. The idea that sand should be sharp (angular) has been disproved by tests which show that there are fewer voids in round than angular sand. There is really little difference in the strengths of concrete made using sharp and round sand except that round sand helps to produce a denser mix. On the other hand, cement paste adheres to the angular sand somewhat better than to the smooth surfaced round sand particles, so either type of sand can be used successfully. The shape of the particles is not nearly so important as their soundness and their being properly graded from fine up to ¼ inch.

Coarse Aggregates

Coarse aggregates for making regular concrete may consist of crushed stone or gravel taken from gravel pits which are found in many parts of the country.

Crushed Stone. Trap rock is the hardest and most durable stone that can be crushed and used for making concrete. This stone is dark, heavy, close-grained, and of igneous origin. Granite makes good crushed stone for concrete, and generally it is less expensive than trap rock. Hard limestone also may be crushed and used to advantage in concrete making, but it is not as strong as granite or trap rock and is affected by fire. Only the hardest grades of sandstone can be used for making concrete.

Concrete: Basic Materials and Mixing

Grading Crushed Stone. In general, stone is crushed in sizes ranging from ¼″ to 2½ inches. When the stone is crushed, some of it will become much smaller than the ¼″ minimum usually considered the smallest usable size. This should be discarded or used as sand. After crushing, the stone should be screened and the different sizes kept separate or mixed together to fit the needs of various concrete mixes. The pieces should be square or triangular in shape. Flat, elongated, or thin pieces should never be used.

In concrete which is to be used for ordinary structural parts such as floors, foundations, footings, etc., the sizes of crushed stone used as coarse aggregate should be a mixture varying from ¼″ to 1¼″, or from ¼″ to 1½ inches. The smaller size is best for thin structural items. Mixtures of even smaller sized crushed stone should be used for concrete which is subject to severe wear.

For reinforced concrete members which are small and have steel bars spaced close together, crushed stone should be graded to include a mixture of pieces varying from ¼″ to ¾″ in size. This size aggregate should also be used for fireproofing structural steel. Where concrete members are larger and the steel not so close together, the crushed stone mixture may vary from ¼″ to 1¼ inches.

For concrete items which are massive, the crushed stone mixture may vary from ¼″ or ½″ up to 2½ inches. Massive items include retaining walls, extra thick foundations, etc.

Gravel. The term *gravel* refers to stone as it occurs naturally in gravel banks. Generally, gravel is small pieces of stone which are somewhat rounded in shape. It makes good coarse aggregate because it is hard and close textured. Often in the past it has been the practice to use the sand and gravel directly from gravel banks for coarse and fine aggregate. This practice should be avoided because in most cases too much sand is present and the pieces of stone are not properly graded. In addition, bank-run material may contain too high a percentage of silt or organic matter. If, however, bank gravel is used, it should be washed. After washing, it should be retested to see that it is suitable for use.

Grading Gravel. When gravel is used as a coarse aggregate, the sizes of individual pebbles making up the various mixes should be approximately the same as outlined for crushed stone.

If, for example, a mixture of gravel ranging from ¼″ to 1½″ in size is required, it is not necessary that exactly equal amounts of the various sizes be used. However, the best grading, and therefore the best

Masonry Simplified

Fig. 7-3. Coarse aggregate for concrete. Top: Good concrete gravel. Bottom: The same gravel which has been screened to show the different particle sizes.

concrete, results when the various sizes are fairly well divided as to the quantity of each. Fig. 7-3 shows good concrete gravel and the variety of sizes. It will be seen that the various sizes of pebbles are nearly, but not exactly, equal in amount.

To secure a suitable mixture of gravel for concrete, it is advisable to screen the gravel as it comes from the bank and then mix the various sizes somewhat as suggested by Fig. 7-3. Good concrete just does not happen. Instead, care must be exercised in the selection of the aggregate. The use of improper aggregate always results in poor concrete and unsatisfactory structural work.

Blast Furnace Slag. Blast furnace slags are composed chiefly of silica, alumina, magnesium, and lime. Any blast furnace slag can be used as coarse aggregate if it weighs a minimum of 70 pounds per cubic foot. Such aggregate, like crushed stone or gravel, should be well graded.

Rubble. The aggregate for rubble concrete is similar to regular concrete except that from 20 to 65 percent of the mass of the concrete is taken up by the large stones. The use of large stones in massive concrete, such as drains, is economical

Concrete: Basic Materials and Mixing

and satisfactory if no voids are left between them in the concrete.

Lightweight Aggregates

Lightweight aggregate for concrete is a commercially manufactured product available under several brand names. It is made by burning certain minerals or blast furnace slag at very high temperature in rotary kilns, causing the minerals to expand, producing a hard, sound aggregate. Concrete made with this type of aggregate called *lightweight concrete* has similar properties of strength and endurance to that made with natural aggregate but is much more expensive. Its main use is in such prefabricated units as large wall panels, especially in tall buildings, where lightness is a definite advantage.

Water

Generally, if water is suitable for drinking (potable), it will be satisfactory for making concrete. Again, silt and organic matter are the two main deleterious materials that may be present in mixing water. Normally, any harmful amount of silt will be visible to the naked eye. If water containing silt must be used, it should be stored in settling basins before use.

Cement

The history and manufacturing process of portland cement were presented in Chapter 4. The American Society for Testing and Materials (ASTM) provides for five types of portland cement: Types I, II, III, IV, and V. Each type has particular properties which suits it for specific purposes.

Type I is a general purpose cement. It may be used in almost any location and in every type of concrete structure where the special properties of the other types are not required. It is also the most economical of all the types.

Type II is a modification of Type I. It has much the same properties as Type I except that it is slightly more resistant to sulfate attack. Sulfates present in groundwater will cause concrete exposed to it to deteriorate. This is a problem only in some of the western states. Type II is usually specified in these areas for below-grade structures such as footings, foundations, basement walls, drainage structures, etc., which are exposed to soil contaminating sulfates.

Sulfate attack is also common in industrial areas due to sulfur dioxide air pollution. Rain water precipitates the sulfur dioxide as a weak sulfuric acid which will attack concrete.

Type II also generates heat at a slightly lower rate than Type I. This property is especially important when concrete is placed during hot weather and the concrete may begin to set too rapidly for proper workability.

Type III is called *high-early strength* cement. This is because it gains its full strength at a much faster rate than the other types. Because of this property, forms may be removed and the structure put in use earlier.

Type IV is a very specialized cement in that it develops its strength at a much slower rate than the other types. Because of this, the rate of heat generation is also much lower. Therefore, Type IV is used almost exclusively in massive structures such as large gravity dams where the rate of temperature rise during setting is a critical factor.

Type V is also a very specialized cement. Its special property is high sulfate resistance. Its use is specified only for concrete exposed to severe sulfate attack.

Cement Distribution. Most cement for job site mixing is packed in paper sacks which hold one cubic foot or 94 pounds. Cement for ready mix concrete plants, which provide most of the concrete used today, is delivered in bulk form and sold by the ton.

Cement Storage. Cement is easily damaged by water and will readily absorb moisture from the atmosphere unless carefully protected. Therefore, prior to use, it should be kept in a dry place.

Damaged Cement. Cement which is allowed to absorb moisture will form into lumps. If these lumps cannot be pulverized by lightly striking them with a shovel, the cement is not fit for use.

Proportioning Concrete Mixtures

The primary objectives in selecting proportions of ingredients for concrete are to achieve a mix that is workable when ready to place; that will, when hardened, have the degrees of strength and durability required for the purpose of the structure, and will be as economical

Concrete: Basic Materials and Mixing

as possible while still retaining the desired properties. No one of these properties is more or less important than the other. In fact, they are so interrelated that the apprentice mason may at first be confused by the several different reasons for varying amounts of materials for particular purposes. However, with continued practice and experience, he will come to understand why and how a mix is proportioned a certain way for a certain job.

Water-Cement Ratio. Cement mixed with water makes up the *cement paste* which binds the aggregate particles into a strong solid mass. The quality of the cement paste is the most important factor in assuring the desired properties of the hardened concrete. A higher proportion of water to cement will afford a more workable mixture, whereas a mixture with less water and more cement will be stronger when hardened. Thus, a basic rule for proportioning concrete is to use as small amount of water in proportion to cement as possible, so long as the mixture remains workable. Workability is difficult to measure, but an experienced mason can readily judge it. Proportioning concrete mixtures, therefore, is an art as well as a science. Specifications for water-cement ratios are given as decimal fractions. For instance, a mixture calling for a 0.45 water-cement ratio means that the mix will consist of 45 percent water and 55 percent cement by *pounds*. In other words, 100 pounds of cement paste using this ratio would consist of 45 pounds of water to 55 pounds of cement. (*Note:* one pound of water is equivalent to one pint liquid measure.) This method is used on large jobs where a slight variance in proportions would multiply the variance and cause large batches to deviate considerably from the prescribed formula. On smaller jobs, the water-cement ratio is given as gallons of water per bag of cement.

Aggregate Proportions. The selection of proper aggregates for the mixture is an important economic factor. The cement is by far the most expensive ingredient in concrete so it is advisable to minimize the cement-water requirement by using as stiff a mixture as possible and by using well graded aggregate. Remember that the aggregate particles must be completely coated with cement paste in order to assure a tightly bonded, strong, durable, and watertight structure. Considerably less paste is required to coat one large particle than to coat several small ones occupying the same amount of space. Thus, for economy as well as for strength, the coarse aggregate should contain the largest size allowable according to specifications. Remember, however, that the smaller particles and the fine

Fig. 7-4. Left: If sand falls apart it is damp. Middle: If sand forms a ball it is wet. Right: If sand sparkles and wets the hand it is very wet.

aggregate fill in the "voids" between the larger particles, improving workability, solidity, and strength. So, as always, the aggregate should be well graded from fine to coarse.

Moisture Content in Sand. Fine aggregate (sand) will always contain a certain amount of water at the time of mixing. This will necessarily affect the amount of water to add in mixing, so it is necessary for the mason to judge the approximate amount of water in the sand he is going to use. For the purpose of proportioning, sand may be classified as *damp*, *wet*, or *very wet*.

A simple method for determining the amount of moisture is by squeezing a handful of the sand to be used. See Fig. 7-4. Table 7-1 gives corrected proportions for water-cement ratios when the sand is damp, wet, or very wet.

Bulking in Sand. Sand will increase greatly in volume as it is handled, particularly if it is very moist. This is called *bulking*. This tendency makes it very difficult to judge the volume accurately. For this reason, sand should be measured by weight rather than volume whenever possible.

TABLE 7–1

IF MIX CALLS FOR:	Use these amounts of mixing water, in gallons, when sand is:		
	DAMP	WET	VERY WET
6 gal. per sack of cement	5 1/2	5	4 1/4
7 gal. per sack of cement	6 1/4	5 1/2	4 3/4

Concrete: Basic Materials and Mixing

Slump. The *consistency*, or degree of stiffness or fluidity, of concrete when it has been mixed and ready to place is called *slump*. Slump is measured in inches. High slump concretes are fluid while low slump concretes are stiff. Wet mixtures contain *too little* coarse aggregate for the amount of cement paste. Stiff mixtures contain *too much* sand and coarse aggregate. Add more sand and coarse aggregate for wet mixes; reduce the amount of sand and coarse aggregate in subsequent batches for stiff mixes.

The field test for slump is shown in Fig. 7-5. The sheet metal cone measures 8 inches in diameter at the bottom and 4 inches at the top and is 12 inches high. It is placed on a clean dry surface and filled approximately 1/3 full (by volume) with the trial mixture. Rod the concrete 25 times with a metal rod. Then another $\frac{1}{3}$ of concrete is added and rodded 25 times. The rod should extend all the way through the second $\frac{1}{3}$ and just into the bottom $\frac{1}{3}$. Next, the cone is completely filled, the top is raked off level, and the rodding repeated as above. Then the cone is removed by lifting it straight up, leaving the fresh concrete. The slump is now measured by placing the rod across the top of the cone and extended over the concrete sample. Then, with a rule, measure the distance between the top of the concrete sample and the bottom edge of the rod. This distance is the slump rating of the entire batch.

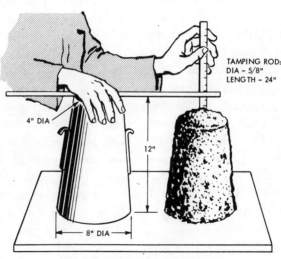

Fig. 7-5. Slump test for concrete.

Masonry Simplified

Trial Proportions

Not many years ago, it was customary to specify the proportions (quantities of cement, sand, and crushed stone or gravel) for regular concrete mixes by such designations as 1:2:4, 1:3:5, etc. The 1:2:4 designation, for example, meant 1 part cement, 2 parts sand, and 4 parts crushed rock or gravel. This method of specification fails to assure satisfactory results for the following reasons.

1. It does not specify the quantity of mixing water which is so essential in making strong and watertight concrete.

2. It does not consider the grading of the aggregates.

3. It does not allow for variation in volume resulting from the tendency toward bulking of moist sands.

Present-day methods of specifying regular concrete mixes are made by paying careful attention to the amounts of water, the moisture content of the sand, and the amounts of aggregates used in the mix.

Recommended trial qualities of concrete and amounts of the ingredients for each, for various classes of work, are shown in Table 7-2. This table can be used as a guide to proportioning concrete materials according to the total amount of water required with each sack of cement.

Determining Suitable Proportions. Suppose, for example, it is necessary to determine the proper mix (proportioning of all ingredients, including damp sand and water) for building a water tank. For this job, the concrete must be watertight and be able to stand severe exposure to weather.

Table 7-2 shows that for a job of this kind, a 6-gallon paste should be used. But, for a trial batch, using one cubic foot (one sack) of cement, $2\frac{1}{4}$ cubic feet of sand, and 3 cubic feet of crushed stone or gravel, only $5\frac{1}{2}$ gallons of water can be added when mixing the ingredients, because (as previously explained) two cubic feet of damp sand will contain $2 \times \frac{1}{4}$, or $\frac{1}{2}$ gallon of moisture. This $\frac{1}{2}$ gallon plus the $5\frac{1}{2}$ gallons makes up the total of 6 gallons required per sack of cement.

Assuming a mechanical mixer is to be used, first place the correct amount of water in the mixer. Add one sack of cement, $2\frac{1}{4}$ cubic feet of sand, and 3 cubic feet of crushed rock or gravel, and run the mixer for at least two minutes. By noting how the resulting mix handles and places, it can be determined readily whether changes in the proportions are necessary to fit the needs of the job. If the concrete is a smooth, plastic, workable mass that will place and finish well, the correct proportions for the job have been determined. Fig. 7-6 shows at (A) a mix which lacks sufficient mortar,

Concrete: Basic Materials and Mixing

TABLE 7-2 RECOMMENDED PROPORTIONS OF WATER TO CEMENT AND SUGGESTED TRIAL MIXES

KIND OF WORK	Add U.S. Gals. Of Water To Each Sack of Cement If Sand Is			Suggested Mixture For Trial Batch		
	Very Wet	Wet	Damp	Cement Sacks	Sand Cu. Ft.	Crushed Rock or Gravel Cu. Ft.

FIVE GALLON PASTE FOR CONCRETE SUBJECTED TO SEVERE WEAR, WEATHER OR WEAK ACID AND ALKALI SOLUTIONS

KIND OF WORK	Very Wet	Wet	Damp	Cement Sacks	Sand Cu. Ft.	Crushed Rock or Gravel Cu. Ft.
Colored or plain topping for heavy wearing surfaces and all two-course work for pavements, tennis courts, floors, etc.	4 1/4	Average sand 4 1/2	4 3/4	1	1	1 1/2
		Maximum size aggregate 3/8 inch				
One-course industrial, floors and all concrete in contact with weak acid or alkali solutions..................	3 3/4	4	4 1/2	1	1 3/4	2
		Maximum size aggregate 3/4 inch				

SIX GALLON PASTE FOR CONCRETE TO BE WATERTIGHT OR SUBJECTED TO MODERATE WEAR AND WEATHER

KIND OF WORK	Very Wet	Wet	Damp	Cement Sacks	Sand Cu. Ft.	Crushed Rock or Gravel Cu. Ft.
Watertight floors such as industrial plant, basement etc........ Watertight foundations............ Concrete subjected to moderate wear or frost action such as driveways, walks, tennis courts, garage floors, etc........................ All watertight concrete for swimming and wading pools, bird baths, fish ponds, septic tanks, storage tanks, etc........................ All base course work such as floors, walks, drives, etc.............. Steps, chimney caps, blocks, concrete masonry, fireplaces, etc..... All reinforced concrete structural beams, columns, lintels, slabs, residence floors, etc............	4 1/4	Average sand 5	5 1/2	1	2 1/4	3
		Maximum size aggregate 1 1/2 inches				

SEVEN GALLON PASTE FOR CONCRETE NOT SUBJECTED TO WEAR, WEATHER OR WATER

KIND OF WORK	Very Wet	Wet	Damp	Cement Sacks	Sand Cu. Ft.	Crushed Rock or Gravel Cu. Ft.
Foundations, walls footings, mass concrete, etc., not subjected to weather, water pressure or other exposure	4 3/4	Average sand 5 1/2	6 1/4	1	2 3/4	4

Masonry Simplified

Fig. 7-6. Top: Concrete mixture which lacks sufficient mortar. Middle: Concrete mixture having excess cement and sand. Bottom: Concrete mixture having correct proportions.

at (B), a mix having excess cement and sand, and at (C) a mix with good proportions. While the mix shown at (C), in Fig. 7-6, is good, it will not satisfy every condition.

For example, it may be too stiff for use in making some concrete objects where the mix must surround reinforcing or run into narrow forms, etc. Thus, the particular job at hand, to some extent, governs the condition of mixes.

Correcting Trial Mixture. If the trial mix is not workable under the conditions of the job, the amounts of aggregate used in the concrete must be changed. *However, the amount of water should not, under any circumstances, be changed.* The trial batch of 1 part cement, 2¼ parts sand, and 3 parts coarse aggregate may, for example, be too stiff or too wet or may lack smoothness and workability.

When the trial proportion gives a mixture that is too wet, add small amounts of sand and coarse aggregate in the proportion of 2¼ parts of sand and 3 parts coarse aggregate until the correct workability is obtained.

If it is necessary to use more sand than is shown in the proportions given in Table 7-2, for instance, an extra ½ cubic foot—it is important to deduct the moisture carried by this additional sand.

If the concrete is too stiff and appears crumbly, succeeding batches can be mixed with less aggregate.

Under ordinary conditions, a concrete mix should be *mushy* but not *soupy*. The mushy mix will hold together while a soupy mix may sepa-

Concrete: Basic Materials and Mixing

rate in handling, with the larger pieces of aggregate sinking in the mass.

In some cases, concrete specifications still call for concrete as a 1:2:4 mix. There may be danger in following such a specification exactly, as explained in the following.

Suppose that a 1:2:4 mix is specified and that the sand available for use is average in regard to moisture. First of all, unless the approximate amount of moisture in the sand is determined, the cement paste will be diluted. In addition, sand of average moisture content is bulked at least 20 percent because the moisture forms a film around each sand particle and thus forces the various particles farther apart. If such bulked sand is used in a 1:2:4 mix, the resulting concrete will be 20 percent short on sand and will detract from the strength and density of the concrete. To overcome this shortage, a mix of $1:2\frac{1}{4}:3$, as recommended in Table 7-2, or $1:2\frac{1}{2}:3\frac{1}{2}$ should be used. This might result possibly in some oversanding, but that condition would be much better than undersanding. As seen, there is a strong possibility for discrepancy between the specification and the actual mix. For this reason, this type of specification should be avoided.

Air Entrainment

One of the most important advances in concrete construction has been the development of air entrainment. Air-entrained concrete contains extremely small air bubbles throughout the cement-water paste. The bubbles are so small that there may be as many as 3000 billion in one cubic yard of concrete.

Although developed and used primarily for its high resistance to freeze-thaw damage, air entrained concrete has been found to have other desirable properties as well. As compared to concrete using regular cement, air entrained concrete is more workable, thus requiring less water and sand, is more watertight, and is highly resistant to chemicals such as sulfates and the various salts used as de-icers on city streets and sidewalks. It is logical, therefore, that air entraining cement or an entraining admixture is specified in concrete formulas for sidewalks, driveways, etc., in areas having severe winters and various salts are used as de-icers. Garage floors in these areas should also be made of air entrained concrete, as vehicles coming in from salted streets and highways will drip enough salty moisture to cause rapid deterioration.

Masonry Simplified

Air entraining portland cements are available in types I-A, II-A, and III-A, which correspond with the specifications for regular Types I, II, and III as described earlier in this chapter. There are also air entraining admixtures available under several brand names. These admixtures will give satisfactory results if the manufacturer's directions on the container are followed.

Admixtures

Admixtures are any materials added to the portland cement-water-aggregate mixture in order to improve or add specific properties such as accelerating or retarding setting time, air entrainment, water repellents, etc. Most specifications or codes discourage the use of admixtures for accelerating or retarding when alternatives, such as modified curing methods, are available. However, the use of air entraining admixtures are advised where air entraining portland cement is not used.

Calcium chloride is sometimes used to accelerate the setting time of fresh concrete in freezing weather. This practice should be avoided if at all possible, as calcium chloride in any appreciable quantity will weaken the concrete. Most building codes, if not actually forbidding the use of calcium chloride, limit the allowable quantity to 2 percent of the total mixture. Recommended alternatives are the using of Type III (high early strength) portland cement, or the warming of materials before mixing or the using of heating devices during curing. Calcium chloride is *never* to be used as an anti-freeze, as the quantity required to be effective is enough to completely ruin the concrete. Calcium chloride also corrodes metal and is prohibited in prestressed concrete.

Ready Mixed Concrete

Most concrete construction jobs use ready mixed concrete. The main reasons for this are that ready mix precludes the time and space involved in storing and handling the raw materials, the labor involved in on-the-job mixing, and the clean-up of waste materials. Purchasers may also be assured of receiving any quantity of concrete of uniform quality at exactly the time it is needed.

Ready mixed concrete is usually prepared in one of two ways: central

Concrete: Basic Materials and Mixing

Fig. 7-7. A typical ready mix concrete truck.

mixed or transit mixed. Central mixing is the complete mixing of the concrete in a central batch plant. The concrete is then loaded into special trucks of the type shown in Fig. 7-7 for delivery to the job site. The concrete is in rotating drums which keep it plastic and workable until time to place.

If a job requires only one truckload or less, the concrete may be transit mixed. In this method, the raw materials, cement, aggregate and water, are placed directly into the drum which revolves and mixes the concrete on the way to the job site.

Job Site Mixing

On some small jobs where ready mix is not available or impractical, concrete may be mixed on the job. Portable mechanical mixers are usually employed in these situations. See Fig. 7-8. These mixers are available with capacities ranging from a few cubic feet up to several cubic yards. The operator must be familiar with the manufacturer's specifications for maximum load and mixing speed. These should *never* be exceeded.

Fig. 7-8. A typical small portable concrete mixer.

The procedure for loading and mixing is as follows. First, the materials are measured according to the specifications. Then, about 10 percent of the amount of water to be used is poured into the drum. The drum is then set in motion, and the cement and aggregate are added gradually along with more water.

The water should be added at such a rate that when all of the cement and aggregate are in the drum, about 90 percent of the specified amount of water has been added. The mixing time is measured from the time all the solid materials are in the drum. Most specifications call for at least one minute of mixing time for mixers of up to 1 cubic yard capacity with an increase of 15 seconds for each additional cubic yard.

After the minimum mixing time has elapsed, the mixture should be tested for stiffness, or slump. If it appears to be too stiff to be workable, additional water may be added up to the maximum amount allowed in the specifications. As the amount of water in the mixture is critical in regard to the ultimate strength of the hardened concrete, the specified amount should never be exceeded. If the mix remains extra stiff, some workability must be sacrificed for the sake of strength.

Hand Mixing

Although hand mixing of concrete is seldom required, the mason should know the correct way of doing it. First, the dry cement and aggregate are mixed thoroughly on a clean, dry, waterproof surface. Then the dry materials are mounded and a depression is made in the middle. Water is gradually added to the depression as the dry material is turned in toward the middle with a shovel. Continue mixing until all the ingredients are thoroughly combined and the aggregate is completely coated with paste.

Concrete: Basic Materials and Mixing

Checking On Your Knowledge

The following questions give you the opportunity to check up on yourself. If you have read the chapter carefully, you should be able to answer the questions. If you have any difficulty, read the chapter over once more so that you have the information well in mind before you go on with your reading.

DO YOU KNOW

1. What the three basic materials in making concrete are?
2. The two classifications of aggregate and their sizes?
3. What is meant by grading of aggregate?
4. What lightweight aggregate is and where it is used?
5. The five types of portland cement and the differences between them?
6. What is meant by a 0.45 water-cement ratio?
7. Why the aggregate proportion in a concrete mix is an important economic factor?
8. How to determine the amount of moisture in sand?
9. What is meant by *bulking* in sand?
10. What is meant by *slump*?
11. The three main factors in modern methods for specifying concrete mixes?
12. How to determine the correct amount of water for a concrete mix?
13. What one of the most important developments in concrete construction is?
14. What admixtures are?
15. Why the use of calcium chloride should be avoided?
16. Why ready mix is so widely used?

Chapter 8
Concrete Construction

Concrete as a building material has almost numberless applications. However, there are certain basic practices in placing and finishing concrete which, when learned by the mason, will enable him to adapt his skills to jobs of any type and magnitude. This chapter will cover the construction of general formwork, including footings, foundations and walls, and will cover the all-important relationship of the mason and carpenter in the construction and use of formwork. Foundations with slab-at-grade construction are also covered. Concrete flatwork will be covered with particular emphasis on the art of finishing concrete. Finally, some of the more specialized uses of concrete will be covered.

The care required in the construction of the footings and foundations of homes cannot be over emphasized. If the footing is not laid correctly on firm earth, cracks will develop in the foundation wall, and it will be very difficult to make the foundation waterproof. If the foundation is not square, if the dimensions are not accurate, and if the top is not exactly according to grade requirements, adjustments must be made in the structure of the building that will affect many aspects of the work to be done later.

The actual forming is done differently in various areas of the country. In rural areas or where only one building is to be built, forms are made of boards and $2'' \times 4''$ members. After the forms are stripped, the lumber is used for rough flooring, sheathing, and structural purposes. Contractors building a number of

Concrete Construction

houses have forms made in modular 4 × 8 foot panels (or other sizes) using 2 × 4 inch frames and sheathing or plywood for face material. These forms are used over again many times. Many contractors use manufactured form panels which are designed for durability and to provide fast, efficient erection. In larger cities, forms may be rented or purchased from companies that make a specialty of concrete products and building forms.

Concrete as a Material

It is important that the carpenter have an understanding of concrete as a material, because he must build forms strong enough to hold it in place until it sets. When forms fail, the time spent in erecting them is lost and much material is expended. On the other hand, labor and forming material can be wasted if the forms are made strong beyond sensible safe limits. The method and equipment used to place the concrete also has a bearing on how the forms should be designed.

Loads on a foundation wall are generally considered as hydraulic loadings that result in lateral pressure. In other words, the concrete acts as though it were in a liquid state, pushing sideways against the forms.

Basic Formwork

The broad surfaces of the forms are generally plywood sheets. These are held the desired distance apart and prevented from spreading further by devices known as ties. Vertical members serve to stiffen the sheathing and horizontal members known as wales (or walers) hold them in line. The ties generally are fastened through some form of holder which transfers the pressure to the wales. Generally the ties, walers and vertical members are spaced uniformly in both the horizontal and vertical direction because they are designed to take the maximum load at all levels in the form. It is important that the builder appreciate the fact that forms must stand a great deal of pressure. The ties do the work of retaining the concrete and spacing the forms to give the required wall thickness. The vertical members and the wales stiffen the forms. Bracing serves the main purpose of keeping the forms in correct position.

Excavation for Basement

In firm soil, shallow excavations up to 5 feet in depth require a clearance of 18 inches outside the building lines for erecting and removing forms. For deeper excavations, 2 feet or more must be allowed for working space between the building lines and the excavation.

To arrive at the depth of the excavation, the builder should study the vertical section view taken

Masonry Simplified

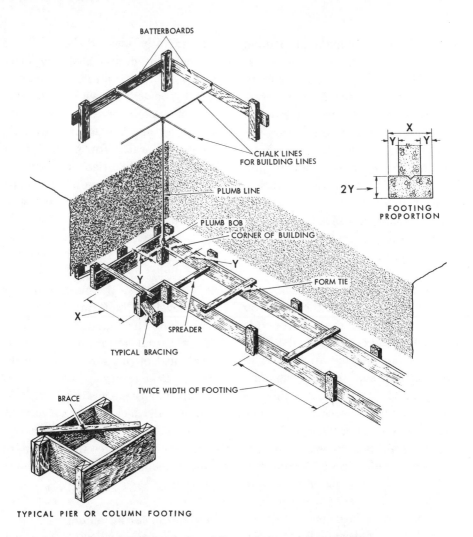

Fig. 8-1. Forming for foundation and pier or column footings.

through the house and calculate the dimensions from the top of the foundation grade to the bottom of the footing. The distance the foundation will project above the finished grade will be marked on the section view.

There must be a minimum exposure of concrete of 8 inches from grade to the lowest wood member as one means of protection from termites. It is also important that surface water be drawn away from the house and,

Concrete Construction

therefore, it becomes necessary that the grade be sloped away in all directions. The excavation should be made so the earth is not disturbed in the bottom of the areas where the footings are to be placed.

The work of excavating should not be done until all stakes have been checked to see that the building is located correctly from the lot lines and has correct dimensions.

Forms for Footing

After all excavations have been made to the correct depths, forms for the footings are laid out and erected. The footings must be straight and level and must rest on undisturbed earth so that the load of the building may be transferred to the ground in a uniform manner. There must be no settling later.

Fig. 8-1 shows a typical method of forming for footings.

"T" Type Footing. A "T" type footing such as shown in Fig. 8-2 provides a starter wall for the foundation and gives the forms a shoulder to rest on. It is used when the foundation is low. Several operations and much time is saved in forming the foundation wall later. The location and the thickness of the foundation wall will not have to be determined. There is no problem in pulling the forms together at the bottom and adjusting them for irregularities in the footing. The manner of making the "T" type footing is shown in Fig. 8-2 and Fig. 8-3.

Some builders in different parts of the country use keys. After the footing has been poured and the concrete has been struck off flush, a key made up of a piece of $2'' \times 4''$ with edges tapered or $2'' \times 2''$ is pressed into the top surface before it has set. The key serves as a tie between

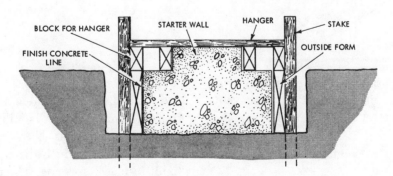

Fig. 8-2. "T" type forms are made by suspending wall form boards accurately above the footing forms.

Masonry Simplified

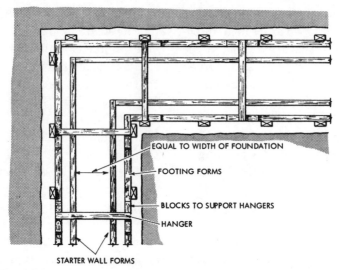

Fig. 8-3. View from above shows how the starter wall forms are spaced.

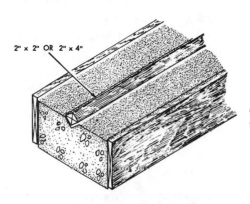

Fig. 8-4. A key is made by pressing a piece of wood into the footing before the concrete sets.

the footing and the foundation wall which will be placed later preventing lateral movement. See Fig. 8-4.

Concrete Foundation Forming Systems

There are several different types of forming systems because of a number of different problems and because of individual preferences. One type of forming, which is still used today though it has been largely replaced, is called "built-up" or "built-in-place" forming. The materials used are 2" × 4" vertical members and 1" × 6" boards. After the

Concrete Construction

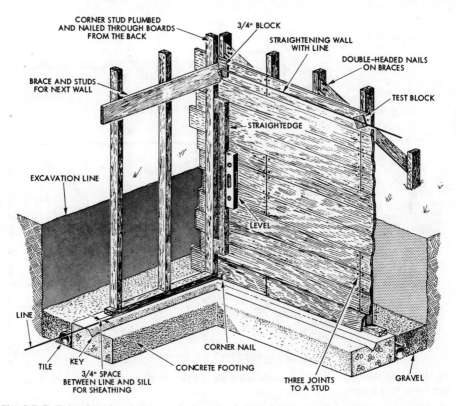

Fig. 8-5. Built-in-place forming is used when the builder wishes to use the form material in the structure of the house.

forms are stripped, the materials used to build the forms are utilized in the building of the house. See Figs. 8-5 and 8-6.

Boards are also used when the form is to be torn apart and the wood used in building the structure. Green lumber should be avoided because exposure to air and sun causes it to shrink. Shrinkage opens joints and allows part of the wet concrete to leak out leaving a concentration of the aggregate. On the other hand, if the wood is too dry it will soak up water from the concrete and will swell. This may bring about distortion in the forms. It is good practice to hose down the forms the day before the concrete is to be placed and to continue hosing up to the time of placing.

Forms for basement foundations

Masonry Simplified

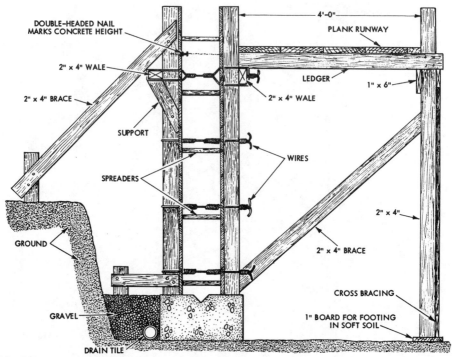

Fig. 8-6. A sectional view through formwork which is built-in-place. Notice the wire ties, wood spreaders, and bracing.

are usually made of 4′ × 8′ plywood panels on frames made of 2″ × 4″s. Templates or jigs are not necessary for these forms, because the sheets of plywood have exact dimensions and have perfectly square corners. The frames are assembled to fit the plywood sheet. The plywood is made for this particular purpose in 5/8 and 3/4 inch thicknesses. Plywood panels are by far the fastest and most economical panels to fabricate. They last longer and give a far better finish to the wall than sheathing, leaving a wall surface which is quite smooth. These panels may be obtained with a plastic surface that is waterproof, abrasion resistant, and easy to clean. Forms with plywood faces can be reused many times. Patented forms with metal frames provide edge protection for the plywood panels. The panels may be replaced when damaged or worn.

Ordinary plywood panels are given a coat of oil so that they will separate from the wall without difficulty. The oil coating also permits

Concrete Construction

them to be cleaned easily. If the walls are to be painted or plastered, however, the oil in the concrete may prevent the finish from bonding. Some other agent may be used, such as a lacquer.

Much of the concrete forming today is done using panel forms made by the builder. Some builders make the forms on the job site, while others build them elsewhere and transport them from job to job. They are made as large as can be conveniently carried and put in place. For low walls, the forms are made by nailing a number of 1 inch boards to evenly spaced 2 × 4 inch uprights. When the forms are stripped, the material may be reused in building the house.

The forms used for houses with basements are modular 4 × 8 foot panels (or other convenient sizes) made with frames of 2 × 4 inch members having intermediate stiffeners and faced with pieces of ¾ inch plywood. Several types of ties are used. Wire ties and band iron are still used but have been largely replaced by snap ties and various patented devices. See Fig. 8-7 for an example of this type of forming.

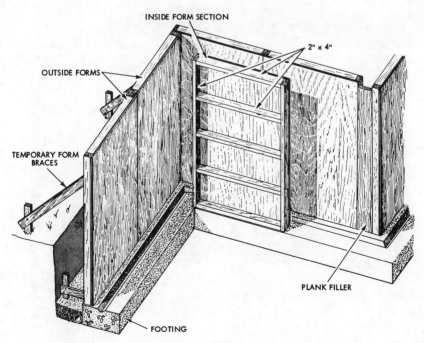

Fig. 8-7. Panels with fillers of various sizes can be adapted to most forming problems and may be reused many times.

Masonry Simplified

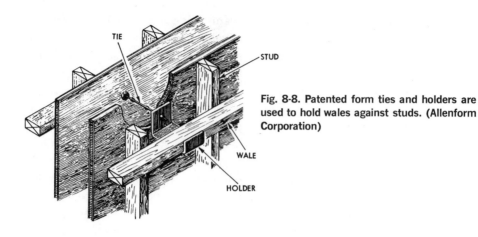

Fig. 8-8. Patented form ties and holders are used to hold wales against studs. (Allenform Corporation)

Fig. 8-9. Some manufactured panels have steel frames and plywood faces. Patented corners and form ties are used to speed up assembly. (Universal Form Clamp Co.)

Concrete Construction

A number of manufacturers have developed several ingenious forming systems to replace, or to supplement, forms built on the site. Some are quite simple, using sheets of plywood and special ties and clamps. Wood wales stiffen the plywood sheets and keep them in line. See Fig. 8-8. Other forming systems use panel units with steel frames and plywood faces. They come in several sizes and require a variety of tying devices to meet special forming problems. See Fig. 8-9. The patented forms have several advantages, the most important of which are that they can be erected quickly by fewer men, they can be reused a great many times, and they are very durable. The companies that manufacture forms rent or sell them to builders. Some provide a service whereby they analyze the forming problems, prepare working drawings, and submit complete material lists. When a builder has much work that must be frequently repeated, it pays him to purchase a set of manufactured forms.

Form Ties. A form tie is a device that passes through both sides of a form, retaining them against the lateral pressure of the concrete. One type of tie that has been used for many years is the wire tie. See Fig. 8-10.

Snap ties are very popular and are used by many builders. The forms are kept the proper distance apart by means of washers which are fixed to the tie. Two places about an inch inside of each washer are weakened by flattening. These are the points where the tie will be broken off after the forms are removed. See Fig. 8-11.

Building Forms for Low Walls

Forms for low walls are generally

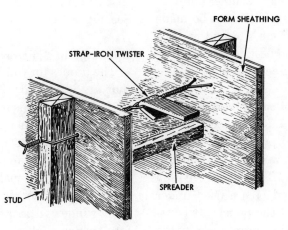

Fig. 8-10. Wire ties are twisted until the studs draw the sheathing tight against the spreaders.

Masonry Simplified

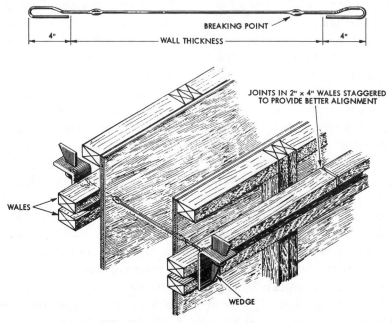

Fig. 8-11. Snap ties provide a means of spacing the walls at the proper distance apart and also clamping the whole assembly together.

made of 2″ × 4″ members and 1″ × 6″ sheathing. They can be made quickly on the job site and may be taken apart and used for other building purposes. See Figs. 8-12 and 8-13.

Alternative Procedure for Building Forms for Low Walls Using "T" Footing. The procedure for building forms for low walls is essentially the same when "T" footings are used except that wedges are used under the sill to level the panels. It is assumed that the forms are made away from the excavation area and that the inner forms will be set first. See Fig. 8-14.

Openings in Concrete Walls

The frames for windows and doors that are located in the foundations are installed in some instances before the pouring of concrete. In the case of some metal basement windows, the frames are put in place and wood forms are used to make the sill and sides of the window opening. For other windows a strip is nailed to the form which makes a recess in the concrete when it is placed. When

Concrete Construction

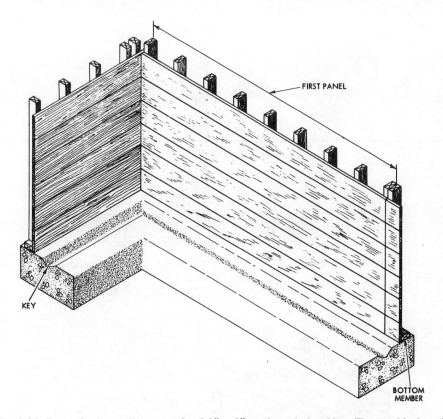

Fig. 8-12. Forms for low walls are made of 2" x 4" studs and sheathing. The outside form is usually erected first.

the form for the window is removed, a recess is left in the wall into which the metal sash is dropped. It is sealed with grout. See Fig. 8-15.

Some finished frames are put in place in the forms and are held by a key strip. The concrete stops against the frame. See Fig. 8-16.

Other openings are made by using a buck which is removed after the concrete is placed, leaving blocks or strips in the wall to which the frame is later fastened. One important thing to remember is that the buck must be made so that it can be removed without too much trouble. The corners of the window buck are mitered, and the bottom piece is cut through, so that it will slip out easily. See Fig. 8-17.

Forms for Steps

There are many different methods of forming for steps. The important

Masonry Simplified

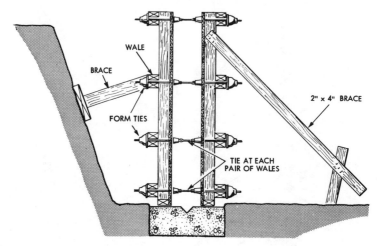

Fig. 8-13. Sectional view through a foundation form shows how the wall is held together and braced.

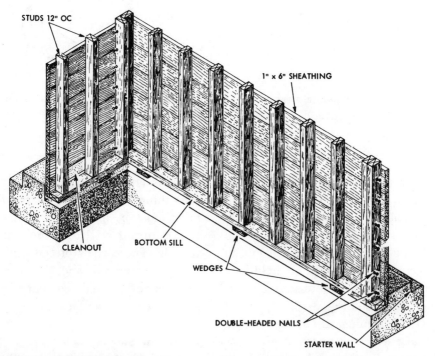

Fig. 8-14. Panel forms used with a "T" footing. Note the wedges used to level the forms. The inside panels are erected first.

Concrete Construction

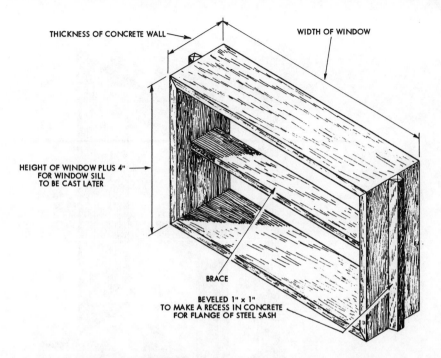

Fig. 8-15. The rough form for metal framed basement windows is removed after the concrete has set, leaving a recess for the window frame.

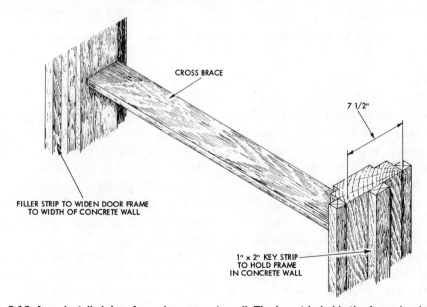

Fig. 8-16. A pre-installed door frame in a concrete wall. The key strip holds the frame in place.

Masonry Simplified

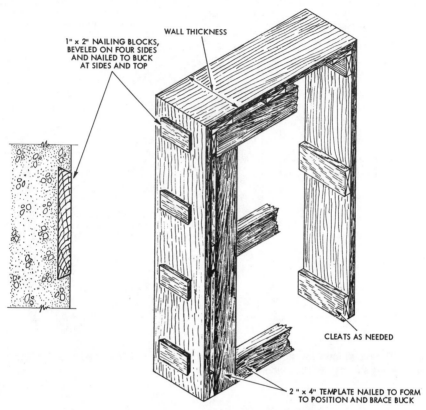

Fig. 8-17. A rough buck is made so that nailing blocks are left in the wall.

considerations are where the steps are located in relation to adjoining structures, and how they are to be supported. In most cases, however, the materials used are the same, and the bracing methods are similar. The riser forms are usually nominal 1 or 2 inch boards. The side forms or braces are made from 2″ × 6″ or 2″ × 8″ stock.

Fig. 8-18 shows a typical forming method for basement steps.

Fig. 8-19 shows two designs for sidewalk steps, one employing curbs. These may also be used as approach steps for building entrances.

Fig. 8-20 shows a simple method of forming for a small porch and steps resting on earth and a foundation.

The steps shown in Fig. 8-21 are self-supporting, reinforced with steel bars, and have a foundation which is not integral with the sidewalk. Such

Concrete Construction

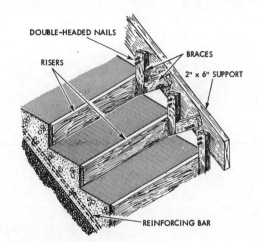

Fig. 8-18. Formwork for basement steps.

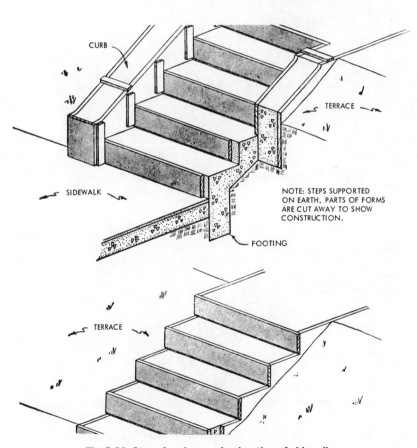

Fig. 8-19. Steps for changes in elevation of sidewalks.

219

Masonry Simplified

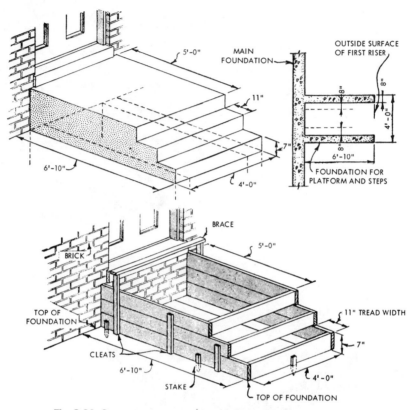

Fig. 8-20. Concrete steps and platform for a residential entrance.

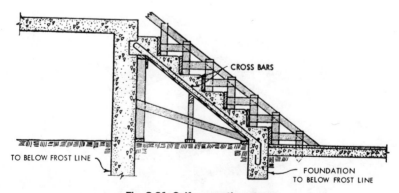

Fig. 8-21. Self supporting steps.

220

Concrete Construction

steps are commonly used where a concrete front porch for a residence is several feet above grade. Note how the top of the steps is notched into the porch foundation wall.

Steel Reinforcement

Reinforcing is frequently used in footings, foundations, and steps to add strength and alleviate stresses. Reinforcing is usually designed by engineers and installed by professional steelworkers. The materials used are either deformed steel bars or welded wire fabric as shown in Fig. 8-22.

Concrete piers, pilasters and columns are also reinforced. Fig. 8-23 shows reinforcing steel in place for a column.

Fig. 8-23. Reinforcing steel for a concrete column.

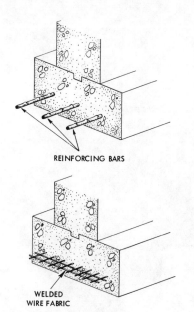

Fig. 8-22. Steel reinforcing in concrete footings.

Placing Concrete in Forms

Proper placing of concrete in forms is one of the most critical operations in building construction. Footings and foundations must carry the weight of the rest of the building. It is essential that the forms be completely filled with concrete and that the aggregate and cement proportions be consistent throughout. Remember that a good concrete mix will be stiff and difficult to work or compact. However, there are certain techniques which, if followed, will assure a sound, solid structure.

In all formed structures, whether they be footings, foundations, walls or others, the first placing should be at the end or ends of the form. Subsequent placing is done continuously

221

until the form is full. Horizontal movement of the concrete after it is placed in the form should be avoided. That is, the concrete should be placed as close as possible to its final position.

During or immediately after placing, the concrete must be compacted tightly against the forms to avoid air pockets or voids both internally and on the surface. This may be accomplished on smaller structures by *lightly* tapping the sides of the form with a hammer. Too heavy tapping, however, will cause the heavier aggregate to segregate and settle to the bottom. This will not only cause an uneven consistence in the mixture, but may also cause the forms to spread and separate at the bottom.

The most common method of compacting on all sizes and types of structures is by the means of mechanical vibrators. Vibrators, used correctly, assure complete consolidation of the concrete. There are no set rules as to the length of time concrete should be vibrated, but in most cases, 10 to 15 seconds for each placing is sufficient. This is another skill that the cement mason

Fig. 8-24. Compacting concrete in the forms with a vibrator. (Adolphi Studio)

Concrete Construction

Fig. 8-25. Float finishing the top surface of a foundation.

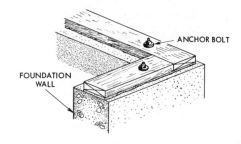

Fig. 8-26. Anchor bolts in a concrete foundation.

must learn by experience. Fig. 8-24 shows mechanical vibrators being used while concrete is being placed.

After the form is filled and compacted to the specified level, the top surface is finished with a hand float. See Fig. 8-25.

At this point, if the foundation is to be used as a base for frame construction, the anchor bolts for the sills should be placed. It is the cement mason's responsibility to check the building plans for the correct locations and to install the anchor bolts. See Fig. 8-26.

Concrete Walls. Concrete walls are placed and compacted in the same way as foundations. It may be impractical, however, to try to place all the concrete up to full height in one operation. In these cases, a few vertical feet are placed and com-

223

Masonry Simplified

pacted as for a foundation. Then succeeding layers, called *lifts*, are placed and compacted until the full height is reached. (From 12 to 15 inches per lift is recommended.) If vibrators are used, the vibrator head is allowed to penetrate through the new lift and at least 6 inches into the preceding lift. This makes for a unified structure with a uniform surface after the forms are removed.

Fig. 8-27 shows a completed foundation-basement wall system for a small residence. Note that the areaways, porch and garage foundations, and other structures and openings were formed in such a way that all the parts became one continuous piece of concrete. This prevents any part from settling away from the rest, insuring a firm and continuous support for the rest of the building.

Removal of Forms

Generally four days should elapse before the forms are removed if time will permit. The panels are usually removed in the reverse order from which they were placed. Form tie clamps are knocked off and the wales removed first. The forms are pried from the walls with care in order not to damage the concrete. The ties pull out of the forms and remain projecting from the wall. It is wise to wait a day or two longer before removing the tie ends. This is done by twisting or pulling on the tie end so that it breaks at the "break back" point within the wall. The hole in the wall is then filled with grout and finished flush with the surface.

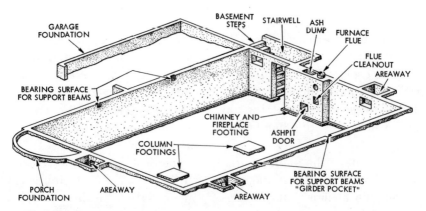

Fig. 8-27. Completed concrete foundation for a small, basement-type residence.

Concrete Construction

Foundations for Homes with Slab-At-Grade

An increasing number of homes are being built in all parts of the country in which a concrete floor is laid directly on the ground. See Fig. 8-28. Certain precautions must be made in order that the floor be satisfactory. The Small Homes Council suggests the following: The earth around the house must be graded so that water will drain away properly. The entire area where the floor will be laid should be covered with 4 inches of washed gravel or crushed rock in order to reduce the capillary rise of moisture. A membrane should be provided over the gravel strong enough to resist puncturing when the concrete is placed. This membrane serves as a vapor barrier to keep moisture from entering the slab from the ground. Polyethylene film, asphaltum board 1/8 inch thick, or reinforced duplex paper with asphaltum center may be used. Overlap paper 4 inches. One additional problem to solve in cold climates is heat loss. The heat loss is primarily around the perimeter of the house,

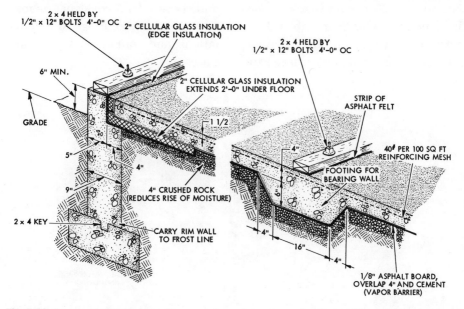

Fig. 8-28. A concrete floor laid on the ground requires a vapor barrier, and in cold climates must have perimeter (edge) insulation.

Masonry Simplified

and to counteract the loss and prevent condensation resulting from the cold floor, edge insulation is required. Two-inch thick rigid waterproof insulation extending 2 feet from the walls is suggested. Where panel heat is used in the floor, the insulation should cover the entire floor area.

In some areas the foundation wall is omitted entirely and a simple perimeter support is constructed instead by merely thickening the edge of the slab. See Fig. 8-29.

In some regions of the South, footings are laid both for the exterior wall and the bearing partition by making trenches in the firm earth. Reinforcing bars are used in the footings and concrete block is used for the foundation wall. A polyethylene film is spread over the earth where the slab is to be laid. The floor slab will be finished with terrazzo and in order to make it as rigid as possible, wire mesh is imbedded in the concrete. See Fig. 8-30.

In some areas where the soil is too unstable to permit the use of conventional footings and foundation walls, grade beams may be used. See Fig. 8-31. (Grade beams are continuous beams running around the house perimeter; they rest on piers.) Holes are dug at the perimeter of the building, a maximum of 8 feet apart, to a depth sufficient to bring them to solid soil well below the frost line. Concrete is placed into the holes, or into shells made for that purpose, to form piles. Forms are made to contain the grade beams which rest on the piles. A steel rod (a dowel) serves to position the grade beam. Horizontal reinforcing bars add strength to the grade beams. This construction may be used with slab-on-ground foundations and for houses with crawl spaces.

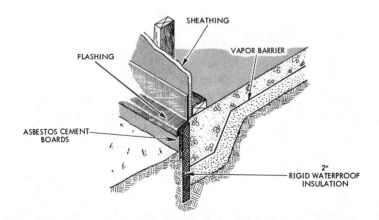

Fig. 8-29. A simple perimeter support for light construction in warm climates.

Concrete Construction

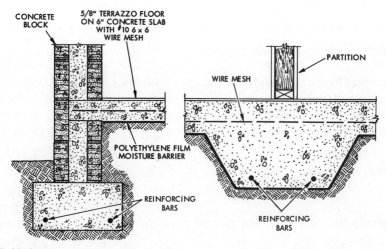

Fig. 8-30. Method used in warm climates for houses made with concrete block exterior walls and terrazzo floors.

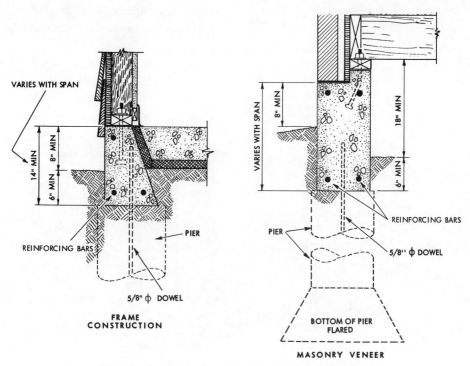

Fig. 8-31. Examples of grade beam and pier construction.

Masonry Simplified

Waterproofing Foundations

There are many methods of waterproofing which may be used successfully. Those discussed here are typical and are recommended by average building codes, the Portland Cement Association, and by other agencies interested in good construction.

Perhaps the simplest form of waterproofing for foundations built in soil which is just damp is to apply hot tar or asphaltum (asphalt) to the outside surfaces. These materials are moistureproof and constitute satisfactory waterproofing. This is shown in (A) of Fig. 8-32.

Where excessive dampness or severe conditions of water in the soil occur, the exterior surfaces of the foundation can be satisfactorily waterproofed as shown in (B) of Fig. 8-32. Two or more plies or layers of membrane (felt, for example) coated with tar or asphaltum are used. Note in Fig. 8-32 that for extraordinary conditions the waterproofing is applied at the joint between the footing and the foundation and that the floor is built in two layers, with the waterproofing placed between the layers. In addition, note that a mix of $1:2\frac{3}{4}:3$ concrete can be used as

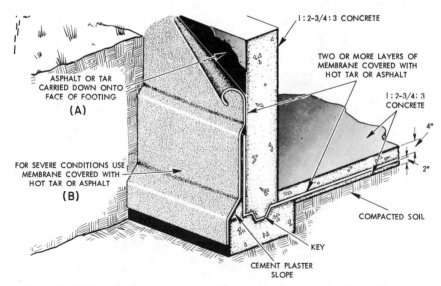

Fig. 8-32. Method of waterproofing foundations against severe groundwater pressure.

Concrete Construction

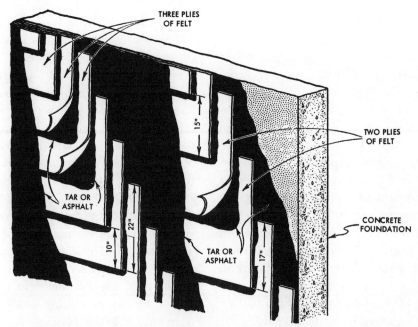

Fig. 8-33. Method of lapping three and two layers of felt waterproofing for foundations.

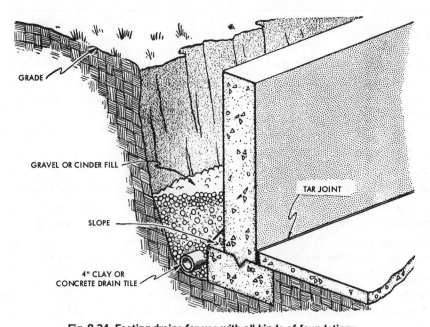

Fig. 8-34. Footing drains for use with all kinds of foundations.

Masonry Simplified

a means of further retarding the flow of water through the foundation. Such concrete becomes dense in setting and, because of this characteristic, tends to prevent the flow or seepage of water through it. The slope at the point where the foundation meets the footing also helps to make that joint waterproof.

Fig. 8-33 shows a slightly different treatment of the membrane and tar or asphaltum treatment.

Another waterproofing method frequently used where there is excessive water in the soil is shown in Fig. 8-34. The clay or concrete tile is laid around all sides of the footing with a gravel or cinder fill covering it to the depth shown. The fill material allows the water to flow directly to the tile where it collects and drains off to some point away from the basement where it cannot do any harm. Occasionally, the water drains into a sump from which it is pumped to the regular sewer. When soil water conditions are really severe, drain tile are used in conjunction with the waterproofing methods explained in Fig. 8-32. Note the tar joint between the floor and foundation in Fig. 8-34. This method of calking is effective and helps materially to keep moisture out of the basement in cases where no waterproofing materials were built into the floor.

Concrete Flatwork

Most of the cement mason's apprenticeship training and his work as a journeyman will be concerned with concrete *flatwork*. Flatwork means the site preparation, placing, and finishing of concrete on a horizontal surface. Virtually all residential and commercial construction involves flatwork in the form of floors, driveways, sidewalks, etc. As these structures are generally in constant use, they must be strong and safe as well as pleasing to the eye. A well finished piece of concrete flatwork is an indication of the skill and pride of the craftsman who constructed it.

The four main steps in concrete flatwork are: site preparation, placing of the concrete, finishing, and curing.

Site Preparation

The forms for concrete slabs are called *screeds*. They are usually constructed of good straight 2" × 4" or 2" × 6" lumber, with wooden stakes at 4' intervals. They are set to grade by the use of a builder's level. See Fig. 8-35.

The next step after setting the screeds is the preparation of the subgrade. This is the surface that the

Concrete Construction

Fig. 8-35. Using a builder's level to set screeds to grade. (Portland Cement Association)

Fig. 8-36. Hand compacting a subgrade. (Portland Cement Association)

concrete will lie upon. It is important that the subgrade be level in order to assure uniform thickness in the finished slab. Low or high spots in the subgrade will cause uneven stress in the slab which will almost invariably lead to cracking. After the subgrade is leveled, it must be compacted in order to provide a firm and relatively non-absorbent support for the concrete. On small jobs, such as sidewalks, compacting may be done with a hand tamper as shown in Fig. 8-36.

Mechanical compactors or rollers are recommended for larger jobs. See Fig. 8-37.

If the subgrade soil is sandy and compacts well, there will probably be no need for additional fill. Other soils, containing organic matter, should be excavated to allow for a

Fig. 8-37. Mechanical compactor. (Portland Cement Association)

fill of cinders or crushed stone. This fill should be from 4 to 6 inches deep, depending on the thickness of the slab. It should be level and well compacted.

Placing Concrete

After the screeds are set and the subgrade leveled and compacted, it is time to place the concrete. Remem-

231

Masonry Simplified

ber that a good concrete mix contains only enough water to make it workable. Therefore, great care must be taken to make sure the fresh concrete is compacted solidly against the forms or screeds to prevent air pockets or voids around the large aggregate. This is done by thoroughly spading and packing along the screeds, as shown in Fig. 8-38. This step is necessary to prevent voids or honeycombs which weaken the edge of the slab. The concrete should be placed *continuously*—that is, beginning at one edge or end of the slab and continuing from there until the form is filled.

If welded wire fabric is to be used as reinforcement in a slab, care should be taken to keep it at the proper level. This may be done by providing supports and avoiding walking on it, or by placing part of the concrete, laying the fabric, and then placing the rest of the concrete. If the steel is placed directly on the subgrade, it becomes useless as reinforcement.

Fig. 8-38. Concrete must be tightly compacted along the screeds. (Portland Cement Association)

Concrete Construction

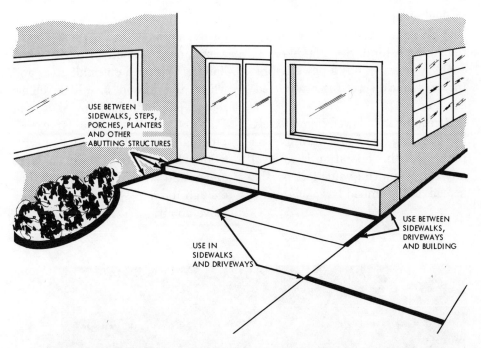

Fig. 8-39. Isolation joints are installed where different parts of the concrete structure meet.

Isolation Joints. Isolation joints are employed where concrete slabs abut against other slabs or structures, and the natural expansion and contraction of each may cause cracking. Examples of this are sidewalk and driveway intersections, slab and wall intersections, etc., as shown in Fig. 8-39. The isolation joint is usually a strip of bituminous fiber material. It is resilient material which absorbs thermal expansion of the concrete.

Finishing the Concrete

The first step in finishing the concrete surface is *screeding*. This is the striking off and compacting of the concrete down to the tops of the forms or screeds. (Note: The term "screed" applies to both the flatwork forms and strike-off tool.) The screed may be made of wood or metal. The important requirements are that the leveling surface be absolutely straight and that it be long enough to completely span the forms. Screeding is done *while* the concrete is being placed. A sawing motion is used back and forth across the surface as the screed is pushed forward along the tops of the forms. There should always be some excess concrete pushed along ahead of the screed so that there are few voids or valleys remaining behind. See Fig. 8-40.

On large scale jobs a power vibrator mechanical screed is often used. See Fig. 8-41.

Screeding is followed immediately by *darbying* or *bull floating*. The purpose of this operation is to level any ridges and fill voids left by the screed and to imbed large aggregate particles just below the surface. The darby, as shown in Fig. 8-42, top, is a long, narrow wood strip with a low handle. It is used on small areas and where the mason has little elbow room. The bull float, as shown in Fig. 8-42, bottom, is a large rectangular metal float equipped with a long handle. It is used on large areas, and where the operator has enough room to stand up and freely move about.

Fig. 8-40. Screeding freshly placed concrete. (Portland Cement Association)

Concrete Construction

Fig. 8-41. A power-vibrated mechanical screed. (Stow Manufacturing Co.)

While screeding and darbying or bull floating are done immediately upon placing, the next step must wait until the concrete has begun to set and all excess surface water (called *bleed*) has evaporated. Under normal weather conditions, this is usually about five hours. Then it is time for jointing and edging.

Jointing is a very important part of concrete flatwork. As discussed previously, all concrete will naturally shrink slightly upon hardening, causing cracks. This is especially true of outdoor structures such as sidewalks and driveways. In order to control the location of the cracks and to hide them from sight, grooves are scored across the partially hardened surface at regular intervals. Special tools called jointers or groovers are used for this operation. Fig. 8-43 shows the correct use of a jointer. Note the use of a straightedged board as a guide. The depth of the joint is usually about $\frac{1}{5}$ of the thickness of the slab. Joints may also be cut at this time with a power saw as shown in Fig. 8-44.

Edging is done at the same time

235

Masonry Simplified

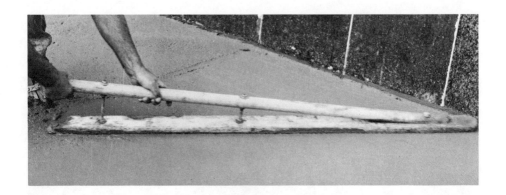

Fig. 8-42. Top: Darbying. Bottom: Bull floating. (Portland Cement Association)

Concrete Construction

Fig. 8-43. Jointing a sidewalk. (Portland Cement Association)

Fig. 8-44. Jointing with a power saw. (Portland Cement Association)

Masonry Simplified

Fig. 8-45. Edging a sidewalk. (Portland Cement Association)

as jointing. The edging tool, as shown in Fig. 8-45, is held flat on the concrete surface and drawn along the edge. The leading edge of the tool should be slightly elevated to facilitate ease of operation. This provides a rounded and compacted edge which prevents chipping when the forms are removed and later when the structure is in use.

Floating is the next step after jointing and edging. This is done with small hand floats as shown in Fig. 8-46. The floats may be made of wood or metal. Most masons prefer a magnesium float because they are light, easy to clean, and have been found to work particularly well with air-entrained concrete. If a coarse texture for skid resistance is desired, floating may be the final finish of the surface. Wood floats, which tend to drag on the surface, are often employed to produce a rough final finish. Floating also is used to remove any remaining irregularities after the previous operations.

Troweling is the final stage of the finishing operation. Steel trowels are used after floating. See Fig. 8-47. This gives a high degree of hardness, density, and smoothness to the finished surface. This operation is usually repeated once or twice, depending upon the degree of smoothness desired. For the first troweling, a fairly large trowel, usually 18 × 4¾ in., is used. The trowel is placed flat on the surface and passed over the surface in a sweeping, arc-like motion with each pass overlapping one half of the previous pass. This way,

238

Concrete Construction

Fig. 8-46. Floating. (Portland Cement Association)

Fig. 8-47. Steel troweling. Note the use of kneeboards. (Portland Cement Association)

each pass covers the surface twice. Smaller trowels are used for each succeeding troweling.

Power trowels, such as shown in Fig. 8-48, are widely used for larger areas.

If joints and edges are disfigured by floating or troweling, they should be rerun.

Although trowelled surfaces are highly impervious to dirt or foreign matter and easily cleaned, they will

Masonry Simplified

Fig. 8-48. A typical power trowel. (Stow Manufacturing Co.)

be slippery when wet. For skid resistance, the surface may be slightly roughened by *brooming,* as shown in Fig. 8-49. The broom is dragged across the surface toward the operator. The stiffness of the bristle of the broom will determine final texture of the surface.

Construction Joints. There will be occasions when work must be halted on a partially placed slab due to the end of the workday, bad weather, lack of materials, etc. In

Fig. 8-49. Brooming a concrete surface. (Portland Cement Association)

Concrete Construction

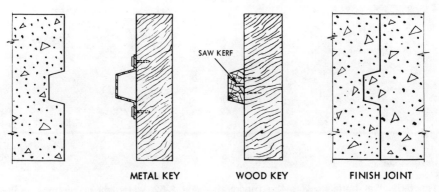

Fig. 8-50. Special forms are used to make a construction joint.

these cases, a construction joint is used. A special form, as shown in Fig. 8-50, is placed between the regular forms and concrete is placed and compacted tightly against it. When work is to be resumed, this form is removed and the fresh concrete is placed directly against that which was previously placed. Note the tongue-and-groove effect. This prevents independent movement of the two placings, making, in effect, a structural whole.

Curing Concrete

Proper curing is of the utmost importance in concrete flatwork. Even an excellent job of placing and finishing may be ruined by hasty or other incorrect curing methods. The chemical reaction of water with cement is called *hydration*. Hydration begins as soon as the concrete is mixed and continues until the concrete is thoroughly hardened and most of the water has evaporated. The essence of good curing is to allow the hydration to proceed at its own natural rate. This is one reason why the use of accelerating or retarding admixtures is generally discouraged unless absolutely necessary. Damp curing is the best, easiest, and most economical method of curing concrete. This is done by simply keeping the surface of the concrete covered with wet or watertight material so that the concrete retains its moisture during the initial setting period. This is usually 3 to 5 days in average climates. The use of burlap or canvas, kept continually wet, is a common method of keeping the surface wet. See Fig. 8-51.

Small areas, such as driveways and sidewalks, may be sprinkled periodically with a garden hose. Another method is to cover with water-

Masonry Simplified

Fig. 8-51. Wet burlap covers the concrete while curing. (Portland Cement Association)

Fig. 8-52. Watertight plastic may be used in curing. (Portland Cement Association)

tight plastic material as shown in Fig. 8-52.

Another method of damp curing is spraying with any one of several patent chemical compounds which form a waterproof membrane on the surface. Membrane compounds containing white pigment are recommended for hot weather curing as the white surface reflects the rays of the sun.

Flatwork Details

Most concrete work in residential construction consists of building sidewalks, driveways, floors, and steps. As these structures are almost constantly in use, great care must be taken in their design and construction. The following examples are more or less typical and are intended only as a general outline for good design and construction. The dimensions given conform to most building codes, but, as always, local codes should be carefully observed.

Sidewalks

Sidewalks, first of all, should be made wide enough to accommodate the expected traffic on them and secondly, to have good proportions relative to the surroundings. In general, sidewalks are divided into two classes, namely, main and secondary walks. Main walks can be further divided into street and approach walks. In Fig. 8-53, the sidewalk along the street and the one leading up to the residence are main walks.

Concrete Construction

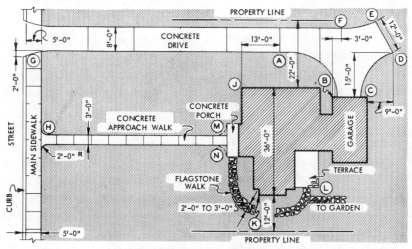

Fig. 8-53. Residence sidewalk and driveway.

The walk along the street is called a street walk and the walk leading to the residence is called an approach walk. The flagstone walk which extends from one part of the residence to another part or, in other cases, from the residence to a garden, etc., is a secondary walk.

Ordinarily, street walks are made at least 5'-0" wide in residential neighborhoods and up to 15'-0" or more in business sections of cities. A 5'-0" walk can accommodate residential district traffic easily and allow people to pass while walking in opposite directions. The wider walks for business districts allow for crowded conditions. Approach walks are generally made at least 3'-0" wide and secondary walks from 2'-0" to 3'-0" wide.

To be in good proportion to the surroundings, a walk, especially approach and secondary walks, must be wide enough to avoid the appearance of extreme narrowness. For example, if the approach walk in Fig. 8-53 were only 2'-0" wide, it would seem out of proportion to the rather large ground area and to the size of the residence.

For ordinary traffic and strength requirements, main and secondary walks should be at least 4" thick and have a rather smooth surface.

Great care must be exercised to see that walks are well aligned or square with intersecting walks and with residences or other buildings except where they are purposely made curving and twisting as are some secondary walks. If the

243

approach walk in Fig. 8-53 were not square with the street walk and with the house, its effect would be annoying and would spoil the general appearance of the lawn. The irregularity of the secondary walk in Fig. 8-53 adds to the charm of the lawn.

Whenever a change of direction (a curve) is required in a sidewalk, care should be taken to make the curve long rather than short. A short curve, such as shown in (A) of Fig. 8-54, does not appear graceful, tends to crowd people when passing, and frequently causes them to break their stride. On the other hand, the curve at (B), being longer, has a much better general appearance and is much easier to negotiate. At points such as shown in (C) of Fig. 8-54 and at H in Fig. 8-53 where an approach walk meets a street walk, the approach walk can be slightly curved or widened to make a better appearing intersection. The radius of such a curve is usually from 2'-0" to 4'-0". In like manner, if desired, curves can be made at M and N in Fig. 8-53, where the approach walk meets the concrete porch.

In some instances, colored sidewalks are desired as a means of ornamentation for borders, etc. In any case, the coloring is used only in the top course and is very carefully and thoroughly mixed before and after water is added to the mix. The following recommendations for producing colored sidewalks are based on tests made by the Portland Cement Association and if observed carefully should result in satisfactory work.

A pigment suitable for use in cement must fulfill the following requirements:

(1) It must be durable under exposure to sunlight and weather.

(2) It must produce intense color.

(3) It must be of such composition that it will not react chemically with the cement to the detriment of either cement or color.

These requirements are best fulfilled with mineral oxide pigments. Other pigments, such as organic dyes, have a tendency to fade and may even reduce the strength of the concrete. There are two kinds of mineral oxides available that are satisfactory—natural oxides that come direct from the mines and manufactured pigments which are prepared especially for concrete work. Ordinarily, natural pigments cost less per pound and may be used where dull colors are satisfactory. However, where bright colors are desired, manufactured pigments produce best results. To achieve a given amount of color, more of the natural pigment is required than the manufactured pigment. It happens frequently that the smaller amount of manufactured pigment actually produces the desired results at a lower cost than the cheaper, natural pigment.

The following recommendations will serve as a guide in determining

Concrete Construction

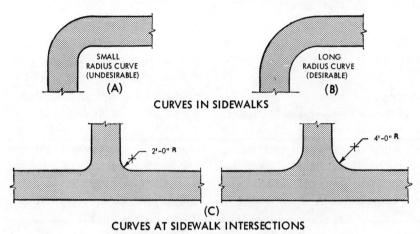

Fig. 8-54. Curves at sidewalk intersections.

proper pigments for use in securing the colors listed:

Buff, yellow, or red — Iron oxide pigments

Green — Chromium oxide pigments

Blue — Cobalt blue

Brown — Iron oxide or iron and manganese oxide pigments

Slate gray and black — Iron oxide or manganese dioxide pigments

The color which is produced in cement work is determined primarily by the proportion of pigment to cement rather than by the proportion of pigment to the total volume of the mix. Because of this, modern color specifications give the weight of pigment to be used per sack of cement. It has been found that pigments may be used in amounts up to about 5 percent of the weight of cement—that is, 2 to 4 pounds of pigment per sack (94 pounds) of cement.

To obtain maximum clearness and brightness in colored finishes, white cement or a mixture of white and gray cement should be used. White finishes are obtained by using all white cement.

Generally, sidewalks should have a smooth finish. However, a rough finish is desirable sometimes to overcome the tendency toward its being slippery. How to obtain various finishes is explained in succeeding pages.

Sidewalks should have their surfaces slightly sloped so that water will run off readily during rains and thawing snow or ice. The slope should be almost unnoticeable to the eye and certainly not enough to constitute a walking annoyance. It is best that the slope be from one side of the walk to the other.

Driveways

A driveway should be wide enough to accommodate an automobile or truck without too much care being required by the driver to keep the wheels on the driveway. This is important because during wet weather or at times when snow and ice are present, even one wheel off the driveway could easily cause trouble in addition to cutting ruts in a lawn. Because of this, driveways are made from 8' to 10' wide. It will be noted that the approach part of the driveway in Fig. 8-53 is 8'-0" wide. If curbs are desirable, they should be beyond a minimum 8'-0" width.

The widths of driveways should be materially increased on curves or places where facilities are provided for turning completely around. For example, the area $ABCDEF$ in Fig. 8-53 allows enough room for backing a car out of the garage, then turning it to head for the street or turning so as to back it into the garage.

A good concrete driveway should be at least 6" thick because of the weight of cars and because at widths of 8' or 10' such a thickness is required for overall strength. The strength can be increased further by using heavy wire mesh in the concrete as reinforcing.

As explained for sidewalks, driveways must be carefully aligned or squared with relation to the street or the shape of the building site. The driveway in Fig. 8-53 would be an eyesore if it did not intersect the street at right angles and was not parallel to the approach sidewalk. Sometimes designers curve such a driveway as a means of relieving the monotony of the severe straightness. This would be all right except that it sometimes makes steering more difficult for the driver. If building sites are situated at an angle to the street, then driveways also may be at an angle.

Floors

Most concrete floors for basements, garages, etc., should be from 4" to 6" thick. In cases where no exceptionally heavy traffic is expected, the 4" thickness is ample. Where machinery or heavy use is expected, the 6" thickness should be used. The thickness of a floor should be uniform over its entire surface. For floors having a 6" thickness, added strength can be provided by putting a heavy wire mesh in the concrete.

Unless floors are placed on sandy soil, it is best to excavate 6" to 8" deeper and fill the added depth with cinders or gravel. This will prevent any possible cracking or buckling because of moisture or freezing. Such practice also helps to waterproof the floor.

Concrete Construction

Floors in basements and garages or barns should have a decided slope or pitch so that laundry water, auto washing water, etc., will run quickly into the drains provided. The slope should be at least $\frac{1}{4}''$ to every foot. Fig. 8-55 shows plan and section views of a typical concrete floor for a two-car garage. Note that all floor areas such as *ABC, ACD, DCE,* and *ECB* all slope toward the drain. Floors used for first floors in residences, in granaries, etc., where no water is involved can be perfectly level throughout.

Floors are not generally exposed to the sun and in most cases are in cool places where appreciable changes in temperature are unlikely. Thus a great deal of expansion need not be expected. However, it would be wise to make some provision for possible expansion. It is recommended that an expansion joint be constructed all around the edges of floors where they meet side walls or foundations. For example, in Fig. 8-55, it would be wise to make a continuous expansion joint where the floor meets the walls along sides

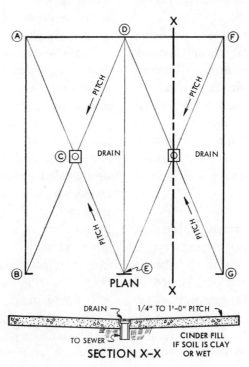

Fig. 8-55. Details of pitch to facilitate draining of concrete garage and basement floors.

BA, *AF*, and *FG*. The cost is not great and the benefit is insurance against the appearance of cracks.

If trenches for sewers and drains are dug, for example, in a residence basement, they should be carefully backfilled and tamped to avoid future settlement. Such settlement under a floor is apt to cause not only cracks but actual breakage as well.

The surfaces of floors should be finished smooth for easy cleaning and ready drainage of water. If carpeting or linoleum is to be laid directly on a concrete floor, the surface should be made somewhat rougher in order to increase the adhesion of the cement between the padding and the concrete floor.

Joints are unnecessary in floors and are not usually made because they constitute a dirt-collecting source which is difficult to clean.

Color may be used in recreation room floors to great advantage. The instructions given for color in connection with sidewalks applies equally well for floors.

Steps

The most important features of steps are the *treads* (the part stepped on) and *risers* (distance between treads). These must be planned to provide comfortable use and safety. The risers should bear a certain relation to the width of the treads and at the same time should be between 6″ and 8″ in height. Low risers and broad treads are generally preferred, especially for exterior steps. High risers and narrow treads should be used only where the horizontal distance into which the stairs must fit is limited. A desirable formula to use for steps is twice the height of the riser plus the width of the tread equals 25. It is a good plan to have the treads project about ½″ beyond the face of the riser.

The width of stairs depends to some extent upon the sidewalks used in connection with them and upon the expected traffic. In general, steps should be at least 3′-0″ wide or wider, depending on what must be taken up and down them. Wide steps seem to cause less mental hazard for the people who use them.

Curbs on either side of steps provide an added means of dressing the stairs and making them more pleasing in appearance. However, there is no other advantage worth contemplating in terms of cost.

Steps should never depend on newly filled ground for support because where steps are placed over such soil, cracking and even breaking are almost sure to occur. If for some reason it is absolutely necessary to place steps over such soil, they should be designed so as to be self-supporting.

The treads of exterior steps should be sloped away from the risers so the pitch is about $\frac{1}{16}$ inch. This practice makes certain that water

Concrete Construction

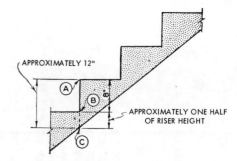

Fig. 8-56. Concrete step thickness.

and ice do not accumulate to become a danger.

As indicated in Fig. 8-56, the total thickness, that is AB plus BC is approximately one-half of the riser height. A dimension for BC less than that specified will result in steps which are structurally inferior.

Color may be incorporated in the steps, as explained for sidewalks.

Special Concrete Uses

Precast Building Units

In residential and small commercial construction, such structures as lintels and beams or joists are often made of precast concrete. They may be made on the job site or ordered in advance from a manufacturer. Figs. 6-48, 6-49, 6-53, and 6-62 illustrate typical installations of precast lintels and joists.

There has been an increase recently of precasting much larger units such as the wall sections shown in Fig. 8-57. Note that the hoisting anchors are cast directly into the unit. The advantage of this type of construction is that the units may be cast at ground level; usually indoors and under ideal conditions.

In "tilt-up" construction, wall panels are precast on the floor slab and tilted into place with a crane and tied together with cast-in-place columns or pilasters. See Fig. 8-58.

Shotcreting

Shotcreting is the most specialized of all methods of placing concrete. In shotcreting, the concrete is conveyed pneumatically through a hose and forced at high velocity through a nozzle directly onto a prepared surface, thus eliminating the need for formwork. See Fig. 8-59.

249

Masonry Simplified

Fig. 8-57. Precast concrete wall units. (Medusa Portland Cement Co.)

Fig. 8-58. Tilt-up concrete construction. (Portland Cement Association)

Concrete Construction

Fig. 8-59. Shotcreting. (American Concrete Institute)

The mix for shotcrete is proportioned according to the specifications for regular mortar and concrete. It usually is about the same as coarse aggregate mortar or fine aggregate concrete (see Chapters 4 and 7).

There are two basic shotcreting processes: dry mix and wet mix. In the dry mix process, the cement and aggregate are pre-mixed and conveyed pneumatically through the hose to the nozzle. Water is fed under pressure into the nozzle and mixed with the cement and aggre-

Masonry Simplified

Fig. 8-60. Preparing wharf structure for shotcrete repairs. (American Concrete Institute)

Fig. 8-61. Deteriorated wharf structure after repair by shotcreting. (American Concrete Institute)

gate. The operator, called the nozzleman, controls the input of water by means of a valve at the nozzle.

In the wet mix process, all ingredients are premixed. The mix is metered into the hose and pneumatically conveyed to the nozzle. Additional air is injected at the nozzle to increase the velocity.

Shotcreting is often used in structures with complex curved surfaces, such as swimming pools, reservoirs, etc., where regular formwork is difficult and expensive. Also, shotcrete has been used for many years in repairing or restoring old or damaged structures as shown in Figs. 8-60 and 8-61.

Concrete Construction

Checking On Your Knowledge

The following questions give you the opportunity to check up on yourself. If you have read the chapter carefully, you should be able to answer the questions. If you have any difficulty, read the chapter over once more so you have the information well in mind before you go on with your reading.

DO YOU KNOW

1. What the most widely used material for the surfaces of concrete formwork is?
2. Why footings must be straight and level and rest on undisturbed soil?
3. The purpose of keys in footings?
4. The purposes of steel reinforcing in concrete?
5. What the most important item from the standpoint of proportioning is in the making of concrete?
6. What happens to a concrete mix when too much water is used per sack of cement?
7. Whether concrete cures best in moist or dry surroundings?
8. What governs the strength of any concrete beyond the quality of aggregates?
9. What the two principal requirements of hardened concrete are?
10. What is meant by the term *segregation* in terms of concrete?
11. If any heat is generated by concrete?
12. What conditions are necessary for the proper hardening of cement paste?
13. Which concrete mix has more water added to it—one to make a water tank or one to make a footing?
14. How the tensile strength of concrete beams is increased?
15. What the colorimetric test is used for in concrete work?
16. How much water a cubic foot of wet sand contains?
17. If it is permissible to add more water when a concrete mix is too stiff?
18. Where the first placing of concrete in forms should be?
19. Two methods of compacting concrete in forms?
20. How the top surface of concrete in forms is finished?
21. When forms should be removed?
22. Where grade beams and piers are used?
23. The four main steps in concrete flatwork?
24. Why leveling the subgrade is important?
25. Where isolation joints are used?
26. What is meant by the term *bleed?*
27. What the purpose of jointing is?
28. What the purpose of troweling is?
29. When construction joints are used?
30. What is meant by the term *hydration?*
31. Three methods of damp curing?
32. The usual dimensions of sidewalks and driveways?
33. Why garage floors should be sloped?
34. The usual dimensions of concrete steps?
35. The main advantage of using precast concrete?
36. Two main uses of shotcrete?

Appendix **A** | **Review of Blueprints, Conventions and Symbols for Masonry**

Buildings are designed by architects and building designers who must be trained in many aspects of structural engineering as well as in artistic fields.

The architect puts his ideas on a drawing which is then duplicated many times by means of blueprints. These blueprints are the source of information to the various building tradesmen as to where and how they will perform the work of their trade. The blueprints are working drawings that each building tradesman must follow if the building is to be built correctly.

Every mason must know how to read and interpret blueprints. He is responsible for following the blueprints as they are drawn, or referring to them with the contractor or architect.

The information included here is not intended to teach you to read blueprints. You should have had a course in blueprint reading before using this text or you should have instruction in blueprint reading along with the explanations given in this chapter. This information should serve to refresh your memory or to emphasize the major points that you, as a mason, will need to know.

The beginner in masonry learns very early how essential it is to be able to read working drawings (blueprints). When he has acquired this skill, he becomes a part of a team, working out a series of construction

Review of Blueprints

problems. Each man must know how to "take off" dimensions accurately so that all of the partitions, windows and doors will be placed exactly according to plan. The beginner must learn to recognize all of the conventions that represent materials, all of the symbols for equipment and fixtures, and all of the abbreviations and notations. He should know about the items which involve other trades so he can make provision for their work as he frames out the house. For instance, he should know where heating ducts occur in the walls and where the plumbing fixtures are located in the bathroom, so he may place the studs and joists to allow for them.

The ability to read blueprints can be acquired in several ways. The trainee should use every opportunity to study plans on the job and to see how the building is progressing step by step. Experienced men will often help to point out special details and show the beginner how to anticipate problems. Where a student does not have an opportunity to get firsthand information, he can learn through home study how to read blueprints using a set of plans like those in this appendix. By means of a systematic study of these plans he will gain a basic knowledge of one building; and by studying more complex structures he will eventually acquire the ability to read all of the blueprints that apply to his work.

The following procedure will be helpful in studying blueprints. When examining a set of plans, forget all of the details at first, and instead try to get a picture of how the house is laid out and what it looks like. (Figs. 1, 2, 3, and 4 are simplified schematic drawings made to bring out this idea.) Look at the first floor plan to see the room arrangement (Fig. 1). The plan view looks as if you were viewing a horizontal slice taken through the building about five or six feet above the floor level. Try to imagine coming into the house through the front door and then going from room to room.

After becoming familiar with the layout of the first floor, study the basement plan (Fig. 2) and then the second floor plan (Fig. 3). Several questions can now be answered. How many bedrooms are there? Can you get to the basement directly from the outside? Where are the bathrooms located? The next thing to do is to look at the front elevation (Fig. 4). Elevation drawings are easy to read because they look like pictures of the house taken from points directly opposite the center of that side of the house. Now more questions can be answered. What do the windows look like? Is the first floor level close to the grade level or several feet above grade? A general idea of what the house is like is essential before beginning to examine it in detail.

Masonry Simplified

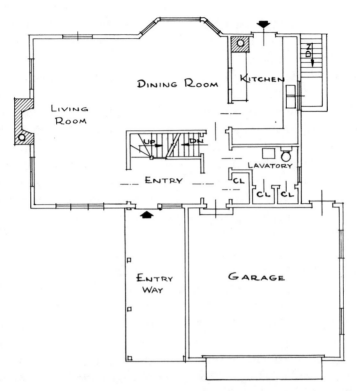

Fig. 1. First floor plan. Trace a path from the front door through the entry and hall to the rear door. Notice the location of the stairs.

Scale

Because of the large size of a house, full-sized drawings would be inconvenient, expensive, and impractical. Therefore, the drawings are made to *scale;* that is, reduced proportionately to a size which can be made and handled conveniently. House drawings usually are drawn to one-fourth inch scale, indicated as ¼″=1′0″. This means that every ¼″ on the drawing will equal one foot on the building, or the building will be 48 times larger than the drawing. To reduce the size of the draw-

ings for large buildings, the ⅛″ scale (⅛″=1′0″) is frequently used. Some parts of a building are more complicated than others and to show the details better these parts are drawn to a larger scale, ½″=1′0″, or ¾″=1′0″, or 1½″=1′0″. Certain complicated details are sometimes drawn full size; for example, the plaster cornice and head of the entrance.

By using these various scales, Fig. 5, the architect makes it possible for the builder to use his own rule to make scaled measurements on draw-

Review of Blueprints

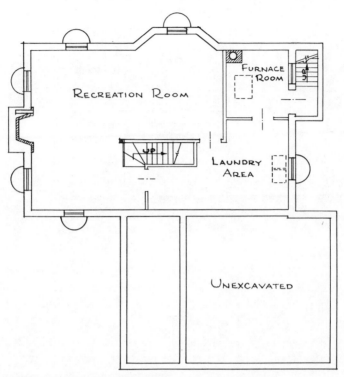

Fig. 2. Basement plan. Trace the path from the exterior stairs to the recreation room. Notice the arrangement of utility and recreation areas.

ings. Therefore, it is essential that the carpenter find out the scale to which the drawings are made before he begins taking any measurements with his rule. Scale is almost always indicated on the blueprint just below the drawing. *Example:* ¾″=1′0″.

Types of Lines

Full or Visible Lines. Border lines and the outline, or visible parts, of the house are always represented by *full*, or *visible lines*, Fig. 6.

Hidden Lines. The outline of hidden, or invisible, parts of a house are shown by *dash lines*. These represent the outline of parts which may be hidden under floors, within walls, or occur beyond or behind elevations.

Center Lines. Fine, alternate long and short lines used to show the center of the axis of an object are called *center lines*. The center of a round object is shown by two intersecting center lines.

Extension Lines. Fine lines which show the extreme limits of a dimension are called *extension lines*.

Masonry Simplified

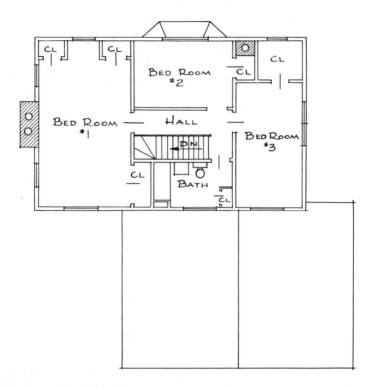

Fig. 3. Second floor plan. Study the room arrangement. Notice how the closets fit in. See how the windows provide cross ventilation.

Fig. 4. Front elevation. Study the general appearance of the house. Note the grade and floor levels, window and door locations, etc.

Review of Blueprints

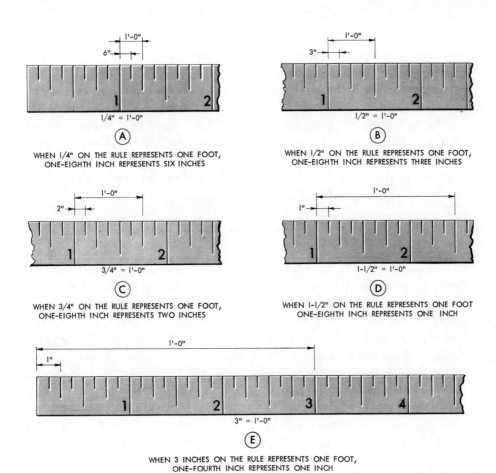

Fig. 5. Scales showing method of reducing dimensions on architectural drawings.

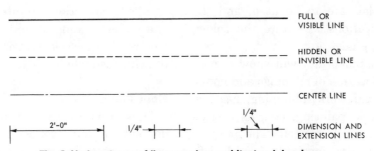

Fig. 6. Various types of lines used on architectural drawings.

Masonry Simplified

Dimension Lines. Fine, solid lines, terminated by arrowheads and used to indicate distances between points and lines, are called *dimension lines*, Fig. 6. The radius of a circle is indicated by a dimension line drawn from the center of the circle and terminating with an arrow at the circumference of the circle. The length of the radius is expressed by the dimension and the letter R; that is, ¾"R means the center of the circle is three-fourths of an inch from the circumference.

The *ceiling lines* and *floor lines* are shown as heavy, alternate long and short lines in the elevation drawings. (The elevation in Fig. 4 illustrates ceiling lines and floor lines.)

Working Drawings

Every construction job except the very smallest cabins and garages has a set of *working drawings*. These are copies made from the original set of architectural drawings done by the architect. Generally a number of sets of working drawings are made so that each building craft can have a set for its own use. The working drawings include all the information necessary to construct the building. They are supplemented with a set of *specifications* which contains, in written form, information that is not shown on the drawings. Thus a blueprint will show where masonry materials are to be used, and the specifications will state the colors, types, grades, etc.

The blueprints will show dimension, location of openings and floors, electrical work, plumbing, heating, millwork, painting, etc., as well as masonry information. The purpose of these drawings is to furnish definite information to every one concerned as to just what the building is to be like. The drawings, and specifications that go with them, allow the contractor, owner, material dealers, and tradesmen to understand just what has been decided by the architect and owner for the building. A well-drawn set of drawings and well-written specifications will prevent disagreements and misunderstandings.

For practice in blueprint reading the mason should study *Building Trades Blueprint Reading, Part I*, by Elmer W. Sundberg. This gives detailed instructions on interpreting and understanding blueprints. Only a representative selection of working drawings is included here.

Floor Plans

Drawings for construction may be divided into four groups: *plans, elevations, sections,* and *details.* The drawings for a building often are called *the plans*. However, this is incorrect as the *plan view* is that part of the drawings which shows the plan views *looking directly down on the flat surface of any particular floor or foundation.* The terms *plan view* and *floor plan* are used interchangeably by architects and builders.

A *plan view* shows the room arrangement, chimneys, fireplaces, stairs, and closets. The plan view also shows the location of various devices, such as plumbing fixtures, lighting outlets, heating apparatus, and other mechanical appliances. The average set of drawings for a house has three floor plans—basement with footing, a first-floor, and a second-floor plan. Sometimes, in addition, a *plot plan* is furnished to show lot lines, location of the house on the lot, trees, and the contour of the grounds. For complicated buildings, special plans are shown of the footings, joists, and rafter layouts. Fig. 7 illustrates a typical first floor plan for a house. In a full set of floor plans for this house a basement plan and a second floor plan would also be included.

Elevations

Elevation drawings show the outside of the building in true proportion. When the architect designs a house he thinks of the elevations in terms of the location of the house on the lot. Therefore, he names them the *south, east, north,* and *west elevation.* Sometimes the front of the building is known as the *front elevation.* As one observes the house from the front, the side to the right of the observer is called the *right elevation;* the side to the left, *left elevation;* and the one showing the back of the house, the *rear elevation.* A typical elevation is illustrated in Fig. 8.

Elevation drawings also show the floor levels and grade lines, story and window heights, and the various materials to be used.

Sections

A *section view* is one in which a part of the building or object has been cut away, exposing the construction, size, and shape of materials which need further clarification. An example of a sectional view for a wall section is shown in Fig. 9. A detailed drawing for a fireplace is shown in Fig. 10.

Construction details of the vertical wall sections provide more information than can be given on floor plans. Fig. 9, for instance, shows how the masonry is joined or fastened to the carpentry framing.

Details

The plans and elevations are usually drawn to a scale which is too small to show accurately the character or construction of certain parts. To show these parts more clearly,

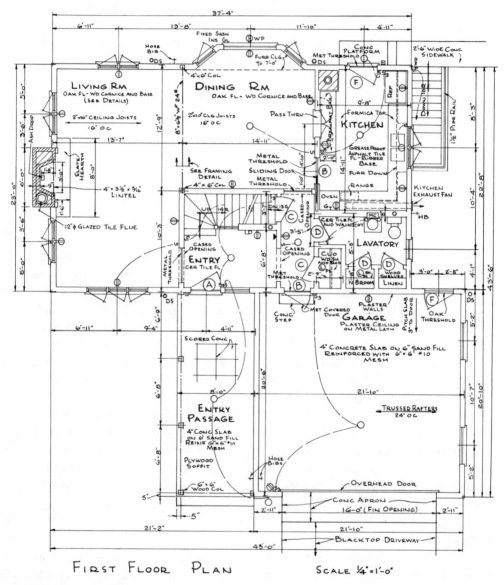

Fig. 7. A typical floor plan.

Review of Blueprints

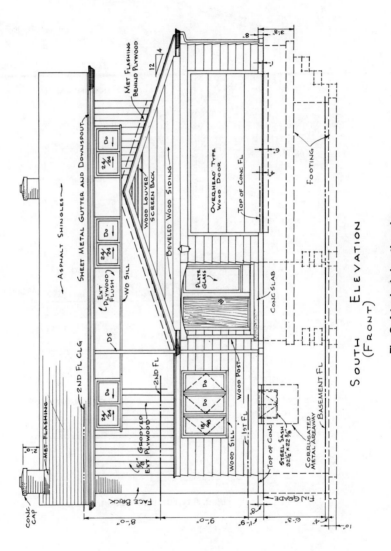

Fig. 8. A typical elevation view.

263

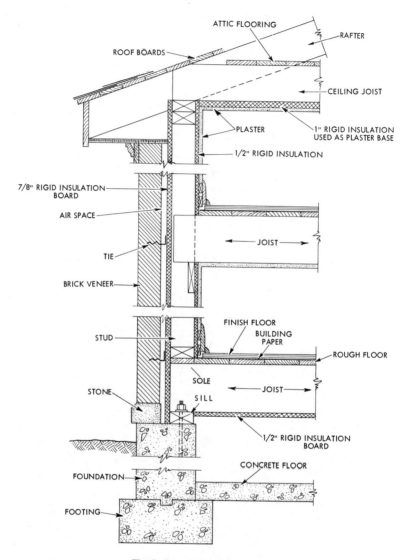

Fig. 9. A typical wall section.

larger scale drawings are made, as in Fig. 11. Details are also made of parts of elevations, floor plans, sections, etc. Fig. 12 shows some typical details that may go on the working drawings. Fig. 9, which is a section, is also a detail drawing; it contributes more information.

Review of Blueprints

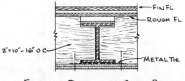

Fig. 11. Beam framing detail.

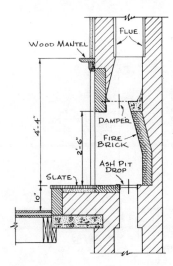

Fig. 10. Typical section through a fireplace.

Transverse or Longitudinal Sections

Transverse or longitudinal sections sometimes are shown. These elevation sections show the interior of a building along a certain line. The transverse section is across the building while the longitudinal section is lengthwise through the building. Different floor levels and interior views of stairs can be illustrated more clearly with this type of sectional view.

Schedules

Separate schedules for doors and windows are shown with the first-floor plans. References to window openings are sometimes indicated by numbers and references to doors by letters. This practice helps to keep the drawings from becoming cluttered with too many details which often make the instructions difficult to read. Fig. 13 illustrates a typical door schedule.

Symbols

Because floor plans are proportionately so small, it is not possible to show all details exactly as they will appear in full size. For example, walls contain many parts and it would be impossible to show all of the parts on such a small scale. Hence, we use symbols; each symbol having a definite meaning either as to structure, or material, or both.

Drawings are simplified by the use of symbols. Various materials, such as wood, stone, brick, and concrete, are represented by certain symbols. Examples of material symbols commonly used in the building trade are shown in Fig. 14. Mechanical devices also are represented by symbols which indicate where

265

Masonry Simplified

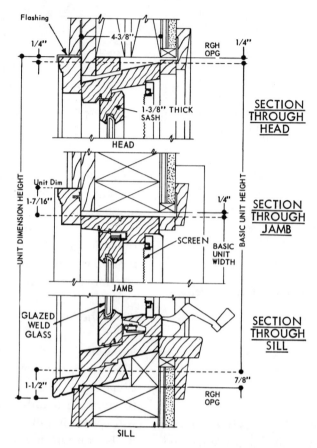

Fig. 12. Typical window section details. (Anderson Corporation)

Door Schedule for Entire House			
Mark	Size	Amt Req'd	Remarks
A	5'-0" x 6'-8" x 1¾"	1	Exterior Flush Door
B	2'-8" x 6'-8" x 1¾"	7	Flush Doors 1-Sliding 1-Metal Covered
C	2'-6" x 6'-8" x 1⅜"	4	Flush Door
C₁	2'-6" x 6'-8" x 1⅜"	2	Louvered
D	2'-4" x 6'-8" x 1⅜"	4	Flush Door
D₁	2'-4" x 6'-8" x 1⅜"	1	Louvered
E	1'-3" x 6'-8" x 1⅜"	1	Bi-Fold Louvered
F	2'-10" x 6'-8" x 1¾"	2	Exterior 2 Lights
G	2'-8" x 6'-8" x 1¾"	1	Exterior 2 Lights

Fig. 13. Typical door schedule.

heating, lighting, and plumbing appliances are to be installed in a new building. Mechanical symbols commonly used are shown in Figs. 15 and 16. The elevation of windows and their common plan symbols and the elevation of doors and their plan symbols are shown in Fig. 17. Fig. 18 illustrates the four symbols used to designate a frame wall indicating a wall and the swing of the door. The

Review of Blueprints

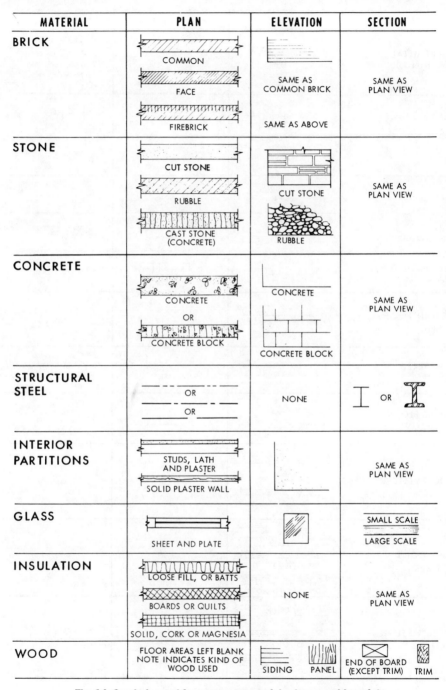

Fig. 14. Symbols used for common materials shown on blueprints.

Masonry Simplified

MATERIAL	PLAN	ELEVATION	SECTION
SHEET METAL FLASHING	INDICATE BY NOTE	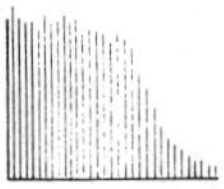	 HEAVY LINE SHAPED TO CONFORM
EARTH	NONE	NONE	
ROCK	NONE	NONE	
SAND	NONE	NONE	
GRAVEL OR CINDERS	NONE	NONE	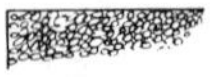
FLOOR AND WALL TILE	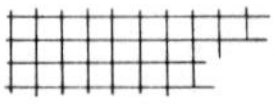	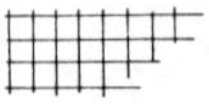	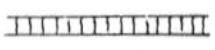
SOUNDPROOF WALL	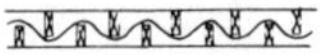	NONE	NONE
PLASTERED ARCH	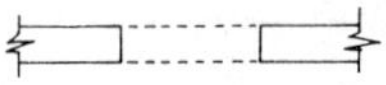	DESIGN VARIES	SAME AS ELEVATION VIEW
GLASS BLOCK IN BRICK WALL	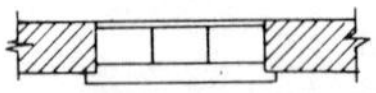	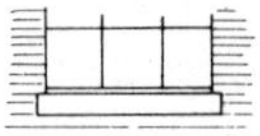	SAME AS ELEVATION VIEW
BRICK VENEER	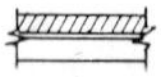 OR 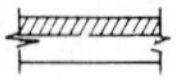 ON FRAME 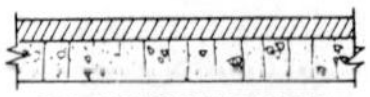 / ON CONCRETE BLOCK	SAME AS BRICK	SAME AS PLAN VIEW
CUT STONE VENEER	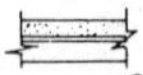 OR 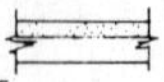 ON FRAME  / ON BRICK 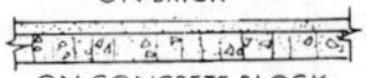 / ON CONCRETE BLOCK	SAME AS CUT STONE	SAME AS PLAN VIEW
RUBBLE STONE VENEER	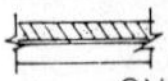 OR 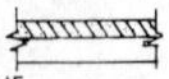 ON FRAME  / ON BRICK  / ON CONCRETE BLOCK	SAME AS RUBBLE	SAME AS PLAN VIEW

Fig. 14. Continued.

Review of Blueprints

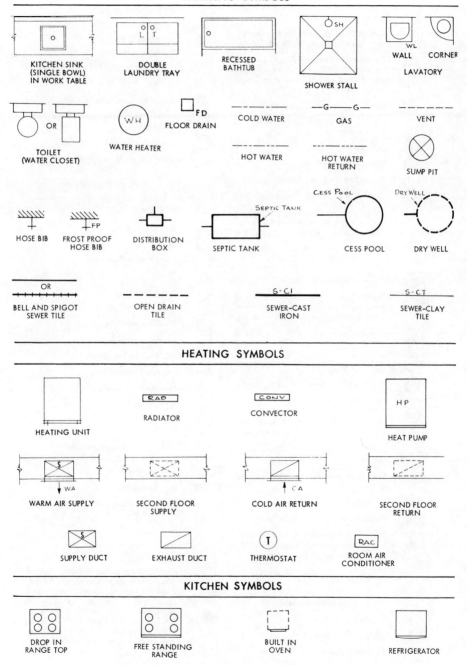

Fig. 15. Common plumbing, heating and kitchen symbols.

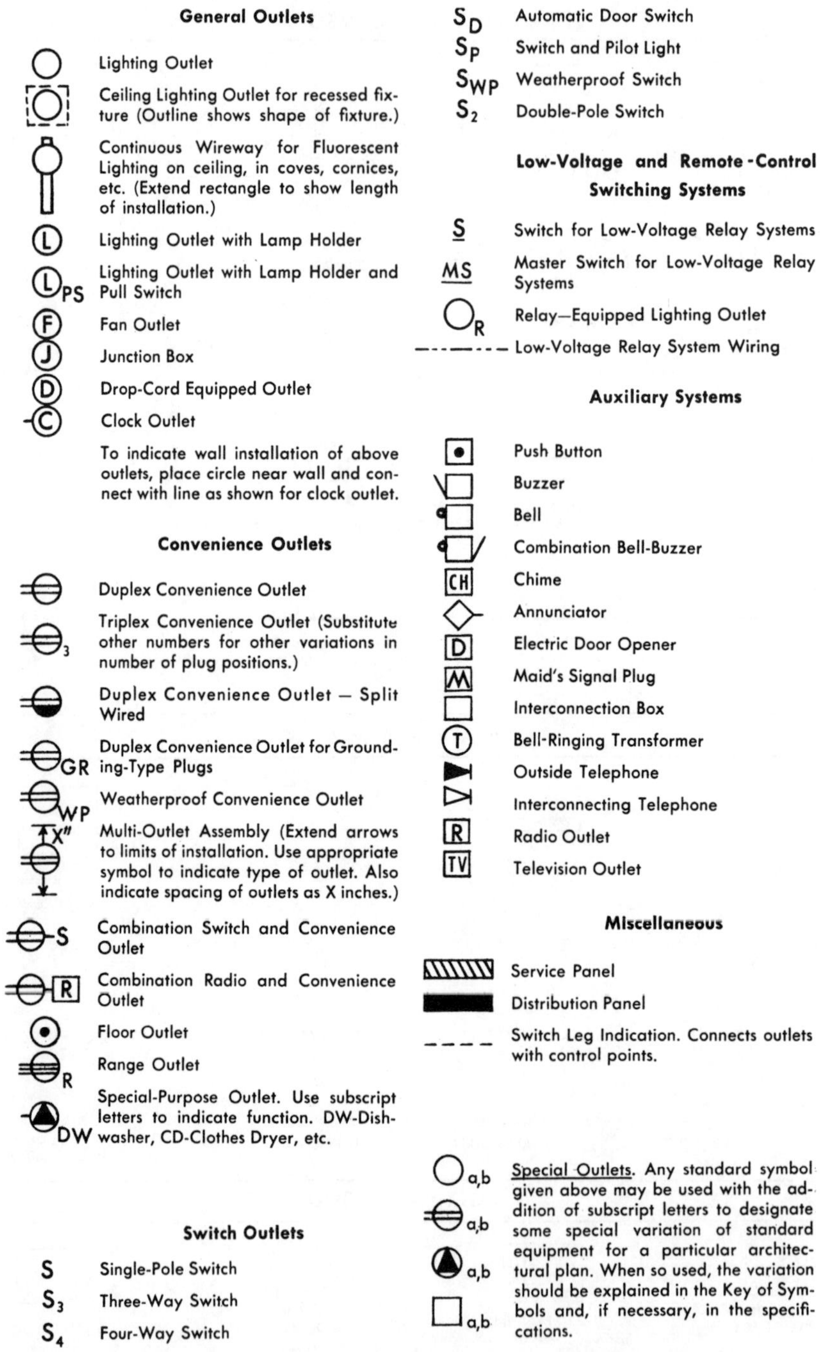

Masonry Simplified

General Outlets

- ○ Lighting Outlet
- Ceiling Lighting Outlet for recessed fixture (Outline shows shape of fixture.)
- Continuous Wireway for Fluorescent Lighting on ceiling, in coves, cornices, etc. (Extend rectangle to show length of installation.)
- ⓛ Lighting Outlet with Lamp Holder
- ⓛ$_{PS}$ Lighting Outlet with Lamp Holder and Pull Switch
- Ⓕ Fan Outlet
- Ⓙ Junction Box
- Ⓓ Drop-Cord Equipped Outlet
- ⓒ Clock Outlet

To indicate wall installation of above outlets, place circle near wall and connect with line as shown for clock outlet.

Convenience Outlets

- Duplex Convenience Outlet
- Triplex Convenience Outlet (Substitute other numbers for other variations in number of plug positions.)
- Duplex Convenience Outlet — Split Wired
- Duplex Convenience Outlet for Grounding-Type Plugs
- Weatherproof Convenience Outlet
- Multi-Outlet Assembly (Extend arrows to limits of installation. Use appropriate symbol to indicate type of outlet. Also indicate spacing of outlets as X inches.)
- Combination Switch and Convenience Outlet
- Combination Radio and Convenience Outlet
- Floor Outlet
- Range Outlet
- Special-Purpose Outlet. Use subscript letters to indicate function. DW-Dishwasher, CD-Clothes Dryer, etc.

Switch Outlets

- S Single-Pole Switch
- S_3 Three-Way Switch
- S_4 Four-Way Switch
- S_D Automatic Door Switch
- S_P Switch and Pilot Light
- S_{WP} Weatherproof Switch
- S_2 Double-Pole Switch

Low-Voltage and Remote-Control Switching Systems

- S Switch for Low-Voltage Relay Systems
- MS Master Switch for Low-Voltage Relay Systems
- ○$_R$ Relay—Equipped Lighting Outlet
- — — — Low-Voltage Relay System Wiring

Auxiliary Systems

- Push Button
- Buzzer
- Bell
- Combination Bell-Buzzer
- CH Chime
- Annunciator
- D Electric Door Opener
- M Maid's Signal Plug
- Interconnection Box
- T Bell-Ringing Transformer
- Outside Telephone
- Interconnecting Telephone
- R Radio Outlet
- TV Television Outlet

Miscellaneous

- Service Panel
- Distribution Panel
- — — — Switch Leg Indication. Connects outlets with control points.

○$_{a,b}$ / $_{a,b}$ / ▲$_{a,b}$ / □$_{a,b}$ Special Outlets. Any standard symbol given above may be used with the addition of subscript letters to designate some special variation of standard equipment for a particular architectural plan. When so used, the variation should be explained in the Key of Symbols and, if necessary, in the specifications.

Fig. 16. Common electrical symbols. (American National Standards)

Review of Blueprints

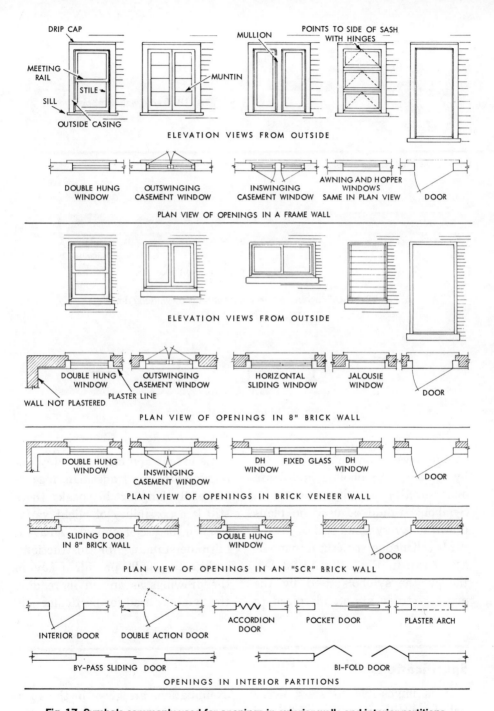

Fig. 17. Symbols commonly used for openings in exterior walls and interior partitions.

Masonry Simplified

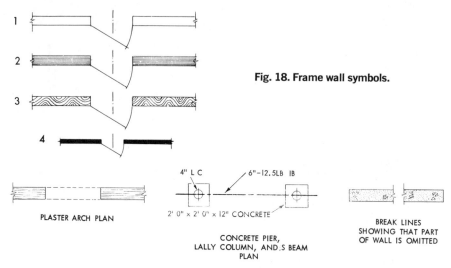

Fig. 18. Frame wall symbols.

Fig. 19. Miscellaneous construction symbols on working drawings.

Fig. 20. Structural steel sections.

standard wall convention is shown by symbol 1. Symbols 2 and 3 are occasionally used by architects. Symbol 4 would be used on plans drawn to a very small scale.

Miscellaneous structural symbols are shown in Figs. 19 and 20. Although the symbols used by the masonry trade are standardized to some extent and can be readily understood by any tradesman, regardless of the language he speaks, there still is a possibility of slight variations in the use of symbols. To avoid misunderstanding of instructions, architects usually provide a *key* on the drawings as an aid in reading the symbols used for particular job.

Specifications

It is impossible to show every detail on drawings, so additional examinations are given in the *specifications* which supplement the drawings. These specifications are written, telling in words what cannot be shown graphically on the drawings. The information furnished by the specifications includes grades of lumber and other materials, and detailed instructions as to how the work is to be performed. These specifications may be defined as *instructions to the builder*, and as such they must be simple and complete. The primary aim of the written specifications is to make perfectly clear to the builder every item that cannot be shown on the *drawings* or *blueprints*.

In addition to their primary purpose, specifications have other important uses. Estimators, including general contractors, subcontractors, manufacturers, and material dealers make use of building specifications as well as the working drawings when calculating cost of materials and labor. If carefully written, specifications make it possible for estimators to price material and labor exactly. The specifications also serve as a guide to all the trades in carrying out their specific parts of a construction job. Well-prepared specifications save time, reduce waste in both material and labor, and assure better workmanship. They also serve as a guide in the purchase of all types and qualities of fixtures, especially millwork and built-in furnishings. Another important use of specifications is the preventing of disputes between the owner and the general contractor; also between the general contractor and the subcontractor. If all necessary items are amply covered by the specifications, there can be no grounds for a dispute. Contracts are made in accordance with specifications. During the process of constructing a new house, whenever the information in the specifications appears to conflict with the instructions shown on the drawings, the carpenter or contractor should consult the architect in order to find out exactly what is wanted before proceeding with his work.

Building specifications also give general information regarding building permits for various trades, contract payments, insurance, liabilities, provisions for changes from original plans, drawings, or specifications, and supervision of construction work.

Appendix

B Math for the Mason

This review can be used by masons and apprentices for quickly finding information to compute areas and volumes, develop geometric shapes, and make practical estimates of materials. For a more detailed study of mathematical problems and formulas, mathematics books should be consulted. Only basic information and items which have general use in the trade can be listed within the scope of this text, which includes:

Operations with Fractions and Decimals
Changing Fractions to Decimals and Decimals to Fractions
Metric-to-English and English-to-Metric Conversions
Computing Areas
Computing Volumes
Elementary Geometry
Estimating Areas, Volumes, and Materials

It is assumed that all readers understand the fundamental processes of adding, subtracting, multiplying and dividing whole numbers and are familiar with common units of measurement such as feet, yards, pounds, gallons, etc.

Operations With Fractions
Addition

To add fractions with the same *denominator* (lower part), such as $1/8 + 2/8$, add the *numerators* (upper parts) and place this sum over the *common denominator* (bottom number). Thus, $1/8 + 2/8 = 3/8$. When the sum

Math for the Mason

of numerators is larger than the common denominator, such as $9/8$, divide this new numerator (top) by the denominator (bottom), giving a *mixed number* (whole number with fractional remainder). Thus, $9/8 = 1\tfrac{1}{8}$.

To add fractions with different denominators, multiply both numerator and denominator of each fraction by a number that will make the denominators equal. Any multiplier may be used without changing the quantity of the fraction. For example, $1/4 = 2/8 = 4/16 = 8/32$, etc. Similarly, $1/4 = 3/12 = 9/36$, etc.

After all fractions to be added have been changed so as to have a common denominator, add the new numerators (top numbers) and place this sum over the common denominator (the bottom number). If the new fraction is larger than 1, it can be changed to a mixed number, as already explained. For example, if $1/2$, $3/8$ and $2/3$ were to be added, the common denominator (the bottom number) would be 24. Thus $1/2$ would become $12/24$, $3/8$ would become $9/24$ and $2/3$ would become $16/24$. To add: $12/24 + 9/24 + 16/24 = 37/24 = 1$ and $13/24$. (Note that $37/24$ breaks down into $24/24 + 13/24$, thus, since $24/24 = 1$, we get 1 and $13/24$, or $1\tfrac{13}{24}$.)

In some cases the various fractions to be added can be changed to have a common denominator by dividing instead of multiplying both numerators and denominators by the same number. Again, the quantities would not be changed. For example, $8/32 = 4/16 = 2/8 = 1/4$.

Although both numerator and denominator of *each individual fraction* must be multiplied or divided by the same number, it is not necessary that the same multiplier or divisor be used for *all fractions* to be added. Thus, to add $1/5 + 2/3$, the first fraction could be multiplied by $3/3$ giving $1/5 = 3/15$. The second fraction could be multiplied by $5/5$, giving $2/3 = 10/15$. The addition would then be $3/15 + 10/15 = 13/15$.

Subtraction

Change the fractions to have a common denominator as in the case of addition. Subtract the smaller of the new numerators from the larger and place this subtracted number over the common denominator. (It is assumed that no negative fractions will be used, such as $-3/8$ or "minus three-eighths", in masonry applications.)

Multiplication

To multiply fractions there is no need to change them first so as to have a common denominator. Simply multiply all the numerators to obtain a new numerator and multiply all denominators to obtain a new denominator. Thus, $2/3 \times 1/8 \times 3/5 = 6/120$. (For the numerator: $2 \times 1 \times 3 = 6$; for the denominator: $3 \times 8 \times 5 = 120$.) This can be simplified to $1/20$ by dividing both numerator and denominator by 6.

275

Another method to simplify calculations is called *cancellation*. In multiplying a series of fractions with no intervening subtractions or additions, it often happens that the same digit appears in a numerator and a denominator. In such cases both numerator and denominator can be cancelled. Thus, in the above case, the digit 3 appears as the denominator in $2/3$ and as the numerator in $3/5$. These 3's cancel each other and disappear in the multiplication. Thus, $2/3 \times 1/8 \times 3/5 = 2/40$. This is readily simplified by dividing both numerator and denominator by 2, giving $1/20$ as the answer, exactly as before.

Division

Division of fractions is very simple but is best understood by example. Consider the problem $1/2 \div 2/3 = ?$ Here the *dividend* (number to be divided) is $1/2$ and the *divisor* (number that divides it) is $2/3$. To do this operation, re-write the dividend without change, change the division sign $\div$ to a multiplication sign $\times$, invert numerator and denominator of the divisor $2/3$ to read $3/2$, and proceed to multiply as the sign directs. Thus $1/2 \div 2/3 = 1/2 \times 3/2 = 3/4$.

It comes as a surprise to some people unused to calculations with fractions that generally the quantities decrease when they are multiplied and increase when they are divided. In common speech, when a person speaks of "half an apple" he is multiplying $1/2 \times 1$ apple, and thus decreasing the quantity.

Operations With Decimals

Decimal numbers, unlike ordinary fractions, have only numerators. Denominators are implied by the place of the last digit to the right of the decimal point. Thus, $0.1 = 1/10$, $0.01 = 1/100$, $0.001 = 1/1000$, etc.

Any number of zeros may follow the significant digits in the decimal number without increasing the quantity. Thus, $0.68 = 0.680 = 0.680000000000$. Zeros to the left of the significant digits and immediately following the decimal point are another matter; the quantity decreases by a factor of 10 for each zero preceding these digits. Thus, $0.75 = 75/100$, $0.075 = 75/1000$, and $0.0075 = 75/10000$.

(Note: It is generally considered good practice to place a zero before the decimal point where no whole number is included with the decimal remainder.)

Operations with decimals differ in no way from those used with whole numbers except in placement of the

Math for the Mason

important decimal point. Because United States money is based on the decimal system, these operations are already familiar to nearly everyone.

Addition

Line up decimal numbers to be added so that the decimal points are directly under each other. Add as with whole numbers and place the decimal point in the sum in exactly the same location as in the numbers added. This is a case where neatness counts.

For example, add 3.236, 0.75, and 107.205. Arrange as:

$$
\begin{array}{r}
3.236 \\
0.750\text{—Adding zero} \\
\underline{107.205} \quad \text{to right}
\end{array}
$$

Adding gives 111.191

This is read as "one hundred eleven and one hundred ninety-one thousandths."

Subtraction

As with addition, arrange the decimal numbers with the decimal points directly underneath each other. Subtract as with whole numbers, and place the decimal point directly under those of the listed numbers.

For example, subtract 0.9 from 2.356. Arrange as:

$$
\begin{array}{r}
2.356 \\
\text{Subtracting} \quad \underline{-0.900}\text{—Add 2 zeros} \\
\text{gives} \quad 1.456 \quad \text{to right}
\end{array}
$$

Multiplication

Multiplication of decimal numbers differs from that of whole numbers only in placement of the decimal point in the product. Multiply the numbers. Then place the decimal point as many places to the left of the significant digits as the sum of such places in the numbers multiplied.

For example, multiply $0.9 \times 0.3 \times 0.5$. The digits in the product will be 135. Since there are three places to the right of the decimal point in the numbers multiplied, the answer is pointed off as 0.135. Here the zero to the left of the decimal point has no meaning other than to assure that there is no whole number. The answer is read as "one hundred thirty-five thousandths."

Division

Division of decimal numbers differs from that of whole numbers only in placement of the decimal point in the *quotient* (the number resulting from the division). Consider the problem $7.835 \div 0.5 = ?$ Here 7.835 is the dividend and has three places of decimals to the right of the decimal point. The divisor, 0.5, has one place to the right of the decimal point. In this particular case the dividend has *two more places* of decimals than the divisor. This difference will determine the placement of the decimal point in the quotient. Dividing gives the digits 1567, the quotient, which is pointed off *two places* as 15.67, read as "fifteen and sixty-seven hundredths".

Changing Fractions to Decimals

A common fraction, such as ⅛, is an instruction to perform an operation. It says "divide 1 by 8". Doing this operation, as shown, converts it to a decimal number.

Set down and divide:

```
      0.125
8 ) 1.000
    8
    ─
    20
    16
    ──
     40
     40
     ──
```

The quotient, 0.125 (read as one hundred twenty-five thousandths), is the conversion. Similarly, ⅜ is an instruction to divide 3 by 8:

```
      0.375
8 ) 3.000
    2 4
    ───
     60
     56
     ──
     40
     40
     ──
```

Thus, ⅜ = 0.375 (three hundred seventy-five thousandths).

Conversions of this type are required so frequently that tables such as the one shown in Table 1 have been compiled for quick reference.

TABLE 1 DECIMAL EQUIVALENTS OF COMMON FRACTIONS OF AN INCH

Common Fractions	Decimal Equivalents	Common Fractions	Decimal Equivalents	Common Fractions	Decimal Equivalents	Common Fractions	Decimal Equivalents
1/64	.015625	17/64	.265625	33/64	.515625	49/64	.765625
1/32	.03125	9/32	.28125	17/32	.53125	25/32	.78125
3/64	.046875	19/64	.296875	35/64	.546875	51/64	.796875
1/16	.0625	5/16	.3125	9/16	.5625	13/16	.8125
5/64	.078125	21/64	.328125	37/64	.578125	53/64	.828125
3/32	.09375	11/32	.34375	19/32	.59375	27/32	.84375
7/64	.109375	23/64	.359375	39/64	.609375	55/64	.859375
1/8	.125	3/8	.375	5/8	.625	7/8	.875
9/64	.140625	25/64	.390625	41/64	.640625	57/64	.890625
5/32	.15625	13/32	.40625	21/32	.65625	29/32	.90625
11/64	.171875	27/64	.421875	43/64	.671875	59/64	.921875
3/16	.1875	7/16	.4375	11/16	.6875	15/16	.9375
13/64	.203125	29/64	.453125	45/64	.703125	61/64	.953125
7/32	.21875	15/32	.46875	23/32	.71875	31/32	.96875
15/64	.234375	31/64	.484375	47/64	.734375	63/64	.984375
1/4	.25	1/2	.50	3/4	.75		

Math for the Mason

Not all conversions are as neat as the ones illustrated. For instance, $1/12 = 0.083333333333333\ldots$ (approx.) As many significant figures are retained in the conversion as practicality requires. For most applications 0.083 would be acceptable.

Changing Decimals to Fractions

Translate the decimal number to fractional form, then reduce this fraction to its simplest form by dividing both numerator and denominator by the same number. For example, $0.84 = 84/100$. This can be reduced to $42/50$ (dividing by 2), and again to $21/25$. No number can evenly divide both 21 and 25, so $21/25$ is the simplest form and the fractional conversion.

Metric-to-English and English-to-Metric Conversions

Rapid expansion of trade and industry on an international basis in the past two decades has increased the need for understanding of both the *metric* or CGS (Centimeter-Gram-Second) system used by nearly all countries of the world and the *English* or FPS (Foot-Pound-Second) system used by the United States and some other English-speaking countries.

If the co-existence of two systems seems inconvenient, as it is, remember that in respect to worldwide agreement we are the exception. In view of the increasing need for a universal system to measure lengths, areas, volumes, weights, temperatures, etc., it now seems likely that the CGS system will ultimately replace the FPS system despite immense costs and problems that will be involved in making the changeover.

Table 2 lists factors for converting units from metric to English, while Table 3 lists factors for converting from English to metric units.

To convert a quantity from *metric* to *English* units:

1. Multiply by the factor shown in Table 2.
2. Use the resulting quantity "rounded off" to the number of decimal digits needed for practical application.
3. Wherever practical in semiprecision measurements, convert the decimal part of the number to the nearest common fraction.

Masonry Simplified

TABLE 2 CONVERSION OF METRIC TO ENGLISH UNITS

LENGTHS:		WEIGHTS:	
1 MILLIMETER (MM)	= 0.03937 IN.	1 GRAM (G)	= 0.03527 OZ (AVDP)
1 CENTIMETER (CM)	= 0.3937 IN.	1 KILOGRAM (KG)	= 2.205 LBS
1 METER (M)	= 3.281 FT OR 1.0937 YDS	1 METRIC TON	= 2205 LBS
1 KILOMETER (KM)	= 0.6214 MILES	LIQUID MEASUREMENTS:	
AREAS:		1 CU CENTIMETER (CC)	= 0.06102 CU IN.
1 SQ MILLIMETER	= 0.00155 SQ IN.	1 LITER (= 1000 CC)	= 1.057 QUARTS OR 2.113 PINTS OR 61.02 CU INS.
1 SQ CENTIMETER	= 0.155 SQ IN.	POWER MEASUREMENTS:	
1 SQ METER	= 10.76 SQ FT OR 1.196 SQ YDS	1 KILOWATT (KW)	= 1.341 HORSEPOWER
VOLUMES:		TEMPERATURE MEASUREMENTS:	
1 CU CENTIMETER	= 0.06102 CU IN.	TO CONVERT DEGREES CENTIGRADE TO DEGREES FAHRENHEIT, USE THE FOLLOWING FORMULA: DEG F = (DEG C X 9/5) + 32	
1 CU METER	= 35.31 CU FT OR 1.308 CU YDS		

SOME IMPORTANT FEATURES OF THE CGS SYSTEM ARE:
1 CC OF PURE WATER = 1 GRAM. PURE WATER FREEZES AT 0 DEGREES C AND BOILS AT 100 DEGREES C.

TABLE 3 CONVERSION OF ENGLISH TO METRIC UNITS

LENGTHS:		WEIGHTS:	
1 INCH	= 2.540 CENTIMETERS	1 OUNCE (AVDP)	= 28.35 GRAMS
1 FOOT	= 30.48 CENTIMETERS	1 POUND	= 453.6 GRAMS OR 0.4536 KILOGRAM
1 YARD	= 91.44 CENTIMETERS OR 0.9144 METERS	1 (SHORT) TON	= 907.2 KILOGRAMS
1 MILE	= 1.609 KILOMETERS	LIQUID MEASUREMENTS:	
AREAS:		1 (FLUID) OUNCE	= 0.02957 LITER OR 28.35 GRAMS
1 SQ IN.	= 6.452 SQ CENTIMETERS	1 PINT	= 473.2 CU CENTIMETERS
1 SQ FT	= 929.0 SQ CENTIMETERS OR 0.0929 SQ METER	1 QUART	= 0.9463 LITER
1 SQ YD	= 0.8361 SQ METER	1 (US) GALLON	= 3785 CU CENTIMETERS OR 3.785 LITERS
VOLUMES:		POWER MEASUREMENTS:	
1 CU IN.	= 16.39 CU CENTIMETERS	1 HORSEPOWER	= 0.7457 KILOWATT
1 CU FT	= 0.02832 CU METER	TEMPERATURE MEASUREMENTS:	
1 CU YD	= 0.7646 CU METER	TO CONVERT DEGREES FAHREHEIT TO DEGREES CENTIGRADE, USE THE FOLLOWING FORMULA: DEG C = 5/9 (DEG F -32)	

To convert a quantity from *English* to *metric* units:

1. If the English measurement is expressed in fractional form, change this to an equivalent decimal form.

2. Multiply this quantity by the factor shown in Table 3.
3. Round off the result to the precision required.

Relatively small measurements, such as 17.3 cm, are generally expressed in equivalent millimeter form. In this example the measurement would be read as 173 mm.

Elementary Geometry

Much of the study of geometry from an academic standpoint has to do with theorems and proofs. From a trade standpoint, however, the properties of lines, figures, etc., are of greater concern.

Lines

The *horizontal line* is a level line. It is the opposite of a vertical line. (Fig. 1.)

A *perpendicular line* is a line at right angles to another line. (Fig. 1.)

A *level line* and a *plumb line* produce a square and are perpendicular to each other. (Fig. 1.)

The *diagonal line* is one joining two opposite angles. (Fig. 1.)

Parallel lines are those having the same direction and are an equal distance from each other at all points. (Fig. 1.)

The Circle

The *circle* is drawn from the center and is a continuous curved line, being of an equal distance from the center at all points. (Fig. 2.)

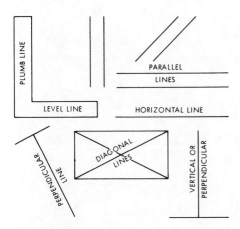

Fig. 1. Straight lines that differ because of their placement.

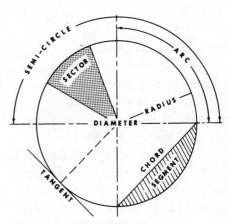

Fig. 2. A circle and its parts.

An *arc* of a circle is any part of its circumference. (Fig. 2.)

The *chord* is a straight line joining two points of the circumference. (Fig. 2.)

The *semi-circle* is one-half of the complete circle. (Fig. 2.)

The *circumference* is the entire distance around the circle.

The *diameter* is the distance across the circle through the center. (Fig. 2.)

The *radius* is half the diameter or the distance from the center to any point of the circumference. (Fig. 2.)

A *sector* is a portion of a circle between two radii and the circumference. (Fig. 2.)

A *segment* is a portion of a circle contained by a straight line and the circumference which it cuts off. (Fig. 2.)

The *tangent* is a straight line which touches a circle or curve but does not cut it and is at right angle to a straight line from the center. (Fig. 2.)

Circle Measurements

Circumference of a circle equals diameter × 3.1416. (3.1416 = Pi or the Greek letter π.)

Area of a circle equals diameter squared (dia.^{2}) × .7854.

Length of arc equals degrees in arc × radius × .01745. *Example:* 45° × 4' radius = 180 × .01745 = 3.141' length of arc.

Degree of arc equals length/radius × .01745. *Example:* 4' radius × .01745 = .0698, 3.141' length ÷ .0698 = 45°.

Radius of arc equals length/degrees × .01745. *Example:* 45° × .01745 = .78525, 3.141' length ÷ .78525 = 4' radius.

To find the *area of a sector* of a circle: 3.1416 × radius squared × degrees of the sector ÷ 360.

To find the *area of a segment* of a circle: Find area of sector and subtract area of included triangle. Fig. 3 illustrates both sector and segment of the circle.

Spherical Measurements

Surface area equals diameter squared × 3.1416. *Example:* dia. 4' sq. = 16' × 3.1416 = 50.26 sq. ft. surface area.

Volume equals diameter cubed × .5236. *Example:* dia. 4' cubed = 64 cu. ft. × .5236 = 33.5104 cu. ft.

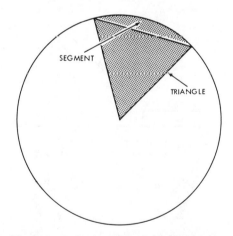

Fig. 3. Area of segment found by subtracting area of triangle from area of sector.

Math for the Mason

Elliptical Measurements

Elliptical *surface area* equals width × height × .7854. *Example:* width 36″ × height 24″ = 864 sq. in. × .7854 = 678.59 sq. in. area.

Triangular Measurements

When two *angles* of a triangle are known, the third can be found by subtracting the sum of the two known angles from 180°. (The sum of the angles of a triangle equals 180°.)

The square of the *hypotenuse* (longest side) of a right triangle is equal to the sum of the squares of the other two sides.

Area of a triangle equals ½ height × base.

Pyramid Measurements

Area equals ½ perimeter of base × slant height + area of base.

Volume equals area of base × ⅓ height or altitude.

Trapezoid Measurements

Area equals height × ½ the sum of its parallel sides.

Rectangular Measurements

Volume = width × length × height. For volume in gallons, divide cubic content *in inches* by 231; for cubic content *in feet* divide by 7.48.

Measurements of Regular Polygons

Regular polygons, also called equi-

TABLE 4 REGULAR POLYGONS: NAME, AREA AND RADIUS OF INCLOSING CIRCLE

NAME OF FIGURE	NUMBER OF SIDES	AREA EQUALS:	RADIUS OF CIRCLE EQUALS:
EQUILATERAL TRIANGLE	3	0.433 X 1 SIDE SQUARED	0.577 X LENGTH OF 1 SIDE
SQUARE	4	1.000 X 1 SIDE SQUARED	0.707 X LENGTH OF 1 SIDE
PENTAGON	5	1.720 X 1 SIDE SQUARED	0.851 X LENGTH OF 1 SIDE
HEXAGON	6	2.598 X 1 SIDE SQUARED	1.000 X LENGTH OF 1 SIDE
HEPTAGON	7	3.634 X 1 SIDE SQUARED	1.152 X LENGTH OF 1 SIDE
OCTAGON	8	4.828 X 1 SIDE SQUARED	1.307 X LENGTH OF 1 SIDE
NONAGON	9	6.182 X 1 SIDE SQUARED	1.462 X LENGTH OF 1 SIDE
DECAGON	10	7.694 X 1 SIDE SQUARED	1.618 X LENGTH OF 1 SIDE
UNDECAGON	11	9.365 X 1 SIDE SQUARED	1.775 X LENGTH OF 1 SIDE
DODECAGON	12	11.196 X 1 SIDE SQUARED	1.932 X LENGTH OF 1 SIDE

lateral polygons, are those having equal sides. All can be inscribed in circles so that all vertices (corners) exactly touch the circle's circumference. It follows that all angles of a regular polygon as well as its sides are equal.

Table 4 lists the more common regular polygons and formulas for calculating their areas.

Estimating Areas, Volumes, and Materials

The most common practical application of the preceding basic math rules is in estimating the amount of materials needed for a given job. Most job estimating is done by professional estimators who are experts in several areas, such as time-labor costs, hardware costs, waste allowances, etc. However, a few examples will indicate to the mason how the estimator arrives at the amount of materials needed for a particular job.

For instance, in estimating for a brick wall, the estimator will study the house plans (working drawings) to determine the *dimensions* (length and height) of the wall. By multiplying the length by the height, he finds the total *area* of the wall. He then calculates the area of openings for doors, windows, etc., and subtracts it from the total area, which gives him the actual area to be laid in brick. Then, by the use of specially prepared tables, he is able to determine the number of bricks required. Table 5 illustrates a typical brick estimating table. Knowing the size of

TABLE 5 QUANTITIES MODULAR BRICK WITHOUT HEADERS

NOMINAL BRICK SIZE Hgt. Th. Lgth.	BRICK Per Sq. Ft. of Wall	CUBIC FEET OF MORTAR Per 100 Sq. Ft. of Wall		
		JOINT THICKNESS		
		1/4 in.	3/8 in.	1/2 in.
2 2/3 x 4 x 8	7.4	4.6	6.6	8.3
3 1/5 x 4 x 8	6.2	4.0	5.8	7.3
4 x 4 x 8	5.0	—	5.0	6.3
5 1/3 x 4 x 8	3.7	—	4.1	5.2
2 x 4 x 8	6.6	—	7.7	9.8
2 2/3 x 4 x 12	5.0	4.2	6.1	7.8
3 x 4 x 12	4.4	—	5.5	7.0
3 1/5 x 4 x 12	4.1	3.7	5.2	6.7
4 x 4 x 12	3.3	3.1	4.4	5.6
5 1/3 x 4 x 12	2.5	—	3.6	4.6
2 2/3 x 6 x 12	5.0	—	9.4	12.2
3 1/5 x 6 x 12	4.1	—	8.3	10.6
4 x 6 x 12	3.3	—	6.9	8.9

Math for the Mason

the brick to be used, the total surface area in square feet and the specified thickness of the mortar, it is easy to determine the number of bricks needed and the amount of mortar required to complete the wall.

In estimating the amount of concrete required for footings and foundations, the estimator, again working from the house plans, takes the vertical surface area (length × height) of each part of the footing or foundation and multiplies it by its horizontal thickness. This gives the estimator the *volume* of the part. If the thickness is less than a foot, it is multiplied as either a fraction of a foot or as its decimal equivalent. If, for instance, the foundation is to be 6" thick, the thickness would be either ½' or 0.5'. The total volume for the complete foundation is determined by adding the volumes of all the parts.

Dimensions in working drawings are given in feet and inches. Thus, at this point, the estimator has the volume in cubic feet. As concrete is usually ordered by the cubic yard, he divides the number of cubic feet by 27, which gives him the number of cubic yards to order.

In estimating for concrete flatwork, the estimator determines the area of the slab by multiplying the length by the width. He then multiplies the area by the thickness in order to determine the volume. If the result is in cubic feet, he again divides by 27 to determine the required number of cubic yards.

Glossary of Trade Terms

A

abrasive: A substance used for wearing away or polishing a surface by friction. A grinding material, such as emery, sand, and diamond. Other abrasives include: crushed garnet and quartz, pumice or powdered lava, also decomposed limestone, known as *tripoli*. There are other abrasives which are made artificially and sold under various trade names.

abrasive paper: Paper, or cloth, covered on one side with a grinding material glued fast to the surface, used for smoothing and polishing. Materials used for this purpose include: crushed flint, garnet, emery or corundum.

abrasive tools: All implements, used for wearing down materials by friction or rubbing, are known as *abrasive tools;* these include: grindstones which are made of pure sandstone, whetstones, emery wheels, sandpaper, emery cloth, and other abrading tools.

abutment: That part of a pier or wall from which an arch is suspended; specifically the support at either end of an arch, beam, or bridge which resists the pressure due to a load.

accelerator: Material added to portland cement concrete during the mixing to hasten its natural development of strength.

acoustical materials: Sound-absorbing materials for covering walls and ceilings; a term applied to special plasters, tile, or any other material for wall coverings composed of mineral, wood, or vegetable fibers; also, cork or metal used to control or deaden sound. See Fig. 1.

acoustics: Science of sound. A study of the effects of sound upon the ear. The sum of the qualities that determine the value of an auditorium as to distinct hearing. The acoustics are said to be *good* or *bad* according to the ease of clearness with which sounds can be heard by the audience. The main factors influencing acoustical conditions are reverberation, extraneous noises, loudness of the original sound, and the size and shape of the auditorium.

acoustic tile: Any tile composed of materials having the property of absorbing sound waves, hence reducing the reflection and reverberation of sound; any tile designed and constructed to absorb sound waves. See Fig. 1.

Glossary

Fig. 1. Nailing acoustic tile to furring.

Fig. 2. Adhesive gun (U.S. Gypsum Co.)

adhesive gun: A mechanical device for applying adhesives to lathing, framework, or other backing material. See Fig. 2.

adhesives: New glues and mastics for construction are being rapidly developed by the industry. The new compounds make it possible to bond almost any like, or unlike, rigid, and some flexible, materials together, for various construction purposes. Improvements are evident in labor and weight saving, economy, strength, and suitability for construction conditions.

adjustable clamp: Any type of clamping device that can be adjusted to suit the work being done, but particularly clamps used for holding column forms while concrete is placed.

admixture: Any ingredient other than cement, aggregate, and water in a concrete or mortar mix, which is added to provide coloring, air-entrainment, or to control setting time.

adobe: An aluminous earth from which unfired brick are made, especially in the western part of the United States; an unfired brick dried in the sun; a house or other structure built of such materials or clay.

adz: A cutting tool resembling an ax. The thin arched blade is set at right angles to the handle. The adz is used for rough-dressing timber.

adze-eye hammer: A claw hammer with the eye extended. This gives a longer bearing on the handle than is the case in hammers not having an extended eye. See Fig. 31.

aerated concrete: A lightweight material made from a specially prepared cement, and used for subfloors. Due to its cellular structure, this material is a retardant to sound transmission.

African mahogany: A large tree remotely related to the mahogany family. The tree is found principally in Africa and produces exceptionally fine figured timber of unusual lengths and widths. The wood is used for fine furniture.

aggregate: A collection of granulated particles of different substances into a compound or conglomerate mass. In mixing concrete, the stone or gravel used as a part of the mix is commonly called the

Masonry Simplified

coarse aggregate, while the sand is called the *fine aggregate.*

aiguille: In masonry, an instrument for boring holes in stone or other masonry material.

air brick: A box of brick size made of metal with grated sides; used to provide ventilation in brick walls. Also called *wall vent, ventilating block,* or *brick.* See Fig. 3.

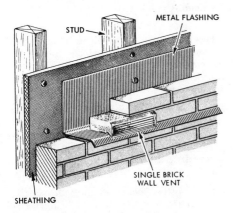

Fig. 3. Air brick.

air conditioning: The process of heating or cooling, cleaning, humidifying or dehumidifying, and circulating air throughout the various rooms of a house or public building. This term has been erroneously used many times and before installing air-conditioning equipment it should be carefully investigated.

air-cooled slag: The product of relatively slow-cooling molten blast-furnace slag, resulting in a solid mass of tough, durable material which is excavated, crushed, and screened for commercial purposes, such as concrete and bituminous aggregate.

air-dried lumber: Any lumber which is seasoned by drying in the air instead of being dried in a kiln or oven.

air entrainment: In concrete or mortar, a process by means of a chemical admixture causing the formation of billions of tiny air bubbles uniformly throughout the mixture. Air entrainment improves the workability of the fresh mix and the frost and deicing salt resistance of the final structure.

air nailer: An automatic air-operated machine which drives large nails at a greater rate of speed than hand nailing. See Fig. 4.

air pocket: An airspace which accidentally occurs in concrete work.

air space: A cavity or space in walls or between various structural members.

air stapler: A staple gun operated by compressed air; capable of driving staples into hard materials at a high rate of speed. See Fig. 5.

aisle: A passageway by which seats may be reached, as the *aisle* of a church.

alcove: Any recess cut in a room. An *alcove* is usually separated from the main room by an archway.

alkali: A water soluble salt sometimes present in soils and water sources. Its presence in concrete or mortar mixes may be detrimental as it may react with some aggregates and cause expansion and deterioration.

all-rowlock wall: In masonry, a wall built so that two courses of stretchers are standing on edge, alternating with one course of headers standing on edge.

aluminum nails: Nails made of this metal are lightweight, stainless, rustless, and sterilized.

American bond: A method of bonding brick in a wall whereby every fifth, sixth, or seventh course consists of headers, the other courses being stretchers. This type of bond is used extensively because it is quickly laid.

anchor blocks: Blocks of wood built into masonry walls, to which partitions and fixtures may be secured.

anchor bolts: Large bolts used for fastening or anchoring a wooden sill to a masonry foundation, floor, or wall. See Fig. 6. Also, any of several types of metal fasteners used

Glossary

Fig. 4. Air-operated nailer. (Duo-Fast Fastener Corp.)

Fig. 5. Air-operated stapler. (Duo-Fast Fastener Corp.)

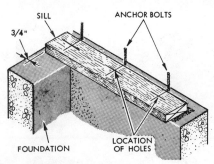

Fig. 6. Anchor bolts.

to secure wood construction to masonry. See Fig. 7.

anchors: In building construction, devices used to give stability to one part of a structure by securing it to another part; metal ties, such as concrete inserts or toggle bolts, used to fasten any structural wood member to a concrete or masonry wall. See Figs. 7 and 8.

angle bead: A molded strip used in an angle, usually where two walls meet at right angles. See *corner bead*.

angle bonds: In masonry work, brick or metal ties used to bind the angles or corners of the walls together.

angle bracket: A type of support which has two faces usually at right angles to each other. To increase the strength, a web is sometimes added.

angle closer: In masonry, a portion of a whole brick which is used to close up the bond of brickwork at corners. See *closer*.

angle dividers: A tool primarily designed for bisecting angles. It can also be used as a try square.

angle gage: A tool used to set off and test angles in work done by carpenters, bricklayers, and masons.

angle iron: A section of a strip of structural iron bent to form a right angle.

anhydrous lime: Unslacked lime which is made from almost pure limestone. Same as *quicklime*. Also called *common lime*.

annual ring: The arrangement of the wood of a tree in concentric rings, or layers, due to the fact that it is formed gradually, one ring being added each year. For this reason the rings are called *annual rings*. The rings can easily be counted in cross section of a

Masonry Simplified

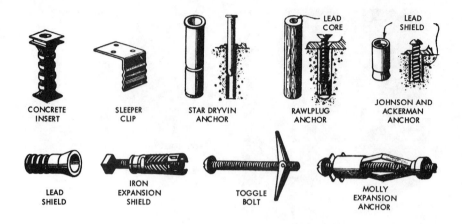

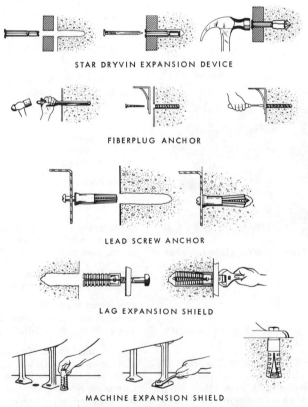

Fig. 7. Anchors used after initial construction.

Glossary

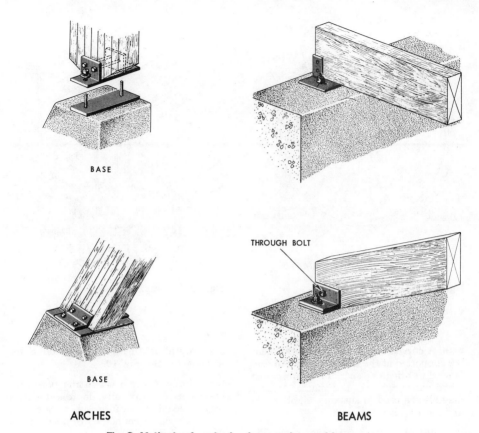

Fig. 8. Methods of anchoring large arches and beams.

tree trunk. If a tree is cut close to the ground, the age of the tree can be estimated by the number of annual growth rings. See Fig. 9.

anta: A rectangular pier or pilaster formed by thickening a wall at its extremity; often furnished with a capital and base; also, a special type of pier formed by thickening a wall at its termination. A pilaster opposite another, as on a door jamb.

apex stone: A triangular stone at the top of a gable wall, often decorated with a carved trefoil. Sometimes called a *saddle stone*.

apron: A plain or molded finish piece below the stool of a window; put on to cover the rough edge of the plastering.

arbor: A type of detached latticework, or an archway of latticework See *trellis*.

arc: Any part of the *circumference* of a circle.

arc welding: Electrical welding process in which intense heat is obtained by arcing between the welding rod and metal. The molten metal from the tip of the electrode is then deposited in the joint and, together with the molten metal of the edges, solidifies to form a sound and uniform connection.

arcade: An arched roof or covered passageway; a series of arches supported either on piers or pillars. An *arcade* may be attached to a wall or detached from the wall.

Masonry Simplified

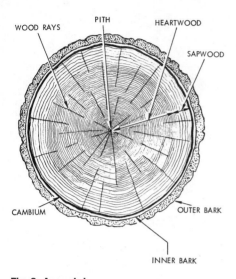

Fig. 9. Annual rings.

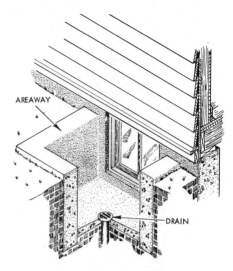

Fig. 10. Areaway.

arch: A curved or pointed structural member supported at the sides or ends. An *arch* is used to bridge or span an opening, usually a passageway or open spaces. An arch may also be used to sustain weight, as the arch of a bridge.

arch bar: A support for a flat arch. The support may be either a strip of iron or a flat bar.

arch brick: Special wedge-shaped brick used in the building of an arch; also suitable for other circular work; a term also applied to brick which have been overburned by being placed in contact with the fire in the arch of the kiln.

architect: One who designs and oversees the construction of a building; anyone skilled in methods of construction and in planning buildings; a professional student of architecture.

arch stone: A stone shaped like a wedge for use in an arch. Same as *voussoir*.

archway: The passageway under an arch.

area: An uncovered space, such as an open court; also, a sunken space around the basement of a building, providing access and natural lighting and ventilation. Same as *areaway*. See Fig. 10.

area drain: A drain set in the floor of a basement areaway, any depressed entryway, a loading platform, or a cemented driveway which cannot be drained otherwise. See Fig. 10.

areaway: An open subsurface space around a basement window or doorway, adjacent to the foundation walls. An *areaway* provides a means of admitting light and air for ventilation, and also affords access to the basement or cellar. See *prefabricated areaway*, Fig. 73.

armored concrete: Concrete which has been strengthened by reinforcing with steel rods or steel plates. See *reinforced concrete*.

arris: An edge or ridge where two surfaces meet. The sharp edge formed where two moldings meet is commonly called an arris.

artificial stone: A special kind of manufactured product resembling a natural stone. A common type is made from pulverized quarry refuse mixed with portland cement (sometimes colored) and water. After being pressed into molds, the mixture is al-

Glossary

lowed to dry out and then is seasoned in the open air for several months before being used.

artisan: A skilled craftsman; an artist; one trained in a special mechanical art or trade; a handicraftsman who manufactures articles of wood or other material.

asbestos cement: A fire-resisting, waterproofing material made by combining portland cement with asbestos fibers.

asbestos shingles: A type of shingle made for fireproof purposes. The principal composition of these shingles is *asbestos,* which is incombustible, nonconducting, and chemically resistant to fire. This makes *asbestos shingles* highly desirable for roof covering.

ashlar: Squared stone used in foundations and for facing of certain types of masonry walls. See Fig. 11.

ashlar brick: A brick that has been roughhackled on the face to make it resemble stone.

ashlar masonry: Masonry work of sawed, dressed, tooled, or quarry-faced stone with proper bond. See Fig. 11.

aspen: A tree common in many parts of the United States. It is especially noted for the trembling of its leaves which are never still. The wood has little commercial value except as pulp for the manufacture of paper. For paper pulp, aspen wood ranks in importance next to spruce and hemlock. The aspen tree grows to a height of 50 feet and the wood weighs 25 pounds per cubic foot.

asphalt cement: A cement prepared by refining petroleum until it is free from water and all foreign material, except the mineral matter naturally contained in the asphalt. It should contain less than one percent of ash.

assize: In masonry, a cylinder-shaped block of stone which forms part of a column, or of a layer of stone in a building.

astragal: A small semicircular molding, either ornamental or plain, used for covering a joint between doors. For decorative purposes it is sometimes cut in the form of a string of beads.

atrium: A large hallway or lobby with galleries at each floor level on three or more sides.

attic: A garret; the room or space directly below the roof of a building. In modern buildings the *attic* is the space between the roof and the ceiling of the upper story. In classical structures the *attic* is the space, or low room, above the entablature or main cornice of a building.

auger: A wood-boring tool used by the carpenter for boring holes larger than can be made with a gimlet. The handle of an *auger* is attached at right angles to the

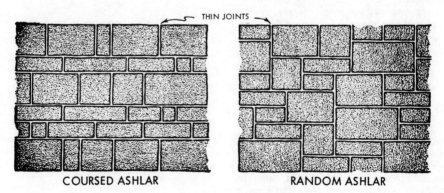

Fig. 11. Ashlar masonry.

Masonry Simplified

tool line. There are several different types of augers made for various purposes.

auger bit: An auger without a handle to be used in a brace. Such a bit has square tapered shanks made to fit in the socket of a common brace. This combination tool is known as a *brace and bit*.

autoclave: An airtight oven-like chamber charged with high pressure steam used to cure freshly molded concrete blocks.

automatic grouter: A pressurized steel form, faced with foam rubber, which forces grout in and around stones, or brick, after it has been poured in from the top of stonework.

avoirdupois weight: A system of weights in common use in English-speaking countries for weighing all commodities except precious stones and metals and precious drugs. In this system 16 ounces equal one pound; 2,000 pounds equal one *short ton*; a *long ton* contains 2,240 pounds.

awl: A small sharp-pointed instrument used by the carpenter for making holes for nails or screws. The carpenter often uses an *awl* to mark lines where pencil marks might become erased.

awl haft: The handle of an *awl*.

awning window: A type of window in which each light opens outward on its own hinges, which are placed at its upper edge. Such windows are often used as ventilators in connection with fixed picture windows. See Fig. 12.

axhammer: A type of cutting tool, or ax, having two cutting edges, or one cutting edge and one hammer face, used for dressing or spalling the rougher kinds of stone.

B

back filling: Coarse dirt, broken stone, or other material used to build up the ground level around the basement or foundation walls of a house to provide a slope for drainage of water away from the foundation. See Fig. 13.

Fig. 12. Awning window.

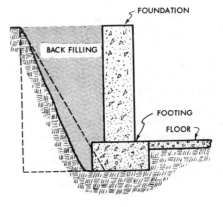

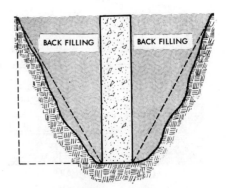

Fig. 13. Back filling.

Glossary

backing hip rafter: The beveling arris as at corners of hip rafters to tie up with adjacent roof surfaces.

backing of a joist or rafter: The blocking used to bring a narrow joist up to the height of the regular width joists. The widths of joists or rafters may vary, and in order to assure even floors or roofs some of the joists or rafters must be blocked up until all the upper surfaces are of the same level.

backing of a wall: The rough inner face of a wall; the material which is used to fill in behind a retaining wall.

backing tier: In masonry, the tier of rough brickwork which backs up the *face tier* of an exterior wall for a residence or other well-built brick structure. This part of a brick wall is often of a cheaper grade of brick than that used for the face tier. See Fig. 14.

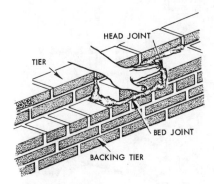

Fig. 14. Backing brick.

back plastering: The application of a ⅜" thick mortar coat on the back of the facing tier for purposes of moisture-proofing and air-proofing; also called *parging*. See Fig. 63.

backsaw: Any saw with its blade stiffened by an additional metal strip along the back. The *backsaw* is commonly used in cabinet work as a bench saw.

badger: An implement used to clean out the excess mortar at the joints of a drain after it has been laid.

badigeon: In building, a kind of cement or paste made by mixing suitable materials for filling holes or covering defects in stones or wood.

balk: A large squared timber, or beam.

balloon framing: A type of building construction in which the studs extend in one piece from the foundation to the roof; in addition to being supported by a ledger board, the second-floor joists are nailed to the studs. See Fig. 15.

ball peen hammer: A hammer having a peen which is hemispherical in shape; used especially by metal workers.

baluster: One of a series of small pillars, or units, of a balustrade; an upright support of the railing for a stairway. See *closed-string stair*, Fig. 83.

balustrade: A railing consisting of a series of small columns connected at the top by a coping; a row of balusters surmounted by a rail.

band saw: A saw in the form of an endless serrated steel belt running on revolving pulleys; the saw is used in cutting woodwork; also used for metal work.

banister: The balustrade of a staircase; a corruption of the word *baluster*.

banker: In masonry, a type of workbench on which bricklayers and stonemasons work when shaping arches or other construction requiring shaped materials.

bar: In masonry, a shortened term for *reinforcing bar*, steel, etc. Also see *rebar, deformed bars*.

bar clamp: A device consisting of a long bar and two clamping jaws, used by woodworkers for clamping large work.

bargeboard: The decorative board covering the projecting portion of a gable roof; the same as a *verge board*; during the late part of the nineteenth century, bargeboards frequently were extremely ornate.

295

Masonry Simplified

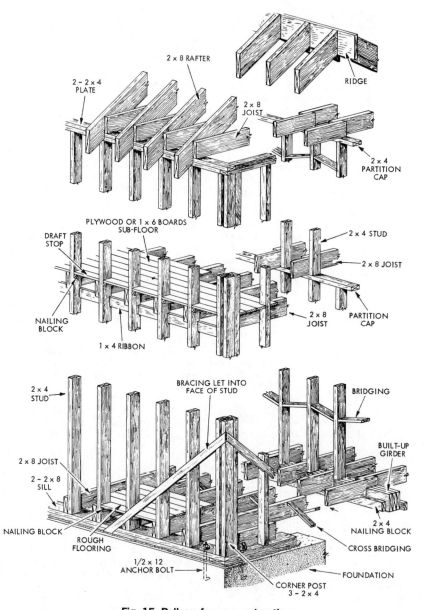

Fig. 15. Balloon frame construction.

barge course: A part of the tiling which usually projects beyond the principal rafters or *bargeboards,* along the sloping edge of a *gable roof;* also, a course of brick laid on edge to form the coping of a wall. See *bargeboard.*

Glossary

bark pocket: A patch of bark nearly, or wholly, enclosed in the wood is known as a *bark pocket*.

base: The lowest part of a wall, pier, monument, or column; the lower part of a complete architectural design.

baseboard: A board forming the base of something; the finishing board covering the edge of the plastered wall where the wall and floor meet; a line of boarding around the interior walls of a room, next to the floor.

base course: A footing course, as the lowest course of masonry of a wall or pier; the foundation course on which the remainder rests.

basement: The story of a building next below the main floor; a story partly or wholly below the ground level; the finished portion of a building below the main floor, or section; also, the lowest division of the walls of a building.

base molding: The molding above the plinth of a wall, pillar, or pedestal; the part between the shaft and the pedestal, or if there is no pedestal, the part between the shaft and the plinth.

base trim: The finish at the base of a piece of work, as a board or molding used for finishing the lower part of an inside wall, such as a *baseboard;* the lower part of a column which may consist of several decorative features, including various members which make up the base as a whole; these may include an ornate pedestal and other decorative parts.

bastard tuck pointing: In masonry, a type of pointing of joints whereby a wider ridge is formed along the center of the joints than in true tuck pointing of mortar joints. See *tuck pointing.*

bat: A piece of brick with one end whole, the other end broken off.

batch: The amount of concrete mixed at one time—it could be a bucketful, a wheelbarrow full, or a mixer full.

batten: A thin, narrow strip of board used for various purposes; a piece of wood nailed across the surface of one or more boards to prevent warping; a narrow strip of board used to cover cracks between boards; a small molding used for covering joints between sheathing boards to keep out moisture. When sheathing is placed on walls in a vertical position and the joints covered by battens, a type of siding is formed known as *boards and battens*. This form of siding is commonly used on small buildings, farm structures, and railroad buildings. A cleat is sometimes called a *batten*. Squared timbers of a special size used for flooring are also known as *battens*. These usually measure 7 inches in width, 2½ inches in thickness, and 6 feet, or more, in length.

batten door: A door made of sheathing boards reinforced with strips of boards nailed crossways and the nails clinched on the opposite side.

batter: A receding upward slope; the backward inclination of a timber or wall which is out of plumb; the upward and backward slope of a retaining wall which inclines away from a person who is standing facing it. A wall is sometimes constructed with a sloping outer face while the inner surface is perpendicular; thus the thickness of the wall diminishes toward the top. See Fig. 16.

batter board: Usually, one of two horizontal boards nailed to a post set up near the proposed corner of an excavation for a new building. The builder cuts notches or drives

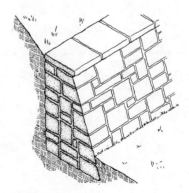

Fig. 16. Batter.

Masonry Simplified

nails in the boards to hold the stretched building cord which marks the outline of the structure. The boards and strings are used for relocating the exact corner of the building at the bottom of the finished excavation. See Fig. 17.

batt insulation: A type of small-sized blanket insulating material, usually composed of mineral fibers and made in relatively small units for convenience in handling and applying. Sometimes spelled *bat*.

bead: A circular or semicircular molding; a beaded molding is known as *beading;* when the beads are flush with the surface and separated by grooves, this type of molding is called *quirk bead*.

bead plane: A special type of plane used for cutting beads.

B & C B: An abbreviation for the term *beaded on the edge and center*.

beam: Any large piece of timber, stone, iron, or other material, used to support a load over an opening, or from post to post; one of the principal horizontal timbers, relatively long, used for supporting the floors of a building.

beam and slab floor construction: A reinforced concrete floor system in which a solid slab is supported by beams or girders of reinforced concrete.

beam ceiling: A type of construction in which the beams of the ceiling, usually placed in a horizontal position, are exposed to view. The beams may be either true or false, but if properly constructed the appearance of the ceiling will be the same, whether the beams are false or true.

beam fill: Masonry or concrete used to fill the spaces between joists; also between a basement or foundation wall and the framework of a structure, to provide fire stops in outside walls for checking fires which start in the basement of a building.

bearing: That portion of a beam or truss which rests upon a support; that part of any member of a building that rests upon its supports.

bearing plate: A plate placed under a heavily loaded truss beam, girder, or column, to distribute the load so the pressure of its weight will not exceed the bearing strength of the supporting member.

bearing wall or partition: A wall which supports the floors and roof in a building; a partition that carries the floor joists and other partitions above it.

bed: In masonry, a layer of cement or mortar in which the stone or brick is embedded, or against which it bears; either of the horizontal surfaces of a stone in position,

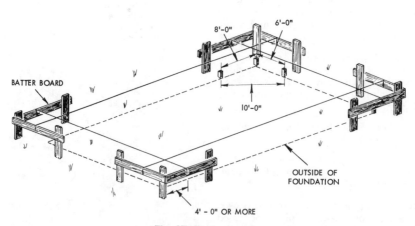

Fig. 17. Batter board.

Glossary

as the *upper* and *lower beds;* the lower surface of a brick, slate, or tile.

bed dowel: A dowel placed in the center of a stone bed.

bed joint: In brickwork, the horizontal joint upon which the bricks rest (Fig. 14); also, the radiating joints of an arch.

bed molding: Finish molding used where the eaves of a building meet the top of the outside walls; the moldings, in any architectural order, used as a finish immediately beneath the corona and above the frieze; any molding in an angle, as between the projection of the overhanging eaves of a building and the sidewalls.

bed of a stone: The under surface of a stone; when the upper surface is prepared to receive another stone, it is called the *top bed,* and the natural stratification of the stone is called the *natural bed.*

bedplate: A foundation plate used as a support for some structural part; a metal plate used as a bed, or rest, for a machine; a foundation framing forming the bottom of a furnace.

bed stone: A large foundation stone, as one used to support a girder.

belt courses: A layer of stone or molded work carried at the same level across or around a building. Also, a decorative feature, as a horizontal band around a building, or around a column. Two types of belt courses are shown in Fig. 18.

bench dog: A wooden or metal peg placed in a hole near the end of a workbench to prevent a piece of work from slipping out of position or off the bench.

bench hook: A hook-shaped device used to prevent a piece of work from slipping on the bench during certain operations; a flat timber or board with cleats nailed on each side and one on each end to hold a piece of work in position and to prevent slipping which might cause injury to the top of the workbench.

bench marks: A basis for computing elevations by means of identification marks or symbols on stone, metal, or other durable

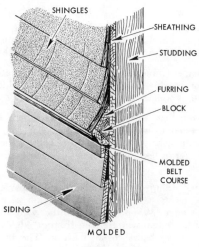
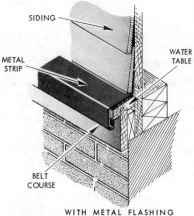

Fig. 18. Belt courses.

matter, permanently fixed in the ground, and from which differences of elevations are measured.

bench plane: Any plane used constantly and kept handy on the bench; a plane used on the bench as a jack plane, a truing plane, or a smoothing plane.

bench stop: An adjustable metal device, usually notched, attached near one end of a workbench, to hold a piece of work while it is being planed.

bench table: A course of projecting stones

Masonry Simplified

forming a stone seat running around the walls at the base of a building such as a large church; a projecting course around the base of a pillar sufficient to form a seat.

bent: A framework transverse to the length of a structure usually designed to carry lateral as well as vertical loads.

bevel: One side of a solid body which is inclined in respect to another side, with the angle between the two sides being either greater or less than a right angle; a sloping edge.

bevel siding: A board used for wall covering, as the shingle, which is thicker along one edge. When placed on the wall the thicker edge overlaps the thinner edge of the siding below to shed water. The face width of the bevel siding is from 3½" to 11¼" wide.

bib: In plumbing, a faucet or tap. A water faucet threaded so a hose may be attached to carry water. Also spelled *bibb*.

bid: An offer to furnish, at a specified price, services, supplies or equipment for performing a designated piece of work; an offer to pay a specified sum for goods sold at auction.

biscuit: A term applied to unglazed tile or ware after first firing in a biscuit oven, but before glazing.

bit brace or bit stock: A curved device used for holding boring or drilling tools; a bit stock, with a curved handle, designed to give greater leverage than is afforded by a boring tool with a straight handle. See Fig. 24.

blade: The longer of the two extending arms of the *framing square*, usually 24 inches long and 2 inches wide. The *tongue* of the square forms a right angle with the *blade*. *Rafter framing tables* and *essex board measure tables* appear on the faces of the blade of the square. Also called *body*.

blank flue: If the space on one side of a fireplace is not needed for a flue, a chamber is built in and closed off at the top in order to conserve material and labor, and to balance the weight. See Fig. 19.

bleed: In concrete work, the exuding of

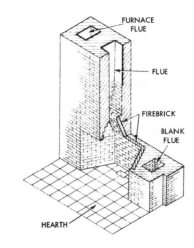

Fig. 19. Blank flue.

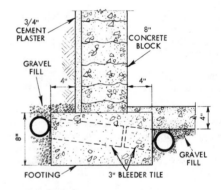

Fig. 20. Bleeder tile.

water from concrete. The water either rises to the surface and weakens the slab, or escapes through the forms, sometimes carrying cement along.

bleeder tile: The pipe placed in the footings of a wall to allow the surface water accumulated by the outside tile drain to pass into the drain provided on the inside of the foundation wall. Sometimes called bleeder pipe. See Fig. 20.

Glossary

blemish: Any imperfection which mars the appearance of wood.

blind header: In masonry work, stones or bricks having the appearance of headers; they really are only short blocks of stone or the ends of bricks.

block: In building construction, a small piece of wood glued into the interior angle of a joint to strengthen and stiffen the joint; a piece of wood placed back of a wainscot for support and to hold it away from the wall; a building unit of terra cotta or cement which differs from brick in being larger and sometimes hollow; also, a small piece of stone which has been cut down, usually for attaching a rope for lifting purposes.

block-in-course: A kind of masonry used for heavy engineering construction, in which the stones are carefully squared and finished to make close joints, and the faces are dressed with a hammer.

block-in-course bond: In masonry, a bond used for uniting the concentric courses of an arch by inserting transverse courses, or *voussoirs,* at intervals.

blocking course: In masonry, a finishing course of stones on top of a cornice, showing above the cornice and crowning the walls, usually serving as a sort of solid parapet, forming a small architectural attic.

block plane: A tool used for working end grain. This type of plane is usually small in size, measuring from 5 to 7 inches in length. The cutting bevel is placed up instead of down, and has no cap iron. Designed to use in one hand when in operation.

bloom: An efflorescence which sometimes appears on masonry walls, especially on a brick wall.

blueprint: A working plan used by tradesmen on a construction job; an architectural drawing made by draftsmen, then transferred to chemically treated paper by exposure to sunlight, or strong artificial light. The sensitized paper, to which the drawing is transferred, turns blue when exposed to light.

blue stain: A discoloration of lumber due to a fungus growth in the unseasoned wood. Although *blue stain* mars the appearance of lumber, it does not seriously affect the strength of the timber.

bluestone: A grayish-blue sandstone quarried near the Hudson River; much used in the East as a building stone, especially for window and door sills; also used for lintels.

board measure: A system of measurement for lumber. The unit of measure being one board foot which is represented by a piece of lumber 1 foot square and 1 inch thick. Quantities of lumber are designated and prices determined in terms of *board feet.*

board rule: A measuring device with various scales for finding the number of board feet in a quantity of lumber without calculation; a graduated scale used in checking lumber to find the cubic contents of a board without mathematical calculation.

boasted work: In masonry, a dressed stone having a finish on the face similar to tooled work. *Boasting* may be done by hand or with a machine tool.

boaster: In stone masonry, a chisel used to smooth the surface of hard stone or to remove tool marks.

body: Same as the *blade* of a *framing square.*

bolster: A crosspiece on an arch centering, running from rib to rib; the bearing place of a truss bridge upon a pier; a top piece on a post used to lengthen the bearing of a beam.

bolt: A fastener, usually consisting of a piece of metal having a head and a threaded body for the receptacle of a nut. For common types of bolts, see Fig. 21.

bond: In masonry and bricklaying, the arrangement of brick or stone in a wall by lapping them upon one another, to prevent vertical joints falling over each other. As the building goes up, an inseparable mass is formed by tying the face and backing together. Various types of bond are shown in Fig. 22.

bond breaker: A material, such as form oil, used to prevent adhesion or sticking of

Masonry Simplified

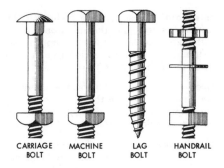

CARRIAGE BOLT   MACHINE BOLT   LAG BOLT   HANDRAIL BOLT

Fig. 21. Bolts.

newly placed concrete and the forms, wall, etc.

bondstones: In masonry, stones running through the thickness of a wall at right angles to its base to bind the wall together.

bossage: In masonry, stones which are roughly dressed, such as corbels and quoins, built in so as to project, and then finish-dressed in position.

Boston hip roof: A method of shingling used to cover the joint, or hip, of a hip roof. To insure a watertight job, a double row of shingles or slate is laid lengthwise along the hip.

boulder wall: In masonry, a type of rustic wall composed of boulders, usually undressed, and mortar.

bow: Any part of a building which projects in the form of an arc or of a polygon.

bow saw: A special type of saw used for making curved cuts. The blade which is thin and narrow is held in tension by the leverage obtained through the twisting of a cord, or by means of rods and turnbuckle.

box column: A type of built-up hollow column used in porch construction; it is usually square in form.

box sill: A header nailed on the ends of joists and resting on a wall plate. It is used in frame-building construction. See Fig. 23.

brace: A piece of wood or other material used to resist weight or pressure of loads; an inclined piece of timber used as a support to stiffen some part of a structure; a support used to help hold parts of furniture in place, giving strength and durability to the entire piece. A term also applied to a tool with which *auger bits* are turned for boring holes in wood. A bit brace is shown in Fig. 24.

brace bit: A tool used for boring holes in wood. An ordinary bit has square, tapered shanks to fit into the socket of a common brace.

brace frame: A type of framework for a building in which the corner posts are braced to sills and plates.

brace jaws: The parts of a bit brace which clamp around the tapered shank of a bit. See Fig. 24.

brace measure: A table which appears on the *tongue* of a *framing square*. This table gives the lengths of common 45° braces plus the length of a brace with a run of 18" to 24" of rise. Limited in use to braces conforming to these specifications.

bracing: The ties and rods used for supporting and strengthening the various parts of a building.

bracket: A projection from the face of a wall used as a support for a cornice, or some ornamental feature; a support for a shelf.

brad: A thin, usually small, nail made of wire with a uniform thickness throughout and a small head.

bradawl: A short straight awl with a chisel or cutting edge at the end; a nontapering awl.

break: A lapse in continuity; in building, any projection from the general surface of a wall; an abrupt change in direction as in a wall.

breaking of joints: A staggering of joints to prevent a straight line of vertical joints. The arrangement of boards so as not to allow vertical joints to come immediately over each other.

break iron: An iron fastened to the top of

Glossary

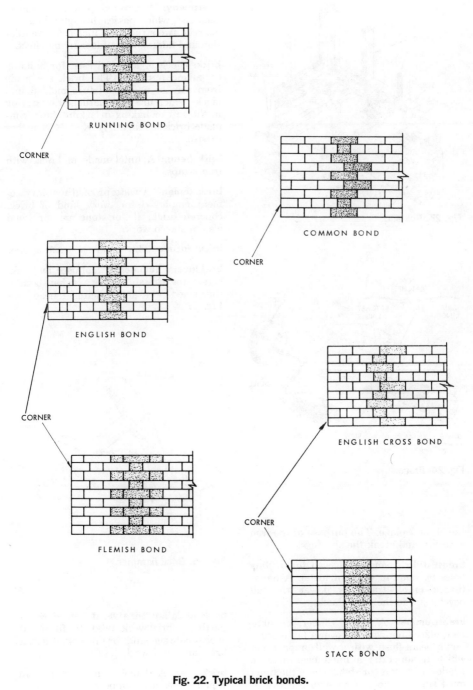

Fig. 22. Typical brick bonds.

303

Masonry Simplified

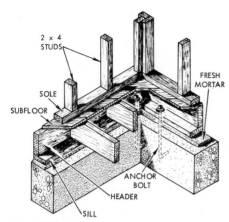

Fig. 23. Box sill.

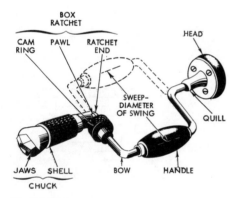

Fig. 24. Brace.

the bit of a plane. The purpose of the iron is to curl and break the shavings.

breast drill: A small tool used for drilling holes by hand in wood or metal. A hand-turned crank transmits power through bevel gears to the drill chuck.

breastsummer: A heavy timber, or summer, placed horizontally over a large opening; a beam flush with a wall or partition which it supports; a lintel over a large window of a store, or shop, where the lintel must support the superstructure above it.

breezeway: A covered passage, open at each end, which passes through a house or between two structures, increasing ventilation and adding an outdoor living effect.

brick: Block of material used for building or paving purposes. The brick are made from clay or a clay mixture molded into blocks which are then hardened by drying in the sun or baking in a kiln. American-made brick average $2\frac{1}{2} \times 4 \times 8$ inches in size.

brick beam: A lintel made of brick, with iron straps.

brick cement: A waterproofed masonry cement employed for every kind of brick, concrete brick, tile, or stone masonry, and also in stucco work.

brick facing: The same as *brick veneer*.

bricklayer's hammer: A tool used by bricklayers for dressing brick. It has both a sharpened peen and a hammer head. See Fig. 25.

Fig. 25. Brick hammer.

brick nogging: In a wood-framed wall or partition, brickwork used to fill in the spaces between studs or timbers; also called brick-and-stud work.

brick pier: A detached mass of masonry which serves as a support.

Glossary

brick set: In masonry, a tool used to cut bricks when exact surfaces are required. The *bricklayer's hammer* is used to force the chisel-like brick set into the brick.

brick trimmer: An arch built of brick between trimmers in the thickness of an upper floor to support a hearth and to guard against fire.

brick trowel: In masonry, a flat triangular-shaped trowel used by bricklayers for picking up mortar and spreading it on a wall. See *buttering trowel*.

brick veneer: A brick facing applied to the surface of the walls of a frame structure, or other types of structures.

bridging: An arrangement of small wooden pieces between timbers, such as joists, to stiffen them and hold them in place; a method of bracing partition studding and floor joists by the use of short strips of wood; cross bridging used between floor joists; usually a piece of 1x3, 2x2, or 2x4. Solid bridging used between partition studs is the same size as the studding.

British thermal unit: The quantity of heat required to raise the temperature of one pound of pure water one degree Fahrenheit at or near the temperature of maximum density of water 39 degrees Fahrenheit. Abbreviation Btu.

broached work: In masonry, broad grooves which give a finish to a building stone, made by dressing the stone with a punch.

brush or spray coat: A waterproofing application of one or more coats of asphalt, or a commercial waterproofing, on the exterior of the foundation, below grade line, with a brush, trowel, or by spraying. May be used where subgrade moisture problems are not severe.

buck: Framing around an opening in a wall. A door buck encloses the opening in which a door is placed.

buggy: A manual or powered vehicle used to transport fresh concrete from the mixer to location where the concrete is to be placed.

builders' tape: Steel measuring tape, usually 50 to 100 feet in length, contained in a circular case. *Builders' tape* is made sometimes of fabricated materials.

building: A structure used especially for a dwelling, barn, factory, store, shop, or warehouse; the art, or work, of assembling materials and putting them together in the form of a structure; the act of one who or that which builds.

building block: Any hollow rectangular block of burned clay, terra cotta, concrete, cement, or glass, manufactured for use as building material.

building brick: A solid masonry unit made primarily for building purposes and not especially treated for texture or color. Formerly called common brick. Its nominal dimensions (which include the thickness of the mortar joint used with it) are 4" x 2⅔" x 8".

building line: The line, or limit, on a city lot beyond which the law forbids the erection of a building; also, a second line on a building site within which the walls of the building must be confined; that is, the outside face of the wall of the building must coincide with this line.

building paper: A form of heavy paper prepared especially for construction work. It is used between rough and finish floors, and between sheathing and siding, as an insulation and to keep out vermin. It is used, also, as an undercovering on roofs as a protection against weather.

building stone: An architectural term applied in general to any kind of stone which may be used in the construction of a building, such as limestone, sandstone, granite, marble, or others.

bulkhead: In building construction, a box-like structure which rises above a roof or floor to cover a stairway or an elevator shaft. Also: a partition blocking fresh concrete from a section of the forms or closing the end of a form, such as at a construction joint.

bull floating: First stage in the final finishing of concrete flatwork; sometimes substituted for darbying. Smooths and levels hills and voids left after screeding.

305

bull header: In masonry, a brick having one rounded corner, usually laid with the short face exposed to form the brick sill under and beyond a window frame; also used as a *quoin* or around doorways.

bull nose: An exterior angle which is rounded to eliminate a sharp or square corner. In masonry, a brick having one rounded corner; in carpentry, a stair step with a rounded end used as a starting step.

bull-nose plane: A small plane which can be used in corners or other places difficult to reach. The mouth can be adjusted for coarse or fine work.

bull's-eye arch: An arch forming a complete circle.

bull stretcher: A brick with one corner rounded and laid with the long face exposed, as a *quoin*.

bungalow: A one-story house with low sweeping lines and a wide veranda; sometimes the attic is finished as a second story. This type of dwelling was developed in India. In the United States, the *bungalow* has become especially popular as a country or seaside residence.

burl: An abnormal growth on the trunks of many trees; an excrescence often in the form of a flattened hemisphere; veneer made from these excrescences, an especially beautiful *burl* veneer is cut from stumps of walnut trees.

Burnett's process: The infusion of timber with chloride of zinc as a preservative.

burnisher: A tool, of hardened steel, used for finishing and polishing metal work by friction. The *burnisher* is held against the revolving metal piece which receives a smooth polished surface due to the compression of the outer layer of the metal. This tool is used, also, to turn the edge of a scraper.

burrs: In brick making, lumps of brick which have fused together during the process of burning; often misshapened and used for rough walling.

butt: A hinge of any type except a strap hinge.

Fig. 26. Butterfly roof.

butterfly roof: A roof constructed so as to appear as two shed roofs connected at the lower edges. See Fig. 26.

buttering: In masonry, the process of spreading mortar on the edges of a brick before laying it.

buttering trowel: In masonry, a flat tool similar to, but smaller than, the brick trowel; used for spreading mortar on a brick before it is placed in position.

butt hinge: A hinge secured to the edge of a door and the face of the jamb it meets when the door is closed, as distinguished from the strap hinge. Usually mortised into the door and jamb.

butt joint: Any joint made by fastening two parts together end to end without overlapping. See Fig. 52.

buttress: A projecting structure built against a wall or building to give it greater strength and stability. See Fig. 27.

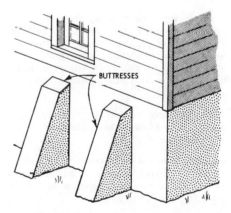

Fig. 27. Buttress.

Glossary

C

cabinet: A piece of furniture, fitted with shelves or drawers, sometimes both, and enclosed with doors, as a *kitchen cabinet* for holding small kitchen equipment; a case with shelves or drawers used as a depository for various articles, such as jewels or precious stones. The doors for such cases are often made of glass, especially when the cases are used for display purposes.

cabinet latch: A name applied to various kinds of catches. These range from the type of catch used on refrigerator doors to the horizontal spring-and-bolt latch operated by turning a knob, as on kitchen cabinets.

cabinet scraper: A tool, made of a flat piece of steel, designed with an edge in such a shape that when the implement is drawn over a surface of wood any irregularities, or uneven places, will be removed, leaving the surface clean and smooth. The *cabinet scraper* is used for final smoothing of surfaces before sandpapering.

cabinetwork: The work of one who makes fine furniture, or beautifully finished woodwork of any kind.

cabin hook: A type of fastener, consisting of a small hook and eye, used on the doors of cabinets.

caisson: A deeply recessed panel sunk in a ceiling or soffit; also, a watertight box used for surrounding work involved in laying a foundation of any structure below water.

caisson pile: A type of pile which has been made watertight by surrounding it with concrete.

calcining: A term applied to the process of producing lime by the heating of limestone to a high temperature. The same as *lime-burning*.

calking: The process of driving tarred oakum or other plastic material into joints or seams between various structural materials such as wood, metal, glass, etc. In masonry, calking is used particularly in *control joints*.

calking tool: A tool used for driving tarred oakum, cotton, and other materials into seams and crevices to make joints watertight and airtight. The *calking tool* is made of steel and in appearance somewhat resembles a chisel.

camber: A slight arching or convexity of a timber, or beam; the amount of upward curve given to an arched bar, beam, or girder to prevent the member from becoming concave due to its own weight or the weight of the load it must carry.

canopy: A rooflike structure projecting from a wall or supported on pillars, as an ornamental feature.

cant: To incline at an angle; to tilt; to set up on a slant, or at an angle; also a molding formed of plain surfaces and angles rather than curves.

cant brick: In masonry, a purpose-made brick with one side beveled. See *splayed brick*.

cant hook: A stout wooden lever with an adjustable steel, or iron, hook near the lever end. The *cant hook* is used for rolling logs and telephone or telegraph poles.

cantilever: A projecting beam supported only at one end; a large bracket, usually ornamental, for supporting a balcony or cornice; two bracketlike arms projecting toward each other from opposite piers or banks to form the span of a bridge making what is known as a *cantilever bridge*.

canting strip: A projecting molding near the bottom of a wall to direct rain water away from the foundation wall; in frame buildings, the same as a *water table*.

cap: The top parts of columns, doors, and moldings; the coping of a wall; a cornice over a door; the lintel over a door or window frame; a top piece. See Fig. 28.

capillary action: In hardened concrete or mortar, the seepage of moisture through the material due to incomplete or faulty surface finishing. Same as *capillary flow*.

capping: The uppermost part on top of a piece of work; a crowning, or topping part. See Fig. 28.

Masonry Simplified

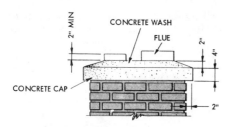

Fig. 28. Concrete capping.

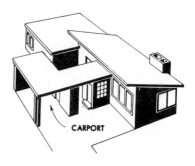

Fig. 29. Carport.

capping brick: In masonry, brick which are specially shaped for capping the exposed top of a wall. Same as *coping brick*.

cap stone: Stone used for the crown or top part of a structure.

Carborundum: A trademark for an abrasive made from a combination of carbon and silicon and sometimes used instead of emery.

carborundum cloth or **paper:** An abrasive cloth or paper made by covering the material with powdered carborundum held in place by some adhesive, such as glue.

carcase: The frame of a house; the unfinished framework, or skeleton, of a building or ship. Also *carcass*.

carpenter: A craftsman who builds with wood; a worker who builds with the heavier forms of wood, as with lumber in the structural work of a building.

carpenter's finish: A term applied to practically all finish work performed by a carpenter, except that classed as *rough finish*. It includes casings, laying finish flooring, building stairs, fitting and hanging doors, fitting and setting windows, putting on baseboards, and installing all other similar finish material.

carpentry: Work which is performed by a craftsman in cutting, framing, and joining pieces of timber in the construction of ships, houses, and other structures of a similar character.

carpet strip: A piece beneath a door attached to the floor.

carport: A shelter, built into a house, which has a roof but only one or two side walls. Generally found in mild climates. See Fig. 29.

carriage: The support for the steps of a wooden stairway; these supports may be wood or steel.

casement window: A window sash that opens on hinges on the sides.

casing: The framework around a window or door.

caster: A wheel, or set of wheels, mounted in a swivel frame attached to the feet or base of a piece of furniture, trucks, and portable machines. Casters help in the moving of furniture without injury to the floor.

cast-in-place: Concrete which is deposited in the place where it is required to harden as part of the structure, as opposed to pre-cast concrete.

catch basin: A cistern, or depression, at the point where a gutter discharges into a sewer to catch any object which would not readily pass through the sewers; a reservoir to catch and retain surface drainage; a receptacle at an opening into a sewer to retain any matter which would not easily pass through the sewer; a trap to catch and hold fats, grease, and oil from kitchen sinks to prevent them from passing into the sewer.

Glossary

caul: A tool used in forming veneer to the shape of a curved surface.

cavetto: A quarter round, concave molding; a concave, ornamental molding opposed in effect to the ovolo—the quarter of a circle called the *quarter round*.

cavil: In masonry, a kind of heavy sledge hammer, having one blunt end and one pointed end, used for rough dressing of stone at the quarry; a term also applied to a small stone ax resembling a *jedding ax*.

cavity wall: A hollow wall, usually consisting of two brick walls erected a few inches apart and joined together with ties of metal or brick. Such walls increase thermal resistance and prevent rain from driving through to the inner face. Also called *hollow wall*.

cellar: A room, or set of rooms, below the surface of the ground, used especially for keeping provisions and other stores; a room beneath the main portion of a building. In modern homes, the heating plant is usually located in the *cellar*.

cement: In building, a material for binding other material or articles together; usually plastic at the time of application but hardens when in place; any substance which causes bodies to adhere to one another, such as portland cement, stucco, and natural cements; also mortar or plaster of paris.

cement-base waterproof coating: Compounds which seal the pores in basement walls or floors, providing protection against water pressure and dampness in the ground. These compounds usually come in a cement base, but some can be purchased in paste, powder, or liquid form to be mixed with cement before application.

cement colors: A special mineral pigment used for coloring cement for floors. In addition to the natural coloring pigment obtained from mineral oxides, there are manufactured pigments produced especially for cement work.

cement gun: A mechanical device used for spraying fine concrete or cement mortar by means of pneumatic pressure. Same as *Cement Gun,* the trade-mark for a machine used to apply *Gunite*. See *shotcrete*.

cement joggle: In masonry construction, a key which is formed between adjacent stones by running mortar into a square-section channel which is cut equally into each of the adjoining faces, thus preventing relative movement of the faces.

cement mortar: A building material composed of portland cement, sand, and water.

center: A fixed point about which the radius of a circle, or of an arc, revolves; the point about which any revolving body rotates or revolves, as the middle, or center, of activity.

centering: The frame on which a brick or stone arch is turned; the false work over which an arch is formed. In concrete work the *centering* is known as the *frames*.

center line: A broken line, usually indicated by a dot and dash, showing the center of an object and providing a convenient line from which to lay off measurements.

centerpiece: An ornament placed in the middle of a ceiling.

centimeter: A measure of length in the metric system equal to the one-hundredth part of a meter, or .3937 inch.

central mixed concrete: Ready mix concrete which is completely mixed in a stationary mixer and then delivered to the site in agitating or fixed trucks. See *ready mix*.

ceramic mosaic: A collective term applied to floor tiles which are marketed in sheet units. The individual units are properly spaced and mounted on sheets of paper. The tiles are produced in a variety of sizes, shapes, and colors and used extensively for shower-bath floors; also for regular bathroom floors.

ceramic tile: A thin, flat piece of fired clay, usually square. These pieces of clay are attached to walls, floors, or counter tops, with cement or other adhesives, creating durable, decorative and dirt-resistant surfaces. Tiles may be plastic process (formed while clay is wet) or dust-pressed process (compressed clay powder). They may be glazed (vitrified coating); unglazed (natu-

Masonry Simplified

ral surface); nonvitrified; semivitrified and vitreous (porous, semiporous, or relatively nonporous).

chain bond: In masonry, the bonding together of a stone wall by the use of a built-in chain or iron bar.

chamfer: A groove, or channel, as in a piece of wood; a bevel edge; an oblique surface formed by cutting away an edge or corner of a piece of timber, or stone. Any piece of work that is cut off at the edges at a 45 degree angle so that two faces meeting form a right angle are said to be *chamfered*.

channel: A concave groove cut in a surface as a decorative feature; a grooved molding used for ornamental purposes; a decorative concave groove on parts of furniture.

channel iron: A rolled iron bar with the sides turned upward forming a rim, making the *channel iron* appear like a channel-shaped trough. In sectional form, the *channel iron* appears like a rectangular box with the top and two ends omitted.

charging: Putting materials into the mixer.

chase: In masonry, a groove or channel cut in the face of a brick wall to allow space for receiving pipes; in building, a trench dug to accommodate a drainpipe; also, a recess in a masonry wall to provide space for pipes and ducts. Also see *sleeve*. See Fig. 30.

chasing: The decorative features produced by grooving or indenting metal.

check: An ornamental design composed of inlaid squares; a blemish in wood caused by the separation of wood tissues.

checkerboard placement: The practice of planning a concrete job so that adjacent slabs of concrete can be placed at separate times. This is handy when special colors or finishes are used and creates a control joint without sawing or jointing.

checking of wood: Blemishes or cracks in timber due to uneven seasoning.

check rail: The middle horizontal member of a double-hung window, forming the

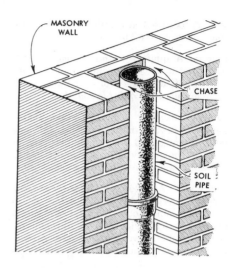

Fig. 30. Chase in masonry wall.

lower rail of the top sash and the top rail of the lower sash.

chevron: The meeting place of rafters at the ridge of a gable roof; a zigzag pattern used as an ornamentation in Romanesque architecture.

chimney: That part of a building which contains the flues for drawing off smoke or fumes from stoves, furnaces, fireplaces, or some other source of smoke and gas.

chimney blocks: Concrete blocks designed to form a continuous flue when placed in position, one on top of another.

chimney bond: In masonry, a form of bond commonly used for the internal division walls of domestic chimneys, as well as for the outer walls. The surface of this type of wall is made up of stretchers which break joints at the center, with a header on each alternate course at the corner.

chimney breast: That part of a chimney which projects from a wall where the chimney passes through a room. When the chimney is a part of a fireplace, the breast of the chimney is usually built much wider than the chimney itself to provide for a

Glossary

mantel or to improve the appearance of the room.

chimney lining: Rectangular or round tiles placed within a chimney for protective purpose. The glazed surface of the tile provides resistance to the deteriorating effects of smoke and gas fumes.

chip ax: Small sharp cutting tool used in the building trades for cutting and shaping structural stone or timbers.

chipping: The process of cutting off small pieces of metal or wood with a cold chisel and a hammer.

chisel: A cutting tool with a wide variety of uses. The cutting edge on the end of the tool usually is transverse to the axis. The cutting principle of the *chisel* is the same as that of the wedge.

chutes: Made of metal, wood, or rubber and used to place fresh concrete in hard-to-reach areas such as in tall structural forms, or over distances where the transporting vehicle can't approach close enough.

cinder blocks: Building blocks in which the principal materials are cement and cinders.

cinder concrete: A type of concrete made from portland cement mixed with clean, well-burned coal cinders which are used as coarse aggregate.

cinder fill: A fill of cinders, from three to six inches deep, under a basement floor as an aid in keeping the basement dry; also, a fill of cinders outside of a basement wall to a depth of twelve inches, over drain tile to facilitate drainage. Pebble gravel is sometimes used instead of cinders.

circular and angular measure: A standard measure expressed in degrees, minutes, and seconds, as follows:

    60 seconds (") = 1 minute (')
    60 minutes   = 1 degree (°)
    90 degrees   = 1 quadrant
    4 quadrants   = 1 circle or
                     circumference

circular saw: A saw with teeth spaced around the edge of a circular plate, or disk, which is rotated at high speed upon a central axis, or spindle, used for cutting lumber or sawing logs.

circumference: The perimeter of a circle; a line that bounds a circular plane surface.

circumscribe: The process of drawing a line to enclose certain portions of an object, figure, or plane; to encircle; to draw boundary lines; to enclose within certain limits.

clamp: A device for holding portions of work together, either wood or metal; an appliance with opposing sides or parts that may be screwed together to hold objects or parts of objects together firmly.

clamping screw: A screw used in a clamp; a screw used to hold pieces of work together in a clamp.

clapboard: A long thin board, graduating in thickness from one end to the other, used for siding, the thick end overlapping the thin portion of the board.

clap post: The upright post of a cupboard where the door *claps* or closes.

classical: Pertaining to a style of architecture in accordance with ancient Greek and Roman models or later styles of architecture modeled upon principles embodied in the early types of Greek and Roman structures.

classic molding: A type of molding similar to that used in classic orders of architecture.

claw hammer: A carpenter's tool having one end curved and split for use in drawing nails by giving leverage under the heads. See Fig. 31.

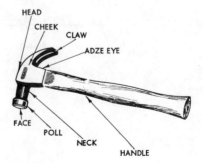

Fig. 31. Claw hammer.

311

Masonry Simplified

claw tool: A stonemason's tool having teeth or claws; used for dressing soft stone. Sometimes called *tooth chisel*.

clay shale: Clay which has a laminated structure; used as one of the ingredients in making bricks.

cleanout: In concrete formwork, an opening left in bottom of the form for the removal of refuse. They are closed before the concrete is placed.

cleat: A strip of wood or metal fastened across a door or other object to give it additional strength; a strip of wood or other material nailed to a wall usually for the purpose of supporting some object or article fastened to it.

clinch: The process of securing a driven nail by bending down the point; to fasten firmly by bending down the ends of protruding nails.

clockwise: Moving in the same direction as the rotation of the hands of a clock; with a right-hand motion.

closed cornice: A cornice which is entirely enclosed by the *roof, fascia,* and the *plancher;* same as *boxed cornice*. See Fig. 32.

closer: In constructing a masonry wall, any portion of a brick used to close up the bond next to the end brick of a course; the last stone, if smaller than the others, in a horizontal course or a piece of brick which finishes a course; also, a piece of brick in each alternate course to enable a bond to be formed by preventing two headers from exactly superimposing on a stretcher; same as *closure*.

coarse aggregate: Crushed stone or gravel used in concrete; the size is regulated by building codes. See *rubble concrete*.

cob: A small mixture of unburned clay, usually with straw as a binder. Used in building walls known as *cob walls*.

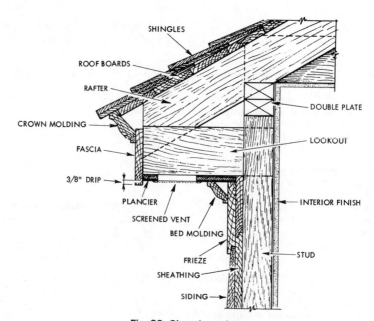

Fig. 32. Closed cornice.

Glossary

cob wall: A wall built of clay blocks made of unburned clay or chalk mixed with straw; also, a wall constructed of *cobs,* such as clay bats.

cocobolo: A tropical wood, which is extremely hard and tough, used for the heads and handles on high-priced tools. It takes a beautiful finish.

code: Any systematic collection or set of rules pertaining to one particular subject, and devised for the purpose of securing uniformity in work or for maintaining proper standards of procedure, as a *building code.*

coffer: An ornamental sunken panel in a ceiling or soffit; a deeply recessed panel in a dome.

cofferdam: A watertight enclosure usually built of piles of clay, within which excavating is done for foundations; also, a watertight enclosure fixed to the side of a ship for making repairs below the water line.

cold chisel: A name applied to a chisel made of tool steel of a strength and temper that will stand up under the hardest usage. A chisel suitable for cutting and chipping cold metal.

cold weather concreting: The process of preventing fresh concrete from freezing during the winter months (40 degrees F or lower).

collar: In carpentry, an encircling band resembling a *collar;* a molding extending around a leg of furniture.

collar beam: A horizontal tie beam, in a roof truss, connecting two opposite rafters at a level considerably above the wall plate.

Colonial: A style of architecture used in America during Colonial times and sometimes used by builders as late as 1840; also, a type of furniture in vogue in early America before the American Revolution, and applied to this type of furniture as late as the nineteenth century.

column: A pillar, usually round; a vertical shaft which receives pressure in the direction of its longitudinal axis; the parts of a column are: the *base* on which the shaft rests, the body, or *shaft,* and the head known as the *capital.*

column footings: Concrete footings, reinforced with steel bars; used as supports for columns which, in turn, carry the load of S beams (formerly called I beams) which serve as supports for the superstructure of a building.

combination pliers: A pincerlike tool, with long, flat, roughened jaws adjustable for size of opening by means of a slip joint. The inner grip is notched for grasping and holding round objects; the outer grip is scored. The tool is used for cutting or bending wire.

combination square: A tool which combines in handy compact form the equivalent of several tools, including an inside try square, outside try square, mitre square, plumb, level, depth gage, marking gage, straightedge, bevel protractor, and center head in addition to square head.

common bond: In masonry, a form of bond in which every sixth course is a header course, and the intervening courses are stretcher courses. Sometimes varied, so a header bond is used every fourth or fifth course. See *typical bonds,* Fig. 22.

common brick: See *building brick.*

common lime: A material produced by the burning of limestone to the proper degree; used for making mortar for plastering and masonry work. Same as *quicklime.*

common rafter: A *rafter* which extends at right angles from the plate line to the *ridge* or *purlin* of a roof. See Fig. 76.

compass brick: In masonry, a curved or tapering brick for use in curved work, such as in arches.

compass plane: A cutting tool used for smoothing concave or convex surfaces; a plane with an adjustable sole.

compass saw: A small handsaw of a special type, with a thin tapering blade designed for cutting a small circle or other small opening, such as a keyhole. Compass saws are often sold in sets called *nests.* (The compass saw is very similar to the *keyhole saw.* See Fig. 53. The keyhole saw, however, is smaller than the compass saw and has smaller blades.)

compound arch: An arch made up of a number of concentric archways placed successively within and behind each other.

compression: A stress or load on a building structure or unit such as the weight of a superstructure on a concrete foundation.

compressive strength: The degree to which a structure can withstand *compression*.

concave: A curved recess; hollowed out like the inner curve of a circle or sphere; the interior of a curved surface or line; a bowl-shaped depression.

concave joint: In masonry, a mortar joint formed with a special tool or a bent iron rod. This type of mortar joint is weather resistive and inexpensive.

concentrated load: The weight localized on, and carried by, a beam, girder, or other supporting structural part.

concrete: In masonry, a mixture of cement, sand, and gravel, with water in varying proportions according to the use which is to be made of the finished product.

concrete-bent construction: A system of construction in which precast concrete-bent framing units are the basic load-bearing members. The principal advantages and problems are similar to those in *post and beam construction*.

concrete blocks: In masonry, precast, hollow, or solid blocks of concrete used in the construction of buildings.

concrete insert: A type of metal anchor used to secure structural wood parts to a concrete or masonry wall. See *anchors*, Fig. 7.

concrete nail: A special nail used to fasten wood or metal to concrete. They are not specified by the penny system, but are available by length. They employ a sinker head, are comparatively thick, and are usually hardened. See Fig. 60.

concrete paint: A specially prepared thin paint, consisting of a mixture of cement and water, applied to the surface of a concrete wall to give it a uniform finish, and to protect the joints against weathering by rain or snow.

concrete wall: In building construction, any wall made of reinforced concrete, such as a basement wall.

condensation: The act or process of changing a substance from a vapor to a liquid state due to cooling. Also, beads or drops of water, and often frost, in cold weather, which accumulate on the inside of the exterior covering of a building, when warm, moisture-laden air, from the interior, reaches a point where the temperature no longer permits the air to sustain the moisture it holds. Use of louvers will reduce moisture condensation in attics.

conduit: A natural or artificial channel for carrying fluids, as water pipes, canals, and aqueducts; a tube, or trough, for receiving and protecting electric wires.

console: In architecture, any bracket, or bracketlike support usually ornamented by a reverse scroll; an ornamental bracketlike support for a cornice or bust; any ornamented bracket-like architectural member used as a support.

construction: The process of assembling material and building a structure; also, that which is built; style of building, as of wood, iron, or steel *construction*.

construction joint: In concrete work, a temporary joint employed when the placing must be interrupted because of weather, time, etc. A rigid, immovable joint where two slabs or parts of a structure are joined firmly to form a solid, continuous unit.

consulting engineer: A person retained to give expert advice in regard to all engineering problems; supposedly an experienced engineer of high rating in his profession.

continuous beam: A timber that rests on more than two supporting members of a structure.

continuous header: The top plate is replaced by 2″ x 6″s turned on edge and running around the entire house. The header is strong enough to act as a lintel over all wall openings, eliminating some cutting and fitting of stud lengths and separate headers over openings. This development is especially important because of the new emphasis on one-story, open-planning houses.

contour: The outline of a figure, as the profile of a molding.

contractor: One who agrees to supply materials and perform certain types of work for a specified sum of money, as a *building contractor* who erects a structure according to a written agreement, or *contract*.

control joint: In concrete flatwork, a groove cut into the top of the slab, usually to about 1/5 the thickness of the slab. The groove predetermines the location of natural cracking as the concrete hardens. In concrete block masonry, a continuous vertical joint without mortar. Such material as rubber, plastic, or building paper is used to key the joint which is then sealed with calk. Control joints are used in very long walls where thermal expansion and concentration may cause cracking in the mortar joints.

coped joint: The seam, or juncture, between molded pieces in which a portion of one piece is cut away to receive the molded part of the other piece.

coping: A covering or top for brick walls. The cap or top course of a wall. The coping frequently is projected out from the wall to afford a decorative as well as protective feature. See Fig. 47.

coping brick: In masonry, brick having special shapes for use in capping the exposed top of a wall. It is sometimes used with a creasing and sometimes without. In the latter case, the brick is wider than the wall and has drips under its lower edges.

coping saw: A saw used for cutting curves and hollowing out moldings. The narrow blade of the *coping saw,* carried on pins set in a steel bow frame, is from $1/16''$ to $1/8''$ wide and $6 1/2''$ long.

corbel: A short piece of wood or stone projecting from the face of a wall to form a support for a timber or other weight; a bracketlike support; a stepping out of courses in a wall to form a ledge; any supporting projection of wood or stone on the face of a wall.

corbeled chimney: A chimney which is supported by a brickwork projection from a wall, forming a sort of bracket or corbel; also, a chimney erected on a bracket constructed of wood members.

corbel out: The building of one, or more, courses of masonry out from the face of a wall to form a support for timbers.

corbel table: A horizontal row of corbels supporting lintels or small arches; a projecting course, as of masonry, which is supported by a series of corbels; a cornice supported by corbels.

corbie stone: Stones used for covering the steps of a crow-stepped gable wall.

cord: Wood cut in four-foot lengths, usually for firewood. A pile of wood measuring four feet in width, four feet in height, and eight feet in length.

corner bead: A small projecting molding, or bead, built into plastered corners to prevent accidental breaking of the plaster; such a *bead* usually is of metal.

corner bit brace: A specially designed *bit brace* for use in positions where it is difficult for a workman to operate the regular bit brace; a corner brace useful for tradesmen who have occasion to work close to perpendicular surfaces and in corners.

cornice: Projection at the top of a wall; a term applied to construction under the eaves or where the roof and side walls meet; the top course, or courses, of a wall when treated as a crowning member. See Fig. 32.

corona: That part of a cornice supported by and projecting beyond the bed molding. The *corona* serves as a protection to the walls by throwing off rain water.

corridor: A passageway in a building into which several apartments open; a gallery, or passage, usually covered into which rooms open, as the *corridor* of a hotel, or of an art gallery.

counterbracing: Diagonal bracing which transmits a strain in an opposite direction

Masonry Simplified

from the main bracing; in a truss or girder, bracing used to give additional support to the beam and to relieve it of transverse stress.

counterclockwise: Motion in the direction opposite to the rotation of the hands of a clock.

countersink: To make a depression in wood or metal for the reception of a plate of iron, the head of a screw, or for a bolt, so that the plate, screw, or bolt will not project beyond the surface of the work; to form a flaring cavity around the top of a hole for receiving the head of a screw or bolt.

course: A continuous level range or row of brick or masonry throughout the face or faces of a building; to arrange in a row. A row of bricks, when laid in a wall, is called a *course*. See Fig. 33.

coursed ashlar: In masonry, a type of ashlar construction in which the various blocks of structural material have been arranged, according to height, to form regular courses in the face of walls. See Fig. 11.

coursed rubble masonry: Masonry composed of roughly shaped stones fitting approximately on level beds and well bonded.

court: An open space surrounded partly or entirely by a building; an open area partly or wholly enclosed by buildings or walls.

cove: A concave molding; an architectural member, as a ceiling, which is curved or arched at its junction with the side walls; also, a large, hollow cornice; a niche.

cove bracketing: The lumber skeleton, or framing for a cove; a term applied chiefly to the *bracketing* of a cove ceiling.

cove ceiling: A ceiling which rises from the walls with a concave curve.

cove molding: A molding called the *cavetto;* a quarter round or concave molding.

coving: The scotia inverted on a large scale; a concave molding often found in the base of a column.

cradling: Lumber work, or framing, for sustaining the lath and plaster of vaulted ceilings.

cramp: In masonry, a contrivance consisting of iron rods or bars with the ends bent to a right angle; used to hold blocks of stone together.

crane: A large hoisting device equipped with a movable boom; used for lifting ma-

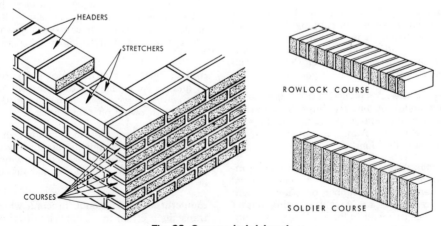

Fig. 33. Courses in brickwork.

Glossary

Fig. 34. Crane. (Adolphi Studios, Inc.)

terials in high level construction. See Fig. 34.

crawl space: In cases where houses have no basements, the space between the first floor and the ground is made large enough for a man to crawl through for repairs and installation of utilities.

crazing: In concrete masonry numerous fine hair cracks in the surface of a newly hardened slab caused by surface shrinkage.

cresting: An ornamental finish of the wall or ridge of a building. The *cresting* of shingle roofs is generally of sheet metal.

crib: A cratelike framing used as a support for a structure above; any of various frameworks, as of logs or timbers, used in construction work; the wooden lining on the inside of a shaft; open-work of horizontally, cross-piled squared timbers, or beams, used as a retaining wall.

cripple: In building construction, any part of a frame which is cut less than full size, as a *cripple studding* over a door or window opening.

cripple rafter: A rafter extending from a hip to a valley rafter.

crosscut saw: A saw made to cut transversely, as across the grain of wood.

cross grain: A section of wood cut at right angles to the longitudinal fiber.

crosslap: A joint where two pieces of timber cross each other. This type of joint is formed by cutting away half the thickness of each piece at the place of joining, so that one piece will fit into the other and both pieces will lie on the same plane. See Fig. 52.

cross section: A transverse section cut at right angles to the longitudinal axis of a piece of wood, drawing, or other work.

crotch veneer: A type of veneer cut from the crotch of a tree forming an unusual grain effect; also, veneer cut from wood of twin trees which have grown together, likewise forming an unusual grain effect.

crown molding: A molding at the top of the cornice and immediately beneath the roof.

cube root: A given number which taken three times as a factor produces a number called its *cube,* as 3×3×3 equals 27, hence *3,* the given number, is the *cube root* of 27.

cubic content: In building construction, the number of cubic feet contained within the walls of a room or combination of rooms and used as a basis for estimating cost of materials and construction; cubic content is also important when estimating cost of installing heating, lighting, and ventilating systems.

cubic measure: The measurement of volume in cubic units, as follows:

    1,728 cubic inches = 1 cubic foot
    27 cubic feet = 1 cubic yard
    231 cubic inches = 1 gallon
    128 cubic feet = 1 cord

cup shake: A defect in wood where annual rings separate from each other, thus forming a semicircular flaw. Such flaws may occur between two or more concentric layers of wood. Because of their appearance, such defects are known as *cup shake,* but, since they are caused by the wind, they are also known as *windshake.* See Fig. 94.

curb edger: In masonry and cement work, a tool specially designed for shaping curved sections which must be finished smooth and true, such as the borders of driveways or pavements.

curb roof: The mansard roof, which takes its name from the architect who designed it. This type of roof has a double slope on each side, with the lower slope almost vertical. Frequently the lower slope contains dormer windows, which make possible the addition of another story to the house.

curing: In concrete work, the drying and hardening process. Also called *setting.* See *hydration.*

curl: A spiral or curved marking in the grain of wood; a feather-form mark in wood.

curtain wall: A thin wall, supported by the structural steel or concrete frame of the building, independent of the wall below.

cushion head: In foundation construction, a capping to protect the head of a pile which is to be sunk into the ground with a *pile driver.* Such a cushion usually consists of a cast-iron cap.

cut nails: Iron nails cut by machines from sheet metal, as distinguished from the more common wire nails now in general use. See Fig. 60.

cutter: In masonry, a brick made soft enough to cut with a trowel to any shape desired, then rubbed to a smooth face. Sometimes called *rubbers.* The same as *seconds.*

cutting gage: A gage similar in construction to the regular marking gages except that it has an adjustable blade for slitting thin stock, instead of the marking pin.

cutting pliers: A type of pliers which, in addition to the flat jaws, has a pair of nippers placed to one side for cutting wire.

cyma: A molding in common use, with a

Glossary

simple waved line concave at one end and convex at the other end, similar in form to an italic *f*. When the concave part is uppermost the molding is called *cyma recta,* but if the convexity appears above and the concavity below the molding is known as *cyma reversa.*

D

dado: The vertical face of an insulated pedestal between the base and surbase or between the base and cornice; a plain, flat surface at the base of a wall as in a room; such a surface is sometimes ornamented.

damper: A device used for regulating the draft in the flue of a furnace; also, a device for checking vibrations.

damp-proofing: The special preparation of a wall to prevent moisture from oozing through it; material used for this purpose must be impervious to moisture.

darby: A flat tool used by plasterers to level the surface of plaster, especially on ceilings. The *darby* is usually about three and one-half inches wide and forty-four inches long, with two handles on the back. See Fig. 35. In concrete flatwork, a smoothing tool used immediately after screeding. See *bullfloating.*

deadening: The use of insulating materials, made for the purpose, to prevent sounds passing through walls and floors.

dead level: An emphatic statement used to indicate an absolute level.

dead load: A permanent, inert load whose pressure on a building or other structure is steady and constant due to the weight of its structural members and the fixed loads they carry, imposing definite stresses and strains upon it; the weight of the structure itself.

deal: A board of fir or pine cut to a specified size, or to one of several specified sizes.

deciduous: Pertaining to trees which shed their leaves annually.

Fig. 35. Plasterer's darby.

decimal: A fractional part of a number, proceeding by tenths, each unit being ten times the unit next smaller.

decimal equivalent: The value of a fraction expressed as a decimal, as $\frac{1}{4}$ equals .25.

deck: In building, any flat horizontal surface on which other materials or structures are placed. See Fig. 36.

defect: In lumber, an irregularity occurring in or on wood that will tend to impair its strength, durability, or utility value.

deflection: A deviation, or turning aside, from a straight line; bending of a beam or any part of a structure under an applied load.

deformation: Act of deforming or changing the shape; alteration in form which a structure undergoes when subjected to the action of a weight or load.

Masonry Simplified

Fig. 36. Placing concrete on a deck. (Republic Steel Corp.)

deformed bars: Reinforcing bars made in irregular shapes to produce a better bond between the bars and the concrete.

degree: One 360th part of a circumference of a circle, or of a round angle.

demarcation: In masonry, a fixed line for marking a boundary limit.

derrick: Any hoisting device used for lifting or moving heavy weights; also, a structure consisting of an upright or fixed framework, with a hinged arm which can be raised and lowered and usually swings around to different positions for handling loads.

design: A drawing showing the plan, elevations, sections, and other features necessary in the construction of a new building.

As used by architects, the term *plan* is restricted to the horizonal projection, while *elevation* applies to the vertical, or exterior, views.

detail: A term in architecture applied to the small parts into which any structure or machine is divided. It is applied generally to moldings or other decorative features and to drawings showing a special feature of construction.

detail drawing: A separate drawing showing a small part of a machine or structure in detail; a drawing showing the separate parts of a machine or other object with complete tabular data, such as dimensions, materials used, number of pieces, and operations to be performed; also, a drawing showing the position of the parts of a ma-

Glossary

chine or tool and the manner in which the various parts are placed in assembling them.

detailer: One who prepares small drawings for shop use; a draftsman who makes detailed drawings.

diagonal bond: In masonry, a form of bond sometimes used in unusually thick walls, or for strengthening the bond in footings carrying heavy loads. The bricks are laid diagonally across the wall, with successive courses crossing each other in respect to rake.

diagram: A figure which gives the outline or general features of an object; a line drawing, as a chart or graph used for scientific purposes; a graphic representation of some feature of a structure.

diameter: A straight line passing through the center of a circle or sphere and terminating in the circumference.

dimension shingles: Shingles cut to a uniform size as distinguished from *random shingles*.

distribution tile: Concrete or clay tile without bell mouths. These tile are laid with a little space at each joint, in lines which fan out from a septic tank distribution box.

dividers: Device for measuring or setting off distances or dividing lines. Also known as *compasses*. Usually in plural only.

dogtooth: In architecture, a toothlike ornament or a molding cut into projecting teeth; a type of early architectural decoration in the form of a four-leafed flower, probably so named from its resemblance to a dogtooth violet.

dome: An inverted cup on a building, as a cupola, especially one on a large scale; the vaulted roof of a rotunda.

door check: A device used to retard the movement of a closing door and to guard against its slamming or banging, but also insures the closing of the door.

doorframe: The case which surrounds a door and into which the door closes and out of which it opens. The frame consists of two upright pieces called *jambs* and the *head* or horizontal piece over the opening for the door.

doorhead: The upper part of the frame of a door.

doorstone: The stone which forms the threshold of a door.

doorstop: A device used to hold a door open to any desired position; a device usually attached near the bottom of a door to hold it open and operated by the pressure of the foot. The *doorstop* may or may not be attached to the door. The strip against which a door closes on the inside face of a door frame is also known as a *doorstop*.

dormer window: A vertical window in a projection built out from a sloping roof; a small window projecting from the slope of a roof.

dormitory: A large sleeping room, or a sleeping apartment containing several rooms; also, a building containing a number of sleeping rooms.

double-acting hinge: A hinge which permits motion in two directions, as on a swinging door, or on folding screens.

double Flemish bond: a bond with two stretchers and one header in succession in each course, with headers centered above and below cross joints of the stretchers.

double-hung window: A window with movable upper and lower sashes. See Fig. 37.

double-pitch skylight: A skylight designed to slope in two directions.

double-pole switch: A switch to connect or break two sides of an electric circuit.

dovetail: In carpentry, an interlocking joint; a joint made by cutting two boards or timbers to fit into each other. A common type of joint used in making boxes or cases.

dovetail cramps: A device, usually of iron bent at the ends, or of dovetail form, used to hold structural timbers or stone together.

dovetail cutter: A tool used for cutting the inner and outer dovetails for joints.

dovetail-halved joint: A joint which is halved by cuts narrowed at the heel, as in a dovetail joint.

Masonry Simplified

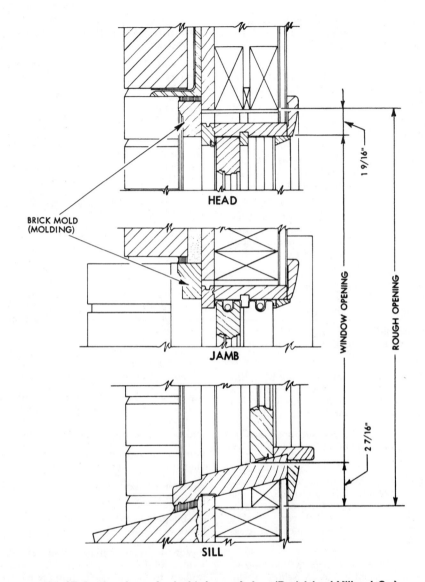

Fig. 37. Section views of a double-hung window. (Rock Island Millwork Co.)

dovetailing: A method of fastening boards or timbers together by fitting one piece into the other as with dovetail joints.

dovetail saw: A small saw similar to a backsaw, with smaller teeth and a different-shaped handle.

dowel: In carpentry, a pin of wood or metal used to strengthen joints between two pieces of timber. See Fig. 52. In masonry, a metal rod used to reinforce the bond in cut stone setting.

downpipe: Same as *downspout*.

Glossary

downspout: Any connector, such as a pipe, for carrying rain water from the roof of a building to the ground or to a sewer connection.

draftsman: One who draws plans or sketches; usually a term applied to one who uses mechanical aids or instruments for preparing drawings for tradesmen.

draftsman's scale: A measuring scale used by draftsmen, usually triangular in shape but sometimes flat. One edge is graduated in $1/16$, $1/8$, $1/4$, $1/2$ and so on, as on a standard scale. Other edges are divided into fractional parts to facilitate reducing measurements.

draft stop or **fire stop:** Any obstruction placed in air passages to block the passing of flames or air currents upward or across a building.

drag: In masonry, a tool which has steel teeth, used for dressing the surface of stone; also called a *comb*.

drain tile: Tubular sections of tile placed around footings to remove excess moisture.

drawer pull: In woodworking, a handle placed on the front of a drawer so that it may be easily opened or pulled out.

drawer slip: A guide or strip on which a drawer moves when it is opened.

drawing: A sketch made with pen, pencil, or crayon representing by lines some figure or object.

drawknife: A woodworking tool with a blade and a handle at each end. The handles are at right angles to the blade which is long and narrow. It is used to smooth a surface by drawing the knife over it.

D & M: An abbreviation for the term *dressed and matched*.

D4S: A symbol used on building plans meaning *dressed on four sides*.

dressed size: The size of lumber after it is planed and seasoned from the unfinished (nominal) size.

dressing: Any decorative finish, as molding around a door; also, in masonry, all those stone or brick parts distinguished from the plain wall of a building, such as ornamental columns, jambs, arches, entablatures, copings, quoins, and string courses. The process of smoothing and squaring lumber or stone for use in a building.

drip: A construction member, wood or metal, which projects to throw off rain water.

drip mold: A molding designed to prevent rain water from running down the face of a wall.

dripstone: In architecture, a stone drip placed over a window to throw off rain water; a label molding sometimes called a *weather molding*.

drive screw or **screw nail:** A type of screw which can be driven in with a hammer but is removed with a screw driver.

driving home: In shopwork, the placing of a part, a nail or screw, in its final position by driving it with the blows of a hammer or screw driver.

drop chute: A special chute recommended to be used when fresh concrete must be dropped more than 3 or 4 feet into narrow forms. See Fig. 38.

drop siding: A special type of weatherboarding used on the exterior surface of frame structures.

drop window: A type of window that is lowered into a pocket below the sill.

dry kiln: An ovenlike chamber in which wood is seasoned artificially, thus hastening the process of drying.

dry masonry: Any type of masonry work laid up without the use of mortar.

dry measure: A system of units of measure used in finding the volume of dry commodities, such as grain, fruit, and vegetables. The units of capacity are as follows:

```
  2 pints (pts.) = 1 quart (qt.)
  8 quarts      = 1 peck (pk.)
  4 pecks       = 1 bushel (bu.)
105 quarts      = 1 barrel (bbl.)
```

dry mortar: In masonry, mortar which con-

323

Masonry Simplified

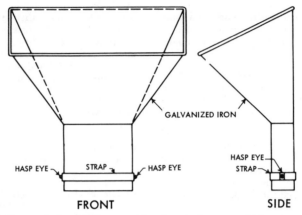

Fig. 38. A drop chute with hopper at top for placing concrete in narrow walls.

tains enough moisture to cause it to set properly, but is not wet enough to cause it to be sticky; also, mortar which still retains a granular consistency.

dry rot: Various types of decay in timber, all of which reduce the wood to a fine powder.

dry-wall construction: A type of construction where the finish material used is other than plaster, such as gypsum panels, wood paneling, plywood, or wallboard.

dry wood: Any timber from which the sap has been removed by seasoning.

dryer: A mechanical drying machine used to take moisture out of veneer.

duplex-headed nail: A specialized wood fastening, designed for use in temporary structures such as formwork for concrete and scaffolding, where ease of removal without damage to the wood is a prime factor. See Fig. 60.

dusting: The appearance of a powdery material at the surface of a newly hardened concrete slab.

Dutch arch: In masonry, a brick arch which is flat at both the top and bottom, constructed with ordinary brick which are not worked to a wedge shape but are laid so as to slope outward from the middle of the arch. See *French arch*.

Dutch bond: In masonry, a bond having the courses made up alternately of headers and stretchers. Same as *English cross bond*. See *typical bonds,* Fig. 22.

dutchman: An odd piece inserted to fill an opening or to cover a defect.

E

easement: In architecture, a curved member used to prevent abrupt changes in direction as in a baseboard or handrail. In stairway construction, a triangular piece to match the inside string and the wall base where these join at the bottom of the stairs.

eaves: That part of a roof which projects over the side wall; a margin, or lower part of a roof hanging over the wall; the edges of the roof which extend beyond the wall.

eaves trough: A gutter at the eaves of a roof for carrying off rain water.

economy brick: Modular brick related to every four-inch module in height, thickness, and length. *Economy brick* are always *cored brick.* Size $3\frac{1}{2}''$ x $3\frac{1}{2}''$ x $7\frac{1}{2}''$.

economy wall: In masonry, a brick wall four inches thick covered with a blanket of back mortaring. The wall is strengthened at intervals with vertical pilasters having brick corbeling which supports the floors

Glossary

Fig. 39. Portable belt sander. (Porter-Cable Machine Co.)

and roof. This provides a four-inch outside reveal for windows and doors, with every window and door frame bricked in.

edging: In concrete flatwork, the finishing of the outside edges of slabs, sidewalks, steps, etc., into a convex arc. Edging compacts and strengthens the edge and prevents chipping.

efflorescence: A whitish, loose powder which forms on the surface of a brick or stone wall.

egg-and-dart molding: A decorative molding with a design composed of an oval-shaped ornament alternating with another ornament in the form of a dart.

electric hand sander: Electrically powered hand sanders with interchangeable abrasive attachments for various on-the-job finishing purposes. See Fig. 39.

electric handsaw: Portable, electrically powered saws are available with interchangeable blades and other attachments for various building jobs, including masonry cutting and tuck pointing, as well as wood cutting, trimming, etc. See Fig. 40.

electronic glue gun: An instantaneous curing glue gun, with its own electric heating

Fig. 40. Ripping fence used with a portable power saw. (Millers Falls Co.)

unit, for synthetic thermosetting of resin adhesives. The gun is lightweight and portable and eliminates nailing and filling of nailheads and hammer marks on many jobs. Makes possible on-the-job gluing of wall paneling, laminates for cabinets or counters, and hardwood flooring.

elevation: A geometrical drawing or projection on a vertical plane showing the external upright parts of a building.

ell: An addition to a building at right angles to one end, or an extension of a building at right angles to the length of the main section.

ellipse: The path of a point the sum of whose distances from two fixed points is a constant; a conic section, the closed intersection of a plane with a right circular cone.

ellipsoid: A solid of which every plane section is an ellipse or a circle.

elliptical arch: An arch which is elliptical in form, described from three or more centers.

embossed: Ornamental designs raised above a surface; figures in relief, as a head on a coin; decorative protuberances.

emery: An impure corundum stone of a blackish or bluish gray color used as an abrasive. The stone is crushed and graded and made into emery paper, emery cloth, and emery wheels. *Emery* is mined in the eastern part of the United States but the best grade comes from Greece and Turkey.

emery cloth: A cloth used for removing file marks and for polishing metallic surfaces. It is prepared by sprinkling powdered emery over a thin cloth coated with glue.

emery wheel: A wheel composed mostly of emery and used for grinding or polishing purposes. It is revolved at high speed.

enameled brick: In masonry, brick with an enamel-like or glazed surface.

encased knot: A defect in a piece of wood, where the growth rings of the knot are not intergrown and homogeneous with the growth rings of the piece in which the knot is encased.

encaustic: Relating to the burning in of colors; applied to painting on glass tiles, brick, and porcelain; any process by which colors are fixed by the application of heat.

end-grain: In woodworking, the face of a piece of timber which is exposed when the fibers are cut transversely.

end-lap joint: A joint formed at a corner where two boards lap. The boards are cut away to half their thickness so that they fit into each other. They are halved to a distance equal to their width, and, when fitted together, the outer surfaces are flush. See Fig. 52.

English bond: In brickwork, a form of bond in which one course is composed entirely of *headers* and the next course is composed entirely of *stretchers,* the header and stretcher courses alternating throughout the wall; a type of bond especially popular for use in a building intended for residential purposes. See *typical bonds,* Fig. 22.

English cross bond: A form of bond similar to Old English bond. It is used where strength and beauty are required. Same as *Dutch bond.* See *typical bonds,* Fig. 22.

engrailed: Indented with curved lines, or small concave scallops.

enrichment: Adornment; embellishing plain work by adding ornamental designs.

erecting: Raising and setting in an upright position, as the final putting together in perpendicular form the structural parts of a building.

escalator: A stairway consisting of a series of movable steps, or corresponding parts, joined in an endless belt and so operated that the steps or treads ascend or descend continuously. Commonly used in large department stores and railway stations.

escutcheon: In architecture, a metal shield placed around a keyhole to protect the wood; also a metal plate to which a door knocker is attached. Such plates are sometimes ornately decorated.

escutcheon pin: A decorative nail having a round head, used in fastening ornamental

Glossary

and/or protective metal plates to wood. See Fig. 60.

espagnolette: A kind of fastening for a French casement window, usually consisting of a long rod with hooks at the top and bottom of the sash and turned by a handle. Also, a decorative feature on corners of Louis XIV furniture.

essex board measure: A method of rapid calculation for finding board feet; the essex board-measure table usually is found on the framing square conveniently located for the carpenter's use.

estimating: A process of judging or calculating the amount of material required for a given piece of work; also the amount of labor necessary to do the work, and finally an approximate evaluation of the finished product.

evaporation: The natural change of liquid into vapor due to temperature and exposure to air.

excavation: A cavity or hole made by digging out of earth to provide room for engineering improvements.

expansion joint: In masonry, a bituminous fiber strip used to separate blocks or units of concrete to prevent cracking due to expansion as a result of temperature changes. See Fig. 41.

expletive: In masonry, a stone which is used to fill a cavity.

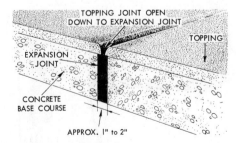

Fig. 41. Expansion joint.

exposed joint: In masonry, a mortar joint on the face of a brick or stone wall above the ground level.

exterior: The outer surface or part, as in building the *exterior* of the structure, or the *exterior* wall.

exterior finish: In building construction, the outside finish which is intended primarily to serve as a protection for the interior of the building and for decorative purposes. It includes the cornice trim, gutters, roof covering, door and window frames, water tables, corner boards, belt courses, and wall covering.

extrados: The exterior curve in an arch or vault.

eyebolt: A bolt which is provided with an eye, or hole, instead of the ordinary head. The eye receives the pin, hook, or stud which takes the pull of the bolt.

F

facade: The entire exterior side of a building, especially the front; the face of a structure; the front elevation, or exterior face, of a building.

facebrick: The better quality of brick such as is used on exposed parts of a building, especially those parts which are prominent in view.

faced wall: A masonry wall in which one or both sides are faced with a different material from the body of the wall, but with the facing and body so bonded that they will exert action as a unit under the weight of a load.

face hammer: A heavy hammer having flat faces, with one blunt end and one cutting end; used for rough dressing of blocks of quarried stone.

face joint: In masonry, a joint between the stones or brick in the face of a building. Since it is visible, the joint is carefully struck or pointed.

face mark: In woodworking, a mark placed on the surface of a piece of wood to indicate that part as the face, according to which all other sides are dressed true.

327

face mix: In masonry, a mixture of stone dust and cement; sometimes used as a facing for concrete blocks in imitation of real stone.

face mold: In architectural drawings, the full-size diagram, scale drawing, or pattern of the curved portions of a sloping handrail. The true dimension and shape of the top of the handrail are given in the drawing.

face of arch: In building, the exposed vertical surface of an arch. See Fig. 42.

facing hammer: In masonry, a special type of stone hammer used for dressing the surface of stone or cast-concrete slabs. See Fig. 43.

faience: Glazed terra-cotta blocks used as facing for buildings or fireplaces. Glazed plastic mosaic is known as *faience mosaic,* and is used for ornamental floors.

false header: In masonry, a half brick that is sometimes used in *Flemish bond;* in framing, a short piece of timber fitted between two floor joists.

false rafter: A short extension added to a main rafter over a cornice, especially where there is a change in the roof line.

falsework: Framework, usually temporary, such as bracing and supports used as an aid in construction but removed when the building is completed.

fascia: The flat outside horizontal member of a cornice placed in a vertical position.

fastening tool: A tool which applies greater force and control to the insertion of fasteners than hand-hammer methods alone. Placing the fastener in a narrow channel or slot, open at one end of the tool, force produced in a larger area is concentrated on the fastener from the other end, thus driving the fastener into the desired surface with speed and accuracy, and leaving no hammer marks. Fastening tools may be activated by a hand-hammer blow, by release of a spring, and by pneumatic (compressed air) hammer action. One of these can be positioned for roof or floor nailing by means of a foot stirrup, thus eliminating all bending. The above are usually called nailing machines and are used for fastening wood. More powerful tools for fastening steel and concrete are activated by exploding powder in a cartridge (as stud drivers); by gasoline (these are used for driving heavy spikes and can be self-contained); and by electricity (as impact wrenches and electric hammers). Tackers of the hammer, gun, and air types operate on the same principles, using a bent wire staple, instead of the usual nail or screw, to fasten thin construction material to walls or other surfaces. Many fastening devices have automatic, magnetic-fastener feed devices. See *anchors.* Fig. 44 illustrates the use of a pneumatic impact hammer used to install anchors for several types of fasteners.

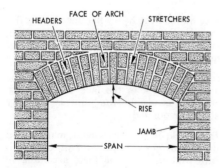

Fig. 42. Face of an arch.

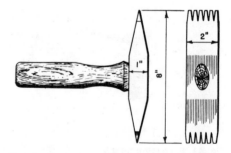

Fig. 43. Facing hammer.

Glossary

①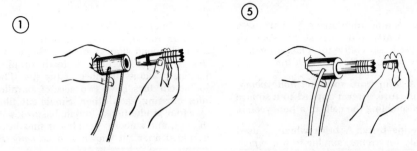

Insert tapered end of snap-off anchor into chuck head attached to any impact hammer.

⑤

Insert hardened steel cone-shaped red expander plug in cutting end of drill.

②

Operate impact hammer to drill into the concrete. Rotate chuck handle while drilling.

⑥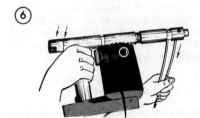

Reinsert the plugged drill in the hole and operate the hammer to expand anchor.

③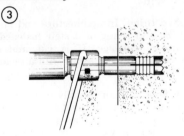

The drill is self-cleaning. Cuttings pass through the core and holes in the chuck head.

⑦

Snap off chucking end of anchor with a quick lateral strain on the hammer.

④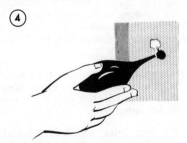

Withdraw the drill and remove grit and cuttings from the drill core and from the hole.

⑧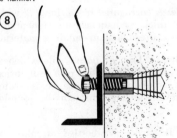

The anchor is now ready to serve as an internally threaded steel bolt hole to support any bolted object.

Fig. 44. Installation of snap-off type anchor. A wide selection of anchor sizes and lengths are available. (Phillips Drill Co.)

fat lime: A quicklime, made by burning a pure or nearly pure limestone, such as chalk; used especially for plastering and masonry work. Also called *rich lime*.

favus: A latin word meaning *honeycomb*; in architecture, a term applied to a square detail resembling the cells in a honeycomb.

featheredge brick: A brick which is used especially for arches, similar to a *compass brick*.

felt papers: Sheathing papers used on roofs and side walls of buildings as protection against dampness; also as insulation against heat, cold, and wind. *Felt papers* applied to roofs are often infused with tar, asphalt, or chemical compounds.

fender: A metal guard before an open fireplace to keep back the live coals.

fender wall: A low brick wall supporting the hearthstone to a ground-floor fireplace.

fenestration: The arrangement of windows and doors in a building; in an architectural composition, the design and proportion of windows as a decorative feature.

ferroconcrete: Concrete work reinforced by steel bars or steel mesh, embedded in the material before it sets, to provide increased strength.

festoon: In architecture and furniture, a decorative feature of carved work representing a garland or wreath of flowers or leaves or both.

fibrotile: A corrugated tile, 4′ long, 3′ wide, and 10″ thick, made of asbestos cement and used as a fire-retardant building material.

field tile: A type of porous tile which is placed around the outside of a foundation wall of a building, to absorb excess water and prevent seepage through the foundation.

figure: In carpentry, mottled, streaked, or wavy grain in wood.

file: A tool, with teeth, used principally for finishing metal surfaces. Common files are from 4 to 14 inches in length. The width and thickness are in proportion to the length. In cross section the *file* may be rectangular, round, square, half round, triangular, diamond shaped, or oval. Single-cut files have parallel lines of teeth running diagonally across the face of the file. The double-cut files have two sets of parallel lines crossing each other. Single-cut files have four graduations—rough, bastard, second cut, and smooth; double-cut files have an added finer cut known as *dead smooth*.

fillet: A narrow concave strip connecting two surfaces meeting at an angle. It adds strength and beauty of design by avoiding sharp angles.

fillister: A plane used for cutting grooves; also, a rabbet, or groove, as the groove on a window sash for holding the putty and glass; in mechanical work, the rounded head of a cap screw slotted to receive a screw driver.

finial: A decorative detail that adorns the uppermost extremity of a pinnacle or gable. Any ornamental device capping a gable or spire.

finished string: In architecture and building, the end string of a stair fastened to the rough carriage. It is cut, mitered, dressed, and often finished with a molding or bead.

finish floor: A floor, usually of high-grade material, laid over the subfloor. The *finish floor* is not laid until all plastering and other finishing work is completed. Also called *finished floor*.

finishing: The final perfecting of workmanship on a building, as the adding of casings, baseboards, and ornamental moldings. In concrete flatwork, the tooling of the surface.

Fink truss: A type of roof truss commonly used for short spans because of the shortness of its struts which makes it economical and prevents waste.

firebrick: Any brick which is especially made to withstand the effects of high heat without fusion; usually made of *fire clay* or other highly siliceous material.

fire clay: Clay which is capable of being subjected to high temperatures without fus-

Glossary

ing or softening perceptibly. It is used extensively for laying *firebrick*.

firecut: The beveled end of a wooden joist resting in a masonry structure. In case of fire, the beveled end will cause the joist to fall *inward* without disturbing the masonry work. Note in Fig. 45 the position of the joist anchor.

fire-division wall: A solid masonry wall for subdividing a building to prevent the spread of fire but one not necessarily extending continuously from the foundation through all the stories and through the roof; also, a wall of reinforced concrete which is more or less fire-resistant, tending to restrict the spread of fire in a building.

fireproof: To build with incombustible materials in order to reduce fire hazards: to cover or treat with an incombustible material; anything constructed of a minimum amount of combustible material. See Fig. 46.

fire-resisting: In building construction, a term applied to any structural material which is relatively immune to the effect of exposure to fire which has a maximum severity for an anticipated duration of time. Same as *fire-resistive*.

fire stops: Blocking, of incombustible material, used to fill air passages through which flames might travel in case the structure were to catch fire; any form of blocking of air passages to prevent the spread of fire through a building. See Fig. 15.

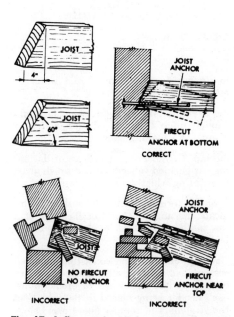

Fig. 45. A fire cut is used on joints in solid masonry walls. A joist anchor prevents the building from falling outward in the event of fire.

Fig. 46. Fireproofing of structural steel building. (Check local codes on the use of fireproofing material.)

Masonry Simplified

firmer tools: In woodworking, the tools commonly used on the workbench, such as the ordinary chisels and gouges.

fished joint: A joint commonly used when a structural piece must be lengthened. The joint is made by placing a second piece end to end with the first one, then covering the juncture with two additional pieces which are nailed or bolted on opposite sides of the joint. These pieces are called *fish plates*, and may be wood or metal. See *fished and keyed joint,* Fig. 52.

fish glue: A type of glue made from specially prepared parts of certain fish, usually the bladderlike portions of hake.

fitment: In carpentry and furniture making, any portion of a wall, room, or built-in furniture which is fitted into place, including chimney pieces, wall paneling, cabinets, and cupboards.

flagstone: A kind of stone that splits easily into flags, or slabs; also, a term applied to irregular pieces of such stone split into slabs from 1 to 3 inches thick and used for walks or terraces. A pavement made of stone slabs is known as *flagging*.

flange: A projecting edge, or rib. Some types of insulation materials are provided with *flanges* for nailing purposes.

flank: In architecture, the side of an arch.

flashing: Metal or plastic material, used around dormers, chimneys, or any rising projection, such as window heads, cornices, and angles between different members, or any place where there is danger of leakage from rain water or snow. These metal pieces are worked in with shingles of the roof or other construction materials used. See Fig. 47.

flat arch or **jack arch:** A type of construction in which both the outside of an arch and the underside of the arch are flat.

flat grain: Lumber sawn parallel with the pith of the log and approximately at right angles to the growth rings.

flat molding: A thin, flat molding used only for finishing work.

flat roof: A roof with just enough pitch

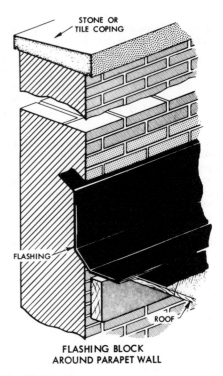

Fig. 47. Flashing.

to provide for drainage of rain water or melting snow.

flat skylight: Any skylight which has only enough pitch to carry off rain water or water from melting snow.

flatting: In veneering, a process of flattening out buckled veneers; also, a finish given to painting which leaves no gloss.

fleam: A term used in woodworking to indicate the angle of bevel of the edge of a saw tooth with respect to the plane of the blade.

Flemish bond: A bond consisting of *headers* and *stretchers,* alternating in every course, so laid as always to break joints, each header being placed in the middle of the stretchers in courses above and below. See *typical bonds,* Fig. 22.

Flemish garden bond: In masonry, a bond consisting of three stretchers, alternating

Glossary

with a header, each header being placed in the center of the stretchers in the course above and below.

flier: Any single one of a flight of stairs whose treads are parallel to each other; a stair tread that is of uniform width throughout its length.

flight of stairs: A series of steps between floors of a building; a single flight of stairs may be broken into two flights by means of a landing.

flint-and-stone work: A system of external ornamentation of buildings used in eastern Europe during the fifteenth century, by means of which inscriptions were produced in stone on a ground of flint which had been split to expose a black surface. Sometimes the stones were sunk about two inches, and the flints were let into it.

flitch girder: A combination beam composed of two or more joists which have between them steel plates ¼ inch or more in thickness. Bolts are used to hold the joists and steel plates together.

float: A tool used to smooth finish concrete flatwork.

floating: The process of spreading plastering, stucco, or cement on the surface of walls to an equal thickness by the use of a board called a *float*. In concrete finishing, an intermediate step usually done after edging and jointing, and before trowelling. If a coarsely textured surface is desired, floating may be the final step.

floating foundation: In building construction, a special type of foundation made to carry the weight of a superstructure which is to be erected on swampy land, or on unstable soil. Such a foundation consists of a large raftlike slab composed of concrete, reinforced with steel bars or fabric.

floatstone: In masonry, a type of stone used by bricklayers to smooth gauged brickwork. See *rubbing stone*.

floor: In architecture and building, different stories of structure are frequently referred to as floors; for example, the *ground floor,* the *second floor,* and the *basement floor;* also, that portion of a building or room on which one walks.

floor chisel: An all-steel chisel with an edge measuring from 2 to 3 inches in width, used especially for removing floorboards.

floor drain: A plumbing fixture used to drain water from floors into the plumbing system. Such drains are usually located in the laundry and near the furnace and are fitted with a deep seal trap.

floor framing: In building construction, the framework for a floor, consisting of sills and joists; also, the method used in constructing the frame for a floor, including the joists, sills, and any floor openings.

floor plan: An architectural drawing showing the length and breadth of a building and the location of the rooms which the building contains. Each floor has a separate plan.

flue: An enclosed passageway, such as a pipe, or chimney, for carrying off smoke, gases, or air.

flue lining: Fire clay or terra-cotta pipe, either round or square, usually obtainable in all ordinary flue sizes and in two-foot lengths. It is used for the inner lining of chimneys with brick or masonry work around the outside. Flue lining should run from the concrete footing to the top of the chimney cap.

flush: The continued surface of two contiguous masses in the same plane; that is, surfaces on the same level.

flush door: A door of any size, not paneled, having two flat faces, frequently of hollow core construction.

flush joint: In masonry, a mortar joint formed by cutting surplus mortar from the face of a wall. If a rough texture is desired, the surface of the joint may be tapped with the end of a rough piece of wood after the mortar has slightly stiffened.

flying bond: In bricklaying, a bond formed by inserting a header course at intervals of from four to seven courses of stretchers.

foaming agent: A chemical which, when foamed by a generator and measured into cement mix, controls the cellular density of the resulting aerated material.

Masonry Simplified

foil: A rounded, leaflike, architectural ornamentation used especially for window decoration, sometimes consisting of three divisions, sometimes of four, and known as *trefoil* and *quatrefoil*.

folding door or partition: Any door or partition in which panels are built to bend back on themselves when opened. It may fold vertically or horizontally toward top or sides of opening. Center-hung types operate on pivoted hangers attached to the center of the same edge of each panel, and running along a track set in the opening. These panels straddle the track when folded. Pair-operated panels operate from hangers placed at opposite corners on the same edge of each of two panels. The hangerless inner edges are usually hinged and project together on only one side of the track, when folded. Light metal collapsible gate frames, covered with fabric, are also hung from tracks and pushed into folds to open. Concealing recesses or passageways are often built to accommodate the folded panels and leave the opening free of obstructions. See *rolling partitions*. All of these may be operated by power or electronic devices for special purposes, as garage doors.

folding stair: A stairway which folds into the ceiling, used for access to areas of a building not in general use. Such a stairway uses only limited space while in use. It is fixed by means of hinges at its uppermost end, while the lower end is provided with any of various catches which fasten the stairway when folded. Quite often the underside of such a stairway is finished so as to match the ceiling area into which it is folded.

foliated: To ornament with foils or foliage; that is, decorated with a leaf design.

footing: A foundation as for a column; spreading courses under a foundation wall; an enlargement at the bottom of a wall to distribute the weight of the superstructure over a greater area and thus prevent settling. *Footings* are usually made of concrete and are used under chimneys and columns as well as under foundation walls. See Figs. 20 and 48.

footing forms: Forms made of wood or metal for shaping and holding concrete for footings which support foundations, walls, columns, chimneys, etc.

footing stop: In concrete work, a term applied to a device consisting of a plank nailed to a 4″ x 4″, placed in the forms to hold the concrete at the close of a day's placing. See Fig. 49.

footstone: A stone placed at the foot of a gable slope to receive and resist the outward thrust of the coping stone above it. A foundation stone.

fore plane: A bench plane 18 inches in length, intermediate between the jointer plane which is larger, and the jack plane, which is smaller.

forming tool: A term frequently applied to

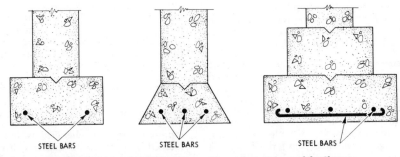

Fig. 48. Steel reinforcing bars used in various types of footings.

Glossary

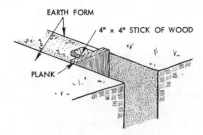

Fig. 49. Form stop.

any device which will facilitate a mechanical operation; a tool especially designed for a particular type of work with its cutting edge shaped like the form to be produced on the work.

forms: In building construction, an enclosure made of boards or metal for holding green concrete to the desired shape until it has set and thoroughly dried.

form stop: Same as *footing stop*. See Fig. 49.

foundation: The lowest division of a wall for a structure intended for permanent use; that part of a wall on which the building is erected; usually that part of a building which is below the surface of the ground and on which the superstructure rests.

foundation bolt: Any bolt or device used to anchor the structural parts of a building to the foundation on which it rests; any bolt used to hold machinery in position on its foundation. Same as *anchor bolt*. See Fig. 6.

four-way switch: A switch used in house wiring when a light (or lights) is to be turned on or off at more than two places. Thus, for three places, use two three-way and one four-way switches; for four places, use two three-way and two four-way switches — an additional four-way switch for each additional place of control.

fracture: In masonry, the breaking apart or separation of the continuous parts of a wall of brick or stone caused by a sudden shock or excessive strain.

frame: In carpentry, the timber work supporting the various structural parts, such as windows, doors, floors, and roofs; the woodwork of doors, windows, and the entire lumber work supporting the floors, walls, roofs, and partitions.

frame high: In masonry, the level at which the lintel or arch of an opening is to be laid; also, the height of the top of window or door frames.

frame of a house: The framework of a house which includes the joists, studs, plates, sills, partitions, and roofing; that is, all parts which together make up the skeleton of the building.

framework: The frame of a building; the various supporting parts of a building fitted together into a skeleton form.

framing: The process of putting together the skeleton parts for a building; the rough lumber work on a house, such as flooring, roofing, and partitions.

framing square: An instrument having at least one right angle and two or more straight edges, used for testing and laying out work for trueness. A good *framing square* will have the following tables stamped on it for the use of the carpenter in laying out his work: *unit length rafter tables; essex board measure; brace measure; octagon scale;* and, a *twelfths scale*. Also called *square* or *steel square*.

free end: The end of a beam which is unsupported, as that end of a cantilever which is not fixed.

freemasons: A term formerly applied to one of a class of skilled stone masons who worked in freestone as distinguished from an unskilled mason who worked only with rough stone.

free-standing vertical sunshades: A wall or fence standing a few feet from a house and placed so as to shade a wall or window opening to help keep the interior of the building cool during warm weather. When louvered or open in style, they do not block out breezes while providing necessary shade. They are excellent against low sun, but can seldom be built high enough to shade walls during the middle of the day;

most effective when painted in a light color and used in combination with tall green plants.

freestone: Any stone, as a sandstone, which can be freely worked or quarried; used for molding, tracery, and other work requiring to be executed with the chisel. Any stone which cuts well in all directions without splitting.

French arch: In masonry, a type of bonded arch which is flat at both top and bottom, having the bricks sloping outward from a common center. Same as *Dutch arch*.

French window: A long double-sash casement window with the sashes hinged at the sides and opening in the middle. The window extends down to the floor and serves as a door to a porch or terrace.

friction: Resistance to relative motion set up between particles of two moving surfaces in contact with each other.

friction catch: A device consisting of a spring and plunger contained in a casing, used on small doors or articles of furniture to keep them tightly closed but not locked.

frilled: An ornamental edging on furniture such as a *frilled* **C** scroll; a term used to refer to any scroll which has added decorative carving along its projecting edges.

frog: A depression, such as a groove or recess, in one or both of the larger sides of a brick or building block, thus providing a key for the mortar at the joints, and also effecting a saving in the weight of the material. A name also applied to a part of a carpenter's plane.

furred: The providing of air space between the walls and plastering or subfloor and finish floor by use of wood strips, such as lath or 1″ x 2″s nailed to the walls or to the subflooring. Walls or floors prepared in this manner are said to be *furred*.

furring: The process of leveling up part of a wall, ceiling, or floor by the use of wood strips; also, a term applied to the strips used to provide air space between a wall and the plastering. See Fig. 50.

furring strips: Flat pieces of lumber used to

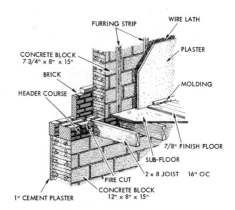

Fig. 50. Furring strips.

build up an irregular framing to an even surface, either the leveling of a part of a wall or ceiling. The term *furring strips* or *furrings* is also applied to strips placed against a brick wall for nailing lath, to provide air space between the wall and plastering to avoid dampness. See Fig. 50.

G

gable: The end of a building as distinguished from the front or rear side; the triangular end of an exterior wall above the eaves; the end of a ridged roof which at its extremity is not returned on itself but is cut off in a vertical plane which above the eaves is triangular in shape due to the slope of the roof.

gable molding: The molding used as a finish for the gable end of a roof.

gable roof: A ridged roof which terminates either at one end or both ends in a gable.

gage: A tool used by carpenters and other woodworkers to make a line parallel to the edge of a board.

gain: The notch or mortise where a piece of wood is cut out to receive the end of another piece of wood.

gallery: An elevated floor or platform

equipped to increase the seating capacity of auditoriums in churches, theaters, and other large audience rooms. Projecting from the interior wall of a building, the *gallery* usually is supported by columns below, but sometimes it is hung on supports from above. In some cases supports are provided both above and below.

gallet: A splinter of stone chipped off by chiseling. Also called a *spall*.

galleting: In rubble work, a term applied to the process of filling in the coarse masonry joints of fresh mortar with small stone chips or gallets. Same as garreting.

gambrel roof: A type of roof which has its slope broken by an obtuse angle, so that the lower slope is steeper than the upper slope; a roof with two pitches.

ganister: In building, a material made of ground quartz and fire clay; used for fireproofing purposes, especially around the hearths of furnaces.

garden bond: A type of masonry construction consisting of three stretchers in each course followed by a header. However, this bond may have from two to five stretchers between headers.

garderobe: A room for keeping articles of clothing, a wardrobe; also, a small private room, as a bedroom.

garnet paper: An abrasive paper used for polishing and finishing surfaces of woodwork. The paper is prepared by covering one side with glue and a reddish abrasive material.

gauged arch: In masonry, an arch constructed with special bricks which have been cut with a bricklayer's saw, then rubbed to the exact shape required for use in an arch where the joints radiate from a common center.

gauging: In masonry, cutting brick or stone to make them uniform in size. In plastering, the mixing of plaster of paris with mortar to effect quick setting.

geometrical stair: A winding stair which returns on itself with winders built around a well. The balustrade follows the curve without newel posts at the turns. It is also known as a *spiral stair*.

geometry: That branch of mathematical science which treats of the properties and relations of lines, surfaces, and solids.

German siding: A type of weatherboarding with the upper part of the exposed face finished with a concave curve and the lower portion of the back face rebated.

gingerbread work: A gaudy type of ornamentation in architecture, especially in the trim of a house.

girder: A large, supporting, built-up, horizontal member used to support walls or joists; a beam, either timber or steel, used for supporting a superstructure. See Fig. 15.

girt: The same as girth; the circumference of round timber.

girt strip: A board attached to studding to carry floor joists; a ledger board.

glass block: A hollow glass building brick having the advantage of admitting light with privacy, insulating against the passage of sound, but not safe to use in a load-bearing wall. Sometimes called *glass brick*.

glazed: Equipped with window panes; the process of placing glass in windows, doors and mirrors is known as *glazing*.

glazed brick: Building brick prepared by fusing on the surface a glazing material; brick having a glassy surface.

glazed doors: Doors which have been fitted with glass and usually having a pattern or lattice of woodwork between the panes.

glazed tile: A type of masonry tile which has a glassy or glossy surface.

glazing: Placing glass in windows, doors, and mirrors. Also the filling up of interstices in the surface of a grindstone or emery wheel with minute abraded particles detached in grinding.

glyph: In architecture, an ornamental channel, or groove; a short, vertical groove.

gooseneck: Something curved like the neck

337

of a goose, such as an iron hook, or other mechanical contrivance bent or shaped like a goose neck; the curved or bent section of the handrail on a stair.

Gothic arch: A type of arch, usually high and narrow, coming to a point at the center at the top, especially one with a joint instead of a keystone at the apex.

gouge: A cutting chisel which has a concave-convex cross section, or cutting surface.

grade: In building trades the term used when referring to the ground level around a building.

grading: Filling in around a building with rubble and earth, so the ground will slope downward from the foundation at an angle sufficient to carry off rain water.

gradual load: The gradual application of a load to the supporting members of a structure, so as to provide the most favorable conditions possible for receiving the stress and strain which these members will be required to carry when the building is completed.

graduate: To mark with degrees of measurement, as the division marks of a scale; the regular dividing of parts into steps or grades.

graduation: The process of separating a unit of measure into equal parts; also, one of the division marks, or one of the equal divisions of a scale.

grain: In woodworking, a term applied to the arrangement of wood fibers; working a piece of wood longitudinally may be either with or against the grain; a cross-section, or transverse, cut of wood is spoken of as cross grain.

granite: An igneous rock composed chiefly of feldspar but containing also some quartz and mica, used extensively in construction work and for monuments. It is extremely hard and will take a high polish.

granulated slag: The product of rapid cooling of molten blast-furnace slag, resulting in a mass of friable, porous grains, most of which are under one-half inch in size. See *air-cooled slag*.

grating: A framework or gratelike arrangement of bars either parallel or crossed, used to cover an opening.

gravel fill: Crushed rock, pebbles, or gravel, deposited in a layer of any desired thickness at the bottom of an excavation, the purpose of which is to insure adequate drainage of any water. See Fig. 20.

green brick: In masonry, a molded clay unit before it has been burned in preparation for building purposes.

green mortar: A term sometimes applied to mortar before it has set firmly.

green wood: A term used by woodworkers when referring to timbers which still contain the moisture, or sap, of the tree from which the wood was cut. Lumber is said to be *seasoned* when the sap has been removed by natural processes of drying or by artificial drying in a kiln.

grille: A grating or openwork barrier, usually of metal but sometimes of wood, used to cover an opening, or as a protection over the glass of a window or door. A *grille* may be plain but often it is of an ornamental or decorative character.

grind: To reduce any substance to powder by friction or crushing; to wear down or sharpen a tool by use of an abrasive, such as a whetstone, emery wheel, or grindstone; to reduce in size by the removal of particles of material by contact with a rotating abrasive wheel.

grinder: Any device used to sharpen tools or remove particles of material by any process of grinding.

grindstone: A flat rotating stone wheel used to sharpen tools or wear down materials by abrading or grinding. *Grindstones* are natural sandstone.

grommet: A metal eyelet used principally in awnings or along the edges of sails.

groove: A channel, usually small, used in woodworking and building for different purposes which sometimes are practical but often merely decorative.

ground: One of the pieces of wood flush with the plastering of a room to which moldings and other similar finish material are nailed. The *ground* acts as a straight edge and thickness gage to which the plasterer works to insure a straight plaster surface of proper thickness.

ground course: A horizontal course, usually of masonry, next to the ground.

ground floor: Usually the main floor of a building; the floor of a house most nearly on a level with the ground; that is, the first floor above the ground level.

ground joist: A joist which is blocked up from the ground.

ground wall: In building, the foundation wall; the wall on which a superstructure rests.

grout: A mortar made so thin by the addition of water it will run into joints and cavities of masonry; a fluid cement mixture used to fill crevices.

Gunite: A construction material composed of cement, sand or crushed slag, and water mixed together and forced through a cement gun by pneumatic pressure. Sold under the trade-mark *Gunite*. Same as *shotcrete*.

gusset: A brace or angle bracket used to stiffen a corner or angular piece of work.

gutter: A channel of wood or metal at the eaves or on the roof of a building for carrying off rain water and water from melting snow.

gypsum: A mineral hydrous sulphate of calcium. In the pure state gypsum is colorless. When part of the water is removed by a slow heating process the product becomes what is known as *plaster of paris*, used extensively for decorative purposes.

gypsum block: A type of building material usually grayish white in color; because of its friable texture it is used only in non-load-bearing partition walls. Gypsum blocks are highly fire resistant and have excellent sound absorbing qualities.

gypsum lath: A plaster base made in sheet form composed of a core of fibered gypsum, faced on both sides with paper.

H

hack saw: A narrow, light-framed saw used for cutting metal; a fine-toothed, narrow-bladed saw stretched in a firm frame. It may be operated either by hand or by electric power.

haft: The handle of any thrusting or cutting tool, such as a dagger, knife, sword, or an awl.

half-back bench saw: A cutting tool in which a stiffening bar extends over only a portion of the blade length, combining the action of both the handsaw and the backsaw.

half bat: A term applied to one-half of a building brick.

half-lap joint: A jointing of two pieces by cutting away half the thickness of each piece so that the pieces fit together with the surfaces flush.

half-round file: A tool which is flat on one side and curved on the other; however, the convexity never equals a semicircle.

half story: An attic in a pitched-roof structure having finished ceiling and floor and some side wall.

half-timbered: A term applied to any building constructed of a timber frame with the spaces filled in, either with masonry or with plaster on laths.

hammer: A tool used for driving nails, pounding metal, or for other purposes. Though there are various types of hammers, used for a variety of purposes, all hammers are similar in having a solid head set crosswise on a handle. See Fig. 31.

hand drill: A hand-operated tool used for drilling holes.

hand file: A tool used in finishing flat surfaces. See *file*.

handiwork: Any work done by the hands; usually refers to work requiring some special skill.

handrail: Any railing which serves as a guard; a rail which is intended to be grasped by the hand to serve as a support,

Masonry Simplified

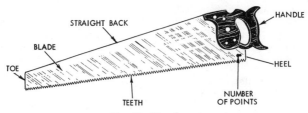

Fig. 51. Handsaw.

as on a stair or along the edge of a gallery. See Fig. 83.

handrail wreath: The curved section of a stair rail. See *wreath*.

handsaw: Any ordinary saw operated with one hand; that is, a one-handled saw, either a ripsaw or a crosscut saw used by woodworkers. See Fig. 51.

hand screw: A clamp with two parallel jaws and two screws used by woodworkers; the clamping action is provided by means of the screws, one operating through each jaw.

hand tools: Any tools which are operated by and guided by hand.

hangar: A shelter or shed for housing aircraft.

hanger: A drop support, made of strap iron or steel, attached to the end of a joist or beam used to support another joist or beam. See *stirrup*.

hanger bolt: A bolt used for attaching hangers to woodwork; it consists of a lag screw at one end with a machine-bolt screw and nut at the other end.

hanging stile: That part of a door to which the hinges are attached; the vertical part of a door or casement window to which the hinges are fixed.

hardboard: A board material manufactured from wood fiber; the wood is broken down into its individual fibers and compressed by hot presses. Lignin, a natural wood substance, bonds the fibers together.

hardpan: A cemented or compacted layer in soils, often containing some proportion of clay, through which it is difficult to dig or excavate.

hardware: In building construction, fastenings which permit movement of parts even when these parts are held together securely. Fastenings, commonly known as *builders' hardware,* include hinges, catches, lifts, locks, and similar devices.

hardwood: The botanical group of broad-leaved trees, such as oak, maple, basswood, poplar, and others. The term has no reference to the actual hardness of the wood.

hasp: A hinged-metal strap designed to pass over a staple and secured by a peg or padlock.

hatching: Parallel lines drawn closely together for the purpose of shading, or to indicate a section of an object shown in a drawing.

hatchway: An opening covered by a hatch, or trap door, to provide easy access to an attic or cellar; any opening in a floor, ceiling, or roof which makes it possible to pass from one story to another; the opening on ships for passage from one deck to another.

haunch: In architecture, either side of an arch between the crown, or vertex, and the impost where the arch rests on the top of a pier or wall; the shoulder of an arch.

header: In building, a brick or stone laid with the end toward the face of the wall.

header bond: A form of bond in which all courses are laid as headers; used for walls or partitions eight inches thick. See *typical bonds,* Fig. 22.

header-high: In masonry, when a portion

Glossary

of a wall has been laid up to the point where headers are necessary, the wall is said to be *header-high*.

header joist: In carpentry, the large beam or timber into which the common joists are fitted when framing around openings for stairs, chimneys, or any openings in a floor or roof; placed so as to fit between two long beams and support the ends of short timbers. See Fig. 23.

head joint: A vertical joint which joins brick at their ends. See Fig. 14.

head room: The vertical space in a doorway; also, the clear space in height between a stair tread and the ceiling or stairs above.

hearth: The floor of a fireplace; also the portion of the floor immediately in front of the fireplace.

heavy joist: In woodworking, a timber measuring between 4 and 6 inches in thickness and 8 inches or over in width.

heel: That part of a timber, beam, rafter, or joist which rests on the wall plate.

height: In reference to an arch, the perpendicular distance between the middle point of the chord and the intrados. Sometimes called the *rise*. See Fig. 42.

helve: The handle of a tool, such as a hammer, hatchet, or ax.

herringbone: In masonry, a pattern used in brickwork where the brick in alternate courses is laid obliquely in opposite directions forming a design similar in appearance to the spine of a herring; a zigzag pattern used in brickwork; in flooring, material arranged diagonally.

herringbone bond: In masonry, the arrangement of bricks in a course in a zigzag fashion, with the end of one brick laid at right angles against the side of a second brick.

hewing: Dealing cutting blows with an ax or other sharp instrument for the purpose of dressing a timber to a desired form or shape.

hexagon: A *polygon* with six sides.

hiearly cement: A contraction of the more generally used term *high early strength portland cement or concrete.*

high early strength cement: A specially prepared portland cement for giving quick strength to concrete work. It is frequently used when the temperature is below freezing.

hinge: A movable joint upon which a door turns; a mechanical device consisting primarily of a pin and two plates which may be attached to a door and the door frame to permit the opening and closing of the door. Hinges are used also on gates and other places where movable joints are desired.

hip rafters: Rafters which form the hip of a roof as distinguished from the common rafters. A *hip rafter* extends diagonally from the corner of the plate to the ridge, and is located at the apex of the outer angle formed by the meeting of two sloping sides of a roof whose wall plates meet at a right angle. See Fig. 76.

hip roof: A roof which rises by inclined planes from all four sides of a building.

hips: Those pieces of timber or lumber placed in an inclined position at the corners or angles of a hip roof. See Fig. 76.

hollow concrete blocks: A type of precast concrete building block having a hollow core.

hollow core door or wall: A faced door or wall with a space between the facings which is occupied by a structure consisting of air or insulation filled cells; made of wood, plastic, or other suitable material. Hollow core constructions have special fire, temperature, and sound insulating properties, as well as being light weight and strong.

hollow masonry unit: A masonry unit whose cross-sectional area in any given plane parallel to the bearing surface is less than 75 percent of its gross cross-sectional area measured in the same plane.

hollow wall: In masonry, a wall constructed of brick, stone, or other materials, having an air space between the outside and inside faces of the wall. Also called *cavity wall.*

341

Masonry Simplified

hollow tile: Clay tile used mainly for decorative purposes.

honeycomb: A cell-like structure. Concrete that is poorly mixed and not adequately puddled having voids or open spaces is known to be *honeycombed*.

honeycomb core: A structure of air cells, resembling a honeycomb, often made of paper, which is placed between plywood panels, sometimes replacing studs. This type of wall construction provides lighter prefabricated walls with excellent insulating properties. See *hollow core door or wall*.

hopper windows: A window in which each sash opens inward on hinges placed at the bottom of each sash.

horizontal: On a level; in a direction parallel to the horizon. For example, the surface of a still body of water is *horizontal*, or level.

horse: In building and woodworking, a trestle; one of the slanting supports of a set of steps to which the treads and risers of a stair are attached; a kind of stool, usually a horizontal piece to which three or four legs are attached, used as a support for work; a braced framework of timbers used to carry a load.

housed string: A stair string with horizontal and vertical grooves cut on the inside to receive the ends of the risers and treads. Wedges covered with glue often are used to hold the risers and treads in place in the grooves.

Howe truss: A type of truss used both in roofs and in construction of bridges; a form of truss especially adapted to wood and steel construction.

hutch: A chest, box, or trough; a small dark room; a storage place; a pen for small animals, as rabbits; also a hut or cabin.

hydrated lime. The material which remains after a chemical reaction due to the contact of quicklime and water. The same as *slaked lime*.

hydration: The chemical process in which water combines with cement to form a hard solid mass, binding the aggregate together to form mortar or concrete.

hydraulic cement: A type of cement which hardens under water.

hydraulic jack: A lifting device operated by a lever from the outside and put into action by means of a small force pump, through the use of a liquid such as water or oil.

hydraulic lime: A lime which will harden under water.

hydraulic limestone: A limestone which contains some silica and alumina, yielding a quicklime that sets or hardens under water.

hydraulic mortar: In masonry, a mortar which will harden under water; used for foundations or any masonry construction under water.

hypotenuse: The side opposite the right angle of any right *triangle*.

I

I beam: A steel beam whose cross section resembles a letter I, used in structural work. Now called **S** *beam*.

impost: The uppermost member of a column, pillar, pier, or wall upon which the end of an arch rests. See Fig. 54.

incinerator: A furnace, or a container, in which waste material and rubbish are burned.

incise: To cut or carve; to cut marks as in the process of engraving.

inclined plane: A surface inclined to the plane of the horizon; the angle which it makes with the horizontal line is known as the *angle of inclination*.

indenture: A contract, or official paper, by means of which an apprentice is legally bound to his employer. Also, a deed, mortgage, or lease.

inglenook: A nook in a corner by a fire, such as a corner by a chimney or fireside.

Glossary

inlay: To decorate with ornamental designs by setting in small pieces of material in the body of a piece of work which is made of different material from the inlaid pieces; also, the designs so made.

inscribe: To write or engrave in any form, especially in a way that will endure. Also, to draw one figure within another as in geometry.

inside calipers: In shopwork, a type of calipers having the points at the ends of the legs turned outward instead of inward so the tool can be used for gauging the inside diameters.

instant grab adhesives: Adhesives which hold their grip on contact without clamping. Plywood wall panels, with adhesived backs, contact studs and hold their position without bracing. These adhesives can be used, with anchor plates, for attaching furring strips, cutting down on nailing damage to masonry walls. Laminating plastics on the job is also possible because of the instant grab property of these glues.

insulated: Any part of a building separated from other parts of the structure to prevent the transfer of heat or electricity is said to be *insulated*.

insulation: Any material used in building construction for the reduction of fire hazard or for protection from heat or cold. Insulation also prevents transfer of electricity.

intake belt course: In building, a belt course with the molded face cut so that it serves as an intake between the varying thicknesses of two walls.

interior finish: A term applied to the total effect produced by the inside finishing of a building, including not only the materials used but also the manner in which the trim and decorative features have been handled.

intersection: The point where two intersecting lines cross each other.

intrados: The interior or bottom curve of an arch.

inverted arch: In masonry, an arch where the keystone is located at the lowest point of the arch.

involute: A curve such as would be described by the unwinding of a string from a cylinder.

ironwork: A term applied to the use of iron for ornamental purposes. Elaborately designed ornamentation in ironwork was used for hinges, door knockers, and escutcheons in the architecture of the Middle Ages.

irregular-coursed: In masonry, rubble walls built up in courses of different heights.

J

jack: A portable machine used for lifting heavy loads through short distances with a minimum expenditure of effort or power.

jack arch: An arch which is flat instead of rounded. This type of arch is sometimes called a *French arch*.

jack plane: A bench plane, appropriately named for a beast of burden often called upon to do the hardest and roughest kind of work. The *jack plane,* likewise, is called upon to do the hardest and roughest work on a piece of timber as it first comes from the saw. This plane is the one used to true up the edges and rapidly prepare the rough surface of a board for the finer work of the smoothing planes.

jack rafter: A short rafter of which there are three kinds: (1) those between the plate and a hip rafter; (2) those between the hip and valley rafters; (3) those between the valley rafters and the ridge board. *Jack rafters* are used especially in hip roofs. See Fig. 76.

jackscrew: A mechanical device operated by a screw, used in lifting weights and for leveling work.

jalousie window: A window consisting of narrow pieces of glass opening outward to admit air but exclude rain. The window appears similar to a venetian blind.

jamb: In building, the lining of an opening, such as the vertical side posts used in the framing of a doorway or window.

343

Masonry Simplified

jambstone: In architecture, a stone which is set in an upright position at the edge of a wall opening, such as for a door or window, so one of the faces of the stone forms a part or all of a jamb.

jedding ax: In masonry, an ax having one flat face and one pointed peen. See *cavil*.

jig saw: In woodworking, a type of saw with a thin, narrow blade to which an up-and-down motion is imparted either by foot power or by mechanical means.

joggle: A projection, or shoulder, to receive the thrust of a brace; also, a key, or projecting pin, set in between two joining surfaces for the purpose of reinforcing the joint.

joggle joint: In masonry, or stonework, a joint in which a projection on one member fits into a recess in another member to prevent lateral movement.

Johnson-Ackerman anchor: A concrete insert, provided with a lead shield, used to fasten wood or metal parts to a concrete or masonry wall. See *anchors,* Fig. 7.

joiner: A craftsman in woodworking who constructs joints; usually a term applied to the workmen in shops who construct doors, windows, and other fitted parts of a house or ship.

joinery: A term used by woodworkers when referring to the various types of joints used in woodworking. Wood joints commonly used in timber framing, in edge joining of boards, and other forms of woodworking are shown in Fig. 52.

joint: In masonry, the mortar bond between individual masonry units.

jointer: In masonry, the tool used in finishing the mortar joints.

jointer plane: A large bench plane used chiefly for long work and for final truing up of wood edges or surfaces for joining two pieces of wood; an iron or wood plane suitable for all kinds of plane work and especially adaptable for truing large surfaces required in furniture making.

jointing: In masonry, the operation of making and finishing the exterior surface of mortar joints between courses of masonry units. In concrete flatwork, the cutting of *control joints*. See *control joints*.

joist: A heavy piece of horizontal timber to which the boards of a floor, or the lath of a ceiling, are nailed. Joists are laid edgewise to form the floor support. See Fig. 15.

K

kellastone: In architecture, a stucco with crushed finish.

kerf: A cut made with a saw.

kerfing: The process of cutting grooves or kerfs across a board so as to make it flexible for bending. *Kerfs* are cut down to about two-thirds of the thickness of the piece to be bent. An example is found in the bullnose of a stair which frequently is bent by the process of kerfing.

kevel: A stonemason's hammer, used for breaking and dressing stone.

key: In building, a wedge for splitting a tenon in a mortise to tighten its hold; a strip of wood inserted in a piece of timber across the grain to prevent casting; also a wedge of metal used to make a dovetail joint in a stone; a hollow in a tile to hold mortar or cement; a groove made in cement footings for tying in the cement foundation of a structure. A footing key is shown in Fig. 90.

keyhole saw: A small cutting tool with a tapered blade used for cutting keyholes, fretwork, and other similar work. A nest of *keyhole saws* is shown in Fig. 53.

keystone: The wedge-shaped piece at the top of an arch which is regarded as the most important member because it binds, or locks, all the other members together. The position of a keystone is shown in Fig. 54.

kick plate: A metal plate, or strip, placed along the lower edge of a door, to prevent the marring of the finish by shoe marks.

kiln: A large oven or heated chamber for the purpose of baking, drying, or hardening, as a *kiln* for drying lumber; a *kiln* for baking brick: a lime *kiln* for burning lime.

Glossary

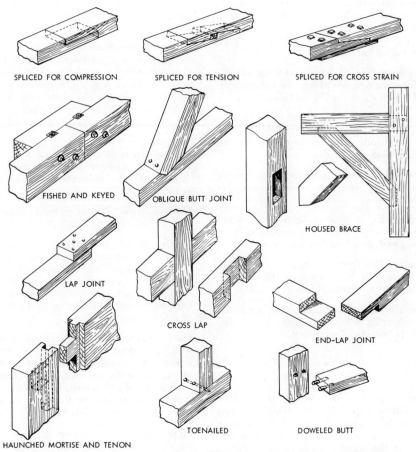

Fig. 52. Wood joints commonly used in timber framing and other woodworking.

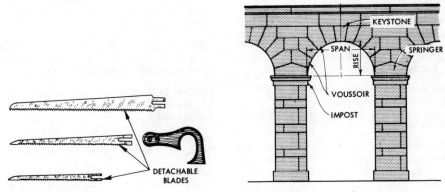

Fig. 53. Keyhole saws. Fig. 54. Parts of an arch.

345

Masonry Simplified

king closer: In masonry, a closer used to fill an opening in a course larger than a half brick. A *king closer* is about three-fourths the size of a regular-sized brick. Also called king *closure*.

king post: In a roof truss, the central upright piece against which the rafters abut and which supports the tie beam.

knee: A piece of lumber bent in an angular shape either naturally or artificially to receive and relieve the strain of a weight on another piece of timber.

kneeler: In masonry, a stone cut to provide a change in direction, as in the curve of an arch.

kraft paper: A type of strong brown paper used extensively for wrapping purposes, and as a building paper.

kyanize: The preparation of wood to prevent decay by a process of infusing the wood with mercuric chloride.

Kyan's process: In woodworking, a method of preserving wood by infusing the timber with bichloride of mercury.

L

label: A molding or dripstone over a door or window, especially one which extends horizontally across the top of the opening and vertically downward for a certain distance at the edges.

lacing course: In masonry, a course of brickwork built into a stone wall for bonding and leveling purposes.

La Farge cement: A cement produced as a by-product during the calcination of hydraulic lime. It is nonstaining and is imported. It develops almost as much strength as portland cement.

lag screw: A heavy wood screw with a square head. Since there is no slot in the head, the screw must be tightened down with a wrench.

lally column: A cylindrically shaped steel member, sometimes filled with concrete, used as a support for girders or other beams.

laminate: The building up with layers of wood, each layer being a lamination or ply; also, the construction of plywood.

laminated construction: Any type of construction where the work is built up by thin layers to secure maximum strength with minimum weight. In pattern making this method is especially desirable since it eliminates cross-grain wood and provides strength, particularly on thin curved members.

landing: In stair construction, a platform introduced at some point to change the direction of the stairway, or to break the run.

landing newel: A post at the landing of a stair supporting the handrail.

landing tread: In building, a term used when referring to the front end of a stair landing. The method of construction usually provides the front edge with a thickness and finish of a stair tread while the back has the same thickness as the flooring of the landing.

lap joint: The overlapping of two pieces of wood or metal. In woodworking, such a uniting of two pieces of board is produced by cutting away one-half the thickness of each piece. When joined, the two pieces fit into each other so that the outer faces are flush. See *end-lap* joint, Fig. 52.

large knot: In woodworking, any sound knot measuring more than $1\frac{1}{2}$ inches in diameter. See *sound knot*.

lath: In building, a term applied to metal screening used as foundation for plaster. Gypsum lath is another type of plaster foundation. See Fig. 55.

lathe: In shopwork, a mechanical device used in the process of producing circular work for wood or metal turning.

lathing: In architecture, the nailing of lath in position; also a term used for the material itself.

lattice: Any open work produced by interlacing of laths or other thin strips.

latticework: Any work in wood or metal made of lattice or a collection of lattices.

Glossary

Fig. 55. Lather screwing gypsum lath to steel partition studs using power screwdriver. (National Gypsum Co.)

lead: In masonry, part of a wall built as a guide for the laying of the balance of the wall; the corner of the wall.

leader: The same as a downspout.

lean-to roof: The sloping roof of a room having its rafters, or supports, pitched against and leaning on the adjoining wall of a building.

ledge: In architecture, any shelflike projection from a wall.

ledger board: In building, the same as a ribbon strip; a support attached to studding for carrying joists; horizontal member of a scaffold.

level: A device (also known as a *spirit level* or *plumb rule*) consisting of a glass tube nearly filled with alcohol or ether, leaving a movable air bubble. This device, protected by a metal or wood casing, is used for determining a point, or adjusting an object, in a line or plane perpendicular to the direction of the force of gravity. When centered, the bubble indicates the line of sight to be truly horizontal. A slight tilting of the *level* at either end will cause the bubble to move away from center, indicating a

347

Masonry Simplified

Fig. 56. Level.

line which is not horizontal. Same as *plumb rule*. An example of a *spirit-level* obtainable in either wood or metal is shown in Fig. 56.

leveling instrument: A leveling device consisting of a spirit level attached to a sighting tube and the whole mounted on a tripod; used for leveling a surface to a horizontal plane. When the bubble in the level is in the center the line of sight is horizontal.

leveling rod or **leveling staff:** A rod, or staff, with graduated marks for measuring heights or vertical distances between given points and the line of sight of a *leveling instrument*. The different types of leveling rods in common use are the *target rods* read only by the rodman and the *self-reading rods*, which are read directly by the men who do the leveling.

level man: The surveyor who has charge of the leveling instrument.

lewis: An iron dovetailed tenon, in sections, designed to fit into a dovetailed mortise in a heavy block of stone; used for attaching a derrick, or other hoisting apparatus, to the stone for lifting it to its proper position in a structure.

lewis bolt: An anchor bolt, with a ragged tapering tail, inserted in masonry and held in place by a lead casing.

light: A window pane; a section of a window sash for a single pane of glass.

lime: A caustic, highly infusible, white substance produced by the action of heat on limestone, shells, or other forms of calcium carbonate. The heat drives out the carbonic acid and moisture, leaving only the quicklime.

lime-burning: The process of producing lime by the burning of limestone. See *calcining*.

limestone: A type of stone used extensively for building purposes, especially in the better grade of structures. *Limestone* is composed largely of calcium carbonate originating usually from an accumulation of organic remains, such as shells, which yield lime when burned. Therefore, this stone is also used extensively as a source of lime.

limewash: Lime slaked in water and applied with a brush or as a spray. Salt is sometimes added to make it adhere better, and bluing may be added to give a white tone. It is used chiefly as a wall covering.

lineal foot: Pertaining to a line one foot in length as distinguished from a square foot or a cubic foot.

linear: Resembling a line, or thread; narrow and elongated; involving measurement in one direction; pertaining to length.

linear measure: A system of measurement in length; also known as *long measure:*

12 inches (in.)	= 1 foot (ft.)
3 feet	= 1 yard (yd.)
16½ feet	= 1 rod (rd.)
320 rods	= 1 mile (mi.)
5280 feet	= 1 mile

lintel: A wood, stone, steel, or reinforced masonry structure placed across the top of door and window openings to support the walls above the openings.

live load: The moving or variable weight to which a building is subjected, due to the weight of the people who occupy it; the furnishings and other movable objects as distinct from the *dead load* or weight of the structural members and other fixed loads.

load-bearing walls: Any wall which bears its own weight as well as other weight; same as a supporting wall. Also called *bearing wall*.

lobby: In architecture, a hall or passage at the entrance to a building; in case of large buildings, the *lobby* is often used as a waiting room.

locking stile: The vertical section of a door to which the lock is fastened.

long-and-short work: In masonry, a method of forming angles of door and window jambs in rubble walls by laying stones horizontally alternating with stones set on end, the upright stones usually being longer than the horizontal stones.

longitudinal section: In shopwork and drawing, a lengthwise cut of any portion of a structure; also, pertaining to a measurement along the axis of a body.

lookouts: Short wooden brackets which support an overhanging portion of a roof (see Fig. 32.); also, a place from which observations are made, as from a watchtower.

loose knot: In woodworking, a term applied to a knot which is not held in position firmly by the surrounding wood fibers; such a knot is a severe blemish in a piece of lumber making the board unfit for first-class work.

louver: An opening for ventilating closed attics or other used spaces; a lantern or turret on a roof for ventilating or lighting purposes, commonly used in medieval buildings; also, a louver board. A slatted opening for ventilation in which the slats are so placed as to exclude rain, light, or vision.

louver boards: In architecture, a series of overlapping sloping boards or slats in an opening so arranged as to admit air but keep out rain or snow.

louvered awning blinds: Adjustable louvers on slanted outside blinds control the amount of shade or sunlight entering windows.

lug sill: In building, a term applied to a window sill in a brick or stone wall, where the sill extends beyond the width of the window opening, with the ends of the sill set in the wall.

lumber: Any material, such as boards, planks, or beams cut from timber to a size and form suitable for marketing.

lump lime: Lime commonly known as *quicklime* produced by burning limestone in a kiln.

M

made ground: In building construction, a portion of land, or ground, formed by filling in natural or artificial pits with rubbish or other material.

magazine: In architecture, a warehouse, a storehouse for merchandise; also a protected building or room such as a depot for storage of military stores, especially explosives and munitions.

magnesite: Carbonate of magnesium obtained from natural deposits.

magnesite flooring: A composition flooring made of calcined *magnesite* and magnesium-chloride solution with a filler of sawdust, wood flour, ground silica, or quartz. It is used to cover concrete floors on which it is floated in a layer about 1½ inches in thickness.

mallet: A small maul, or hammer, usually made of wood used for driving another tool, such as a chisel.

malm: In brickmaking, an artificial marl produced by mixing clay and chalk in a wash mill. The product is used as clay in the manufacture of brick.

malm rubber: In brickwork, a soft form of malm brick which is capable of being worked by cutting or rubbing into special shapes.

mansard roof: A roof with two slopes on all four sides, the lower slope very steep, the upper slope almost flat; frequently used as a convenient method of adding another story to a building.

mantel: The ornamental facing around a fireplace, including the shelf which is usually attached to the breast of the chimney above the fireplace.

marble: Any limestone capable of taking a high polish; used extensively for both interior and exterior finish of buildings; because of the wide range of colors from white to dark gray and brown, *marble* is much used in architectural work for decorative purposes.

marezzo marble: An artificial marble produced by mixing cement with fiber; specially designed for interior decoration.

margin draft: In ashlar work, a smooth surface surrounding a joint.

marine glue: In woodworking, an adhesive substance composed of crude rubber, pitch, and shellac; the proportions are: 1 part rubber, 2 parts shellac, and 3 parts pitch.

mason: A workman skilled in laying brick or stone, as a *bricklayer,* a *stonemason.*

masonry: A term applied to anything constructed of stone, brick, tiles, concrete, and similar materials; also, the work done by a mason who works in stone, brick, tiles, or concrete.

masonry arches: Arches, usually curved, thrown across openings in masonry walls for providing support for the superimposed structure. The arches may be constructed of such material as stone blocks or brick put together in a particular arrangement, so a completed masonry arch will resist the pressure of the load it carries by a balancing of certain thrusts and counterthrusts. See Fig. 42.

masonry nail: A hardened-steel nail of specialized design, used for fastening wood, etc., to *masonry work.* See Fig. 60.

masonry saw: Portable, electrically powered hand saw, similar to all-purpose *electric hand saw.* Designed specifically for cutting masonry, this saw has a variety of masonry blade choices, including diamond blades and abrasive blades. See Fig. 57.

masonry wall: Any wall constructed of such material as stone, brick, tile, cement blocks, or concrete put in place by a mason.

matched boards: Boards which have been finished so as to hold a tongue-and-groove joint securely in place; also boards finished with a rebated edge for close fitting.

maul: A heavy hammer or club used for

Fig. 57. Masonry saw. (Skil Corp.)

Glossary

driving stakes or piles; also, a heavy mallet or mace; any of various types of heavy hammers used for driving wedges, piles, or stakes.

meager lime: In building, a lime in which the impurities are in excess of 6 percent. See *poor lime*.

mechanic: Pertaining to a handicraft, or one skilled in some manual art; also, a skilled workman who makes repairs or assembles machines; a skilled worker with tools or machines.

mechanic arts: In school-shop training, a term applied to craftsmanship and the use of tools and machines.

meeting rail: The strip of wood or metal forming the horizontal bar which separates the upper and lower sash of a window.

mensuration: The process of measuring, especially that branch of mathematics which deals with the determining of length, area, and volume; that is, finding the length of a line, the area of a surface, and the volume of a solid.

metal-edged gypsum plank: A gypsum plank with tongue-and-groove metal edges. Its advantage over ordinary gypsum board is that it results in cleaner, better fitting joints when used for wall, floor, or ceiling construction.

metal strip: A term sometimes applied to metal flashing; used on water tables or around chimneys to prevent water seeping into the roof or walls. See *belt course*, Fig. 18.

metal ties: In masonry, a type of steel tie which is either plain or galvanized, and is used to bond two separate wall sections together in cavity-type walls. Typical metal ties, commonly used, are shown in Fig. 58. Metal ties are also used to secure siding to masonry walls. Metal ties are used to hold the masonry wall and the frame superstructure together. Joists are sometimes fastened to walls by a metal strip at the center of the span. See Fig. 59.

mezzanine: A low story between two higher stories, usually a gallerylike floor midway between the main floor and the next floor above it.

millwork: In woodworking, any work which has been finished, machined, and partly assembled at the mill.

millwright: A workman who designs and sets up mills or mill machinery; also, a mechanic who installs machinery in a mill or workshop.

mineral aggregate: In masonry work, an aggregate consisting of a mixture of broken stone, broken slag, crushed or uncrushed gravel, sand, stone, screenings, and mineral dust.

mineral wool: A type of material used for insulating buildings, produced by sending a blast of steam through molten slag or rock; common types now in use include: rock wool, glass wool, slag wool, and others.

minute of arc: A measure used by architects to find the proportion of a column; one sixtieth of a degree. See *module*.

miter: In carpentry, the ends of any two pieces of board of corresponding form, cut off at an angle and fitted together in an angular shape.

miter box: A device used by a carpenter for guiding a handsaw at the proper angle for cutting a miter joint in wood. The carpenter usually makes his own *miter box* on the job.

miter cut: In carpentry, a cut made at an angle for joining two pieces of board so cut that they will form an angle.

mitering: The joining of two pieces of board at an evenly divided angle; joining two boards by using a miter joint.

miter plane: A tool used for any type of utility work where a joint is made without overlapping of the boards, as in butt or miter joints.

miter-saw cut or **miter-sawing board:** A device used to guide a saw at a desired angle.

miter square: A square similar to the try square, but with one edge of the handle having a 45-degree angle, so it can be used for laying out miter joints.

351

Masonry Simplified

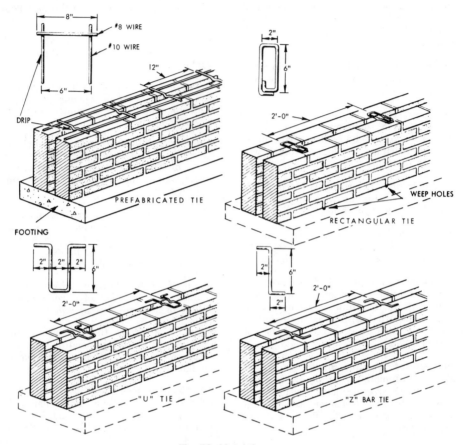

Fig. 58. Metal ties.

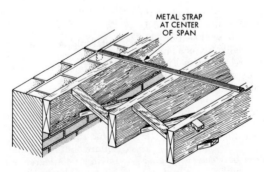

Fig. 59. Metal strap connects floor framing.

Glossary

modular brick: Brick which are designed for use in walls built in accordance with the modular dimensional standards.

modular dimensional standards: Dimensional standards approved by the American National Standards Institute for all building material and equipment, based upon a common unit of measure of 4 inches, known as the *module*. This module is used as a basis for the *grid* which is essential for dimensional coordination of two or more different materials.

modular masonry: Masonry construction in which the size of the building material used, such as brick or tile, is based upon common units of measure, known as the *modular dimensional standards*.

module: A measure used by architects when designing columns; for example, taking the size of some part, as the semidiameter of a column at the base of the shaft, as a unit of measure for regulating the proportions of the entire column.

moisture barrier: A waterproofed material used to retard passage of vapor or moisture through an insulator placed on the warm side.

molded-intake belt course: In building, such a course usually is an elaboration of a plain-band course of masonry or cut-stone work located at a point where the thickness of the upper wall is less than the thickness of the wall below it.

molding: A strip of material, either plane or curved, formed into long regular channels or projections; used for finishing and decorative purposes. *Molding* can be bought in many different sizes and shapes.

molding plane: A small tool used in furniture making for cutting molding into various sizes, shapes, and widths.

molly expansion anchor: A type of metal fastener, consisting of a bolt encased in a shell which expands, wedging itself into a hole drilled to receive it; used principally for securing structural wood parts to a concrete or masonry wall. See *anchors*, Fig. 7.

monolithic: Pertaining to a hollow foundation piece constructed of masonry, with a number of open wells passing through it. The wells are finally filled with concrete to form a solid foundation; a term applied to any concrete structure made of a continuous mass of material or cast as a single piece.

mortar: In masonry, a pasty building material, composed of sand, lime, and cement mixed with water, which gradually hardens when exposed to the air. *Mortar* is used as a joining medium in brick and stone construction.

mortar bed: A thick layer of plastic mortar in which is seated any structural member, the purpose of which is to provide a sound, contour-formed base.

mortar board: In plastering, a small square board, with a handle underneath, on which a plasterer holds his mortar. Same as *hawk*. In masonry, 3′ square platform on which the mortar is placed.

mortar mixer: A machine which mixes mortar by means of rotating paddles in a drum.

mortise: In woodworking, a cavity cut in a piece of wood, or timber, to receive a tenon, or tongue, projecting from another piece; for example, a mortise-and-tenon joint.

mortise chisel: A tool used in woodworking for cutting mortises; a heavy-bodied chisel with a narrow face.

mortise gage: A carpenter's tool consisting of a head and bar containing two scratch pins which may be adjusted, for scribing parallel lines for cutting mortises to whatever width may be desired.

mortising machine: A carpenter's tool used for cutting mortises in wood, either by using a chisel or a circular cutting bit.

mosaic: A combination of small colored stones, glass, or other material so arranged as to form a decorative surface design. The various pieces are inlaid usually in a ground of cement or stucco.

mudsill: The lowest sill of a structure, as a foundation timber placed directly on the ground or foundation.

mullion: The slender bars between the lights or panes of windows.

muntin: Small strips of wood, or metal,

which separate the glass in a window frame; sometimes less correctly a *mullion*.

N

nail: A slender piece of metal pointed at one end for driving into wood, and flat or rounded at the other end for striking with a hammer; used as a wood fastener by carpenters and other construction workers. The sizes of nails are indicated by the term *penny,* which originally indicated the price per hundred, but now refers to the length. Although the sizes of nails may vary as much as $\frac{1}{8}''$ to $\frac{1}{4}''$ from that indicated, the approximate lengths as sold on the market are:

    4-penny nail $= 1\frac{1}{4}''$
    6-penny nail $= 2''$
    8-penny nail $= 2\frac{1}{2}''$
   10-penny nail $= 3''$
   20-penny nail $= 4''$
   60-penny nail $= 6''$

For different types of nails, nailheads, nail points, and nail shanks, see Figs. 60 and 61.

nail-glued roof truss: A glued truss with plywood gusset plates that uses nails to hold it together only until the glue dries. A few hours after assembly, the strength of the truss bonds depends on the glue alone. Grade A casein glue is applied with a specially designed spreader to members whose design specifications must be followed exactly to get proper stresses. This truss requires no special jointing or cutting, and is stronger than conventional truss designs, showing less deflection under test loads.

nail puller: Any small punch bar suitable for prying purposes, with a **V**-shaped, or forked, end which can be slipped under the head of a nail for prying it loose from the wood; also, a mechanical device provided with two jaws, one of which serves as a leverage heel for gripping a nail and prying it loose from a board.

nail set: A tool usually made from a solid bar of high-grade tool steel, measuring about 4 inches in length, used to set the heads of nails below the surface of wood. One end of the tool is drawn to a taper and the head is so shaped there is slight possibility of the device slipping off the head of a nail. Both ends are polished, body machine knurled.

natural beds: The surface of stone as it lies in the quarry. In stratified rocks, if the walls are not laid in their natural bed, the laminae, or scales, separate.

natural cement: A cement made from a natural earth requiring but little preparation; similar to hydraulic lime.

neat cement: In masonry, a pure cement uncut by a sand admixture.

neat work: In masonry, the brickwork above the footings.

nest of saws: A set of saw blades intended for use in the same handle, which is detachable. Such a collection of thin, narrow-bladed saws usually consists of one or more compass saws and a keyhole saw designed primarily for cutting out small holes, such as keyholes. See Fig. 53.

newel: In architecture, an upright post supporting the handrail at the top and bottom of a stairway, or at the turn on a landing; also, the main post about which a circular staircase winds; sometimes called the *newel post.* See Fig. 83.

nogging: In masonry, the filling-in with bricks of the spaces between timbers, such as studding in walls and partitions.

nonbearing partition: A term used in the building trade when referring to a dividing wall which merely separates space into rooms, but does not carry overhead partitions or floor joists.

nosing: The rounded edge of a stair tread projecting over the riser; also, the projecting part of a buttress. See Fig. 83.

O

obelisk: A four-sided shaft of stone, usually monolithic, tapering as it rises, and terminating in a pyramid at the apex.

O.C.: In working drawings, an abbreviation for *on center,* as in the spacing of joists, studs, or other structural parts.

octagon: A *polygon* with eight sides.

Glossary

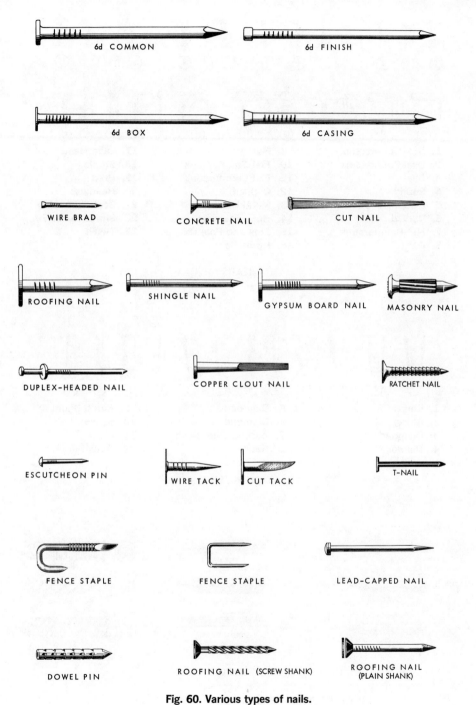

Fig. 60. Various types of nails.

Masonry Simplified

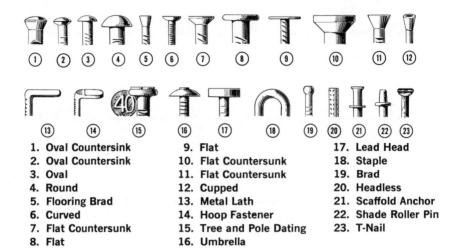

1. Oval Countersink
2. Oval Countersink
3. Oval
4. Round
5. Flooring Brad
6. Curved
7. Flat Countersunk
8. Flat
9. Flat
10. Flat Countersunk
11. Flat Countersunk
12. Cupped
13. Metal Lath
14. Hoop Fastener
15. Tree and Pole Dating
16. Umbrella
17. Lead Head
18. Staple
19. Brad
20. Headless
21. Scaffold Anchor
22. Shade Roller Pin
23. T-Nail

HEADS

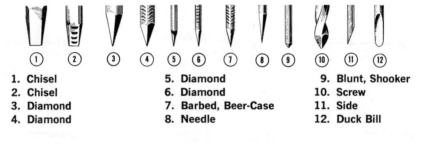

1. Chisel
2. Chisel
3. Diamond
4. Diamond
5. Diamond
6. Diamond
7. Barbed, Beer-Case
8. Needle
9. Blunt, Shooker
10. Screw
11. Side
12. Duck Bill

POINTS

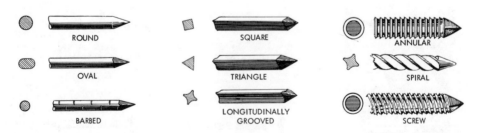

SHANKS

Fig. 61. Types of nail heads, points, and shanks.

Glossary

octagon scale: A scale which appears on the *tongue* of a *framing square;* used for laying out figures with eight equal sides.

odeum: In architecture, a small gallery, or hall, used for musical or dramatic performances.

offset: A term used in building when referring to a set-off, such as a sunken panel in a wall, or a recess of any kind; also, a horizontal ledge on a wall formed by the diminishing of the thickness of the wall at that point.

ogee: A molding with an S-shaped curve formed by the union of a concave and convex line; that is, a cyma recta or cyma reversa.

oilslip: A term used by woodworkers when referring to a small unmounted oilstone held in the hand while they sharpen the cutting edges of gouges.

oilstone: A fine-grained whetstone whose rubbing surface is moistened with oil when used for sharpening the cutting edges of tools.

Old English bond: In masonry work, a bond consisting of alternating courses of stretchers and headers, with a *closer* laid next to the corner bricks in every course of headers. See *typical bonds*, Fig. 22.

oölitic limestone: A variety of rock formation composed of rounded concretions, usually carbonate of lime, resembling the roe of fish, cemented together.

open-corner fireplace: A fireplace of which two adjacent sides are open. Important factors which differ from conventional fireplace construction considerations are flue capacity and cross draft. A $1/4''$ steel plate usually supports the corner overhang extending back beyond the corbeled abutment. Angle iron and plate rest on the cap of a steel column.

open planning millwork: Some millwork stock includes special profiles for use in large window areas and open modular planning constructions, where framing members are often used as finish members.

open-string stairs: In the building trade, a term applied to a stairway with a wall on one side and the other side open, so that a protective balustrade or hand rail is necessary on the open side. The balustrade is supported at top and bottom by upright posts known as *newel posts*. The construction is such that the treads and risers are visible from the room or hallway into which the stairs lead. See Fig. 83.

openwork: Any type of construction which shows openings through the substance of which the surface is formed, especially ornamental designs of wood, metal, stone, or other materials.

oriel: In architecture, a window projecting from the outer face of a wall, especially an upper story, and supported by brackets or corbels.

orifice: A small opening as at the end of a vent pipe, or any similar mouthlike aperture.

ornament: In architecture and furniture making, any decorative detail added to enhance the beauty or elegance of the design.

ornamentation: In masonry, a design formed by the laying of stone, brick, or tile so as to produce a decorative effect.

orthography: In the building trade, a geometrical elevation of a structure which is represented as it actually exists and in perspective as it would appear to the observer.

out of true: In shopworking and the building trade, a term used when there is a twist or any other irregularity in the alignment of a form; also, a varying from exactness in a structural part.

outrigger: A projecting beam used in connection with overhanging roofs. A support for rafters in cases where roofs extend two or more feet beyond the walls of a house.

outside gouge: In woodworking, a type of gouge where the bevel is ground on the convex, or outside, face.

overhand work: In masonry, work performed on the outside of a wall from a scaffold constructed on the inside of the wall.

overhead door: A door which may either be mounted on a sliding track or a pivoted

357

Masonry Simplified

canopy frame, which moves upward to an overhead position when opened. Such doors may be manually operated or may be impelled by a variety of power mechanisms and are commonly used as garage doors. See *roll-up doors, tilt-up doors.*

oversize brick: Modular brick related to the 4-inch module, every 12 inches in height; size 2½" x 3½" x 7½".

ovolo: A convex molding, forming or approximating in section a quarter of a circle; a quarter-round molding.

P

packing: In masonry, the process or operation of filling in a double or hollow wall; also, any material used in the operation of filling or closing up a hollow space, as in a wall. See *furring.*

pad saw: A small compass saw with a detachable handle, which also serves as a socket or holder for the narrow tapering blade when not in use.

pad stone: In building, a stone *template;* a stone placed in a wall under a girder or other beam to distribute the weight or pressure of the load above; also, a lintel of stone spanning a doorway and supporting joists. See *template.*

paint for concrete: Paints made of zinc oxide or barium sulphate, mixed with tung oil, are frequently used for the painting of concrete work.

pan: In half-timbered work, a panel of brickwork or lath and plaster; any large division of an exterior wall, such as the space between upright and horizontal timbers in a frame structure where the surface is to be filled with boards, brickwork, or lath and plaster; also, in carpentry, a recess bed for the leaf of a hinge.

pan forms: pan-like metal or fiberglass structures used as forms for the bottom side of concrete floors. Reinforcing bars are placed in the recesses between the pans, which, when filled with concrete, become, in effect, floor joists. See Fig. 62.

panel: In architecture, a section or portion of a wall, ceiling or other surface, either sunken or raised, enclosed by a framelike border; a term especially applied to woodwork. In masonry, prefabricated wall sections using special high strength mortars so that the panels may be transported and lifted into place in the building.

panel saw: A carpenter's handsaw with fine teeth, making it especially suitable for cutting thin wood.

panel strip: A term used in the building trade when referring to a molded strip of wood or metal used to cover a joint between

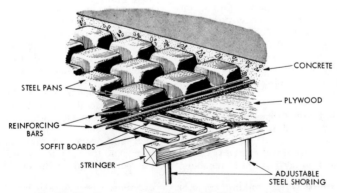

Fig. 62. Metal pans used for concrete floor construction.

Glossary

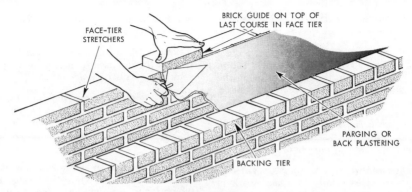

Fig. 63. Parging.

two sheathing boards forming a panel; also, a strip of any kind of material used in the framing of panels.

pantile: A type of roofing tile with straight lengthwise lines, but curved in cross section, laid so that the joint between two concave tiles is covered by a convex tile; also, a type of tile in which there is both a concave and convex portion; this tile is laid so that the convex portion overlaps the rim of the concave portion of an adjoining tile; also, a gutter tile.

parapet: In architecture, a protective railing or low wall along the edge of a roof, balcony, bridge, or terrace.

pargeting: A term used by architects when referring to the decoration of a room with plaster work, or stucco, in relief, such as raised ornamental figures; also, plastering on the inside of flues which gives a smooth surface and improves the draft.

parging: A horizontal layer of mortar between the tiers of a multi-wythe wall. See Fig. 63.

paring: A term used by wood turners when referring to a method of wood turning which is opposed to the scraping method commonly employed by patternmakers.

paring chisel: A type of long chisel employed by patternmakers for slicing, or paring, cuts in wood so as to make a smooth surface which is difficult to obtain when cutting directly across the grain.

paring gouge: A woodworker's bench tool with its cutting edge beveled on the inside, or concave face, of the blade.

parquetry: An inlaid pattern of various designs in wood, used especially for floors, and for decorative features in furniture.

parting tool: A narrow-bladed turning tool used by woodworkers for cutting recesses, grooves, or channels.

partition: An interior wall separating one portion of a house from another; usually a permanent inside wall which divides a house into various rooms. In residences, partitions often are constructed of studding covered with lath and plaster; in factories, the partitions are made of more durable materials, such as concrete blocks, hollow tile, brick, or heavy glass.

partition plate: A term applied by builders to the horizontal member which serves as a cap for the partition studs and also supports the joists, rafters, and studding.

party wall: In architecture, a term used when referring to a wall on the line between adjoining buildings, in which each of the respective owners of the adjacent buildings share the rights and enjoyment of the common wall.

pavilion: A partially enclosed structure, usually roofed, for shelter purposes at the seaside, in parks, or other places where people gather for amusement or pleasure. A *pavilion* sometimes is adorned with orna-

Masonry Simplified

mental designs intended to add a decorative feature to a landscaped park or garden.

Payne's process: A method of fireproofing wood by first treating it with an injection of sulphate of iron, then later infusing the wood with a solution of sulphate of lime or soda.

pebble dash: A term used for finishing the exterior walls of a structure by dashing pebbles against the plaster or cement.

peen hammer: A hammer of various designs, used especially by metal workers and by stonemasons. This hammer sometimes has two opposite cutting edges and is roughly toothed to facilitate the cutting of stone or the breaking of brick.

pegboard: A board with holes evenly spaced over its entire surface. May be cut to desired size and used to line closets or hang on walls. Hooks placed in the holes at convenient intervals provide facilities for hanging household objects of almost any size or shape, simplifying storage problems.

pent roof: A roof like that of a penthouse, attached to and sloping from a wall of a building in one direction only.

perch: A solid measure used for stone work, commonly $16\frac{1}{2}$" x $1\frac{1}{2}$" x 1', or $24\frac{3}{4}$ cu. ft. However, the measure for stone varies according to locality and custom, sometimes $16\frac{2}{3}$ cu. ft. being used for solid work.

perpend: In masonry, a header brick or large stone extending through a wall so that one end appears on each side of the wall and acts as a binder.

perron: An architectural term referring to an out-of-door stairway leading to the first floor of a building; a name sometimes applied to the platform upon which an entrance door opens, together with the flight of steps leading up to it; also, a flight of stairs, as in a garden, leading to a terrace or upper story.

perspective drawing: The representation of an object on a plane surface, so presented as to have the same appearance as when seen from a particular viewpoint.

pew: A name commonly applied to long benches with backs used for seating the audience in a church.

picket: A stake, or narrow board, sharpened at the top used in making fences; also, sometimes called a *pale*.

picture mold: A narrow molding, fastened to an interior wall, used for hanging pictures, which are suspended from the molding by means of fine wire and a metal hook.

picture window: A large window whose bottom ledge is not more than waist high, which includes a dominant fixed sash area, though movable sashes may also be enclosed by the frame. The fixed sash area is usually wider than it is high. See Fig. 64.

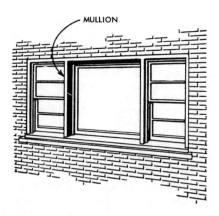

Fig. 64. Picture window.

pier: One of the pillars supporting an arch; also, a supporting section of wall between two openings; a masonry structure used as an auxiliary to stiffen a wall.

pier glass: In building, a term applied to a large mirror between two windows.

pike pole: An implement, equipped with a sharp metal point, used for holding poles, such as telephone or telegraph poles, in an upright position while planting or removing them.

Glossary

pilaster: A rectangular column attached to a wall or pier; structurally a pier, but treated architecturally as a column with a capital, shaft, and base.

pile: A large timber, steel member, or precast concrete shaft with a steel casing driven into the ground for the support of a structure or vertical load.

pile driver: A machine for driving piles; usually a high vertical framework with appliances attached for raising a heavy mass of iron which, after being lifted to the top of the framework, is allowed to fall, by the force of gravity, on the head of the pile, thus driving it into the ground.

pillar: An upright shaft or column of stone, marble, brick, or other materials, relatively slender in comparison to its height. Used principally for supporting superstructures but may stand alone as for a monument.

pin: In carpentry, a piece of wood used to hold structural parts together, as a small peg or wooden nail.

pincers: A jointed tool, with two handles and a pair of jaws used for gripping and holding an object.

pinchbar: A type of crowbar, or lever, on one end of which a pointed projection serves as a kind of fulcrum; used especially for rolling heavy wheels. See *wrecking bar*.

pin knot: A term used by woodworkers when referring to a blemish in boards, consisting of a small knot of $\frac{1}{2}$ inch or less in diameter.

pinnacle: In architecture, a tall, pointed, relatively slender, upright member usually terminating in a cone-shaped spire, used as a decorative feature on a buttress or in an angle of a pier; also, a slender ornament as on a parapet or any turretlike decoration.

pitch board: In building, a thin piece of board, cut in the shape of a right-angled triangle, used as a guide in forming work. When making cuts for stairs, the *pitch board* serves as a pattern for marking cuts; the shortest side is the height of the riser cut and the next longer side is the width of the tread.

pitch of a roof: The angle, or degree, of slope of a roof from the plate to the ridge. The pitch can be found by dividing the height, or rise, by the span. For example, if the height is 8 feet and the span 16 feet, the pitch is $\frac{8}{16}$ equals $\frac{1}{2}$. Then the angle of pitch is 45 degrees.

pith knot: In woodworking, a term used when referring to a blemish in boards, consisting of a small knot with a pith hole not more than $\frac{1}{4}$ inch in diameter.

pivoted casement: A casement window which has its upper and lower edges pivoted.

plancier: The underside of an eave or cornice. See Fig. 32.

placing concrete: A more appropriate term than *pouring*. All good concrete is too stiff to pour; thus the term used in the trade is *placing*.

plain sawing: Cutting wood so the saw cuts are parallel to the squared side of a log. See *flat grain*.

plan: In architecture, a diagram showing a horizontal view of a building, such as floor plans and sectional plans.

plane: In woodworking, a flat surface where any line joining two points will lie entirely in the surface; also, a carpenter's tool used for smoothing boards or other wood surfaces.

planing mill: An establishment equipped with woodworking machinery for smoothing rough wood surfaces, cutting, fitting and matching boards with tongue-and-grooved joints; a wood-working mill.

plank: A long, flat, heavy piece of timber thicker than a board; a term commonly applied to a piece of construction material 6 inches or more in width and from $1\frac{1}{2}$ to 6 inches or more in thickness.

plank truss: Any truss work constructed of heavy timbers such as planking in a roof truss or in a bridge truss.

plan shape: A plan shape is the basic pattern on which a house is laid out. Most

Masonry Simplified

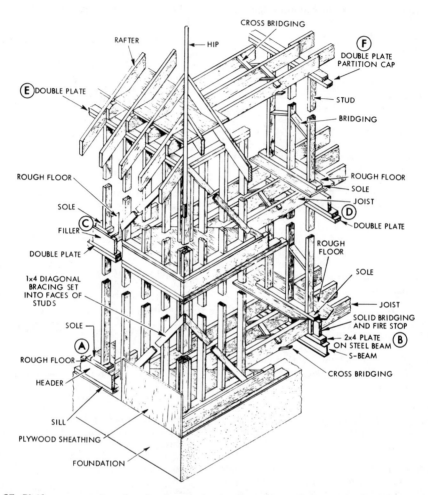

Fig. 65. Platform or western framing is popular because it compensates for shrinkage and can be erected quickly and easily. The rough floor provides a platform for the workmen.

commonly used plan shapes are the square, rectangular, **T, L, H, U,** and split-level patterns. **T, L, H,** and **U** plan shapes roughly follow the shape of the alphabet letter by which they are indicated. See *split-level*.

plaster: Any pasty material of a mortar-like consistency used for covering walls and ceilings of buildings. Formerly, a widely used type of plastering composed of a mixture of lime, sand, hair, and water. A more durable and popular plastering is now made of portland cement mixed with sand and water.

plaster board: A rigid insulating board made of plastering material covered on both sides with heavy paper.

plaster lath: Thin, narrow strips of wood nailed to ceiling joists, studding, or rafters, as a groundwork for receiving plastering.

Glossary

plastic wood: A manufactured product useful to the building industry. In making wood compounds, choice softwoods, such as white pine, spruce, and fir, which have been converted into sawdust, are employed. The sawdust, when ground into wood flour and mixed with the proper adhesives, forms a plastic material used extensively for filling cracks, defects in wood, and for other purposes. Since it dries quickly and hardens upon exposure to air, *plastic wood* can be painted almost immediately after being applied.

plate: A term usually applied to a 2" x 4" placed on top of studs in frame walls. It serves as the top horizontal timber upon which the attic joists and roof rafters rest, and to which these members are fastened. See Figs. 15, 23, and 65. Also, a flat piece of steel used in conjunction with angle irons, channels, or beams in the construction of lintels.

plate glass: A polished, high-grade glass cast in the form of a plate, or sheet, used principally in high-priced structures. A sheet of glass usually thicker and of a better quality than ordinary window glass, also with a smoother surface free from blemishes.

plate rail: A narrow railing usually placed at the edge of a drain for holding plates and other dishes; also, a narrow shelflike molding attached to the interior of a wall for supporting decorated pieces of chinaware, especially plates.

platform framing: A type of construction in which the floor platforms are framed independently; also, the second and third floors are supported by studs of only one story in height. Also called *western framing*. See Fig. 65.

pliers: A small pincerlike tool having a pair of long, relatively broad jaws which are roughened for gripping and bending wire or for holding small objects. Pliers are sometimes made with nippers at the side of the jaws for cutting off wire. See *cutting pliers*.

plinth: The lowest squared-shaped part of a column; a course of stones, as at the base of a wall. See *plinth block*, Fig. 66.

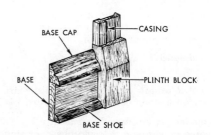

Fig. 66. Plinth block.

plinth block: A small block slightly thicker and wider than the casing for interior trim of a door. It is placed at the bottom of the door trim against which the baseboard or mopboard is butted. See Fig. 66.

plot plan: A plan showing the size of the lot on which the building is to be erected, with all data necessary before excavation for the foundation is begun.

plum: In masonry, a large, undressed stone which, together with other similar stones, is used in mass concrete to form footings for walls. When plum stones are used, less concrete is required.

plumb and level: A well-finished hardwood or metal case containing glass tube with bubble set lengthwise for testing accuracy of horizontal planes and lines, also containing a second glass tube with bubble set crosswise for testing accuracy of vertical lines and perpendicular walls.

plumb bob: A weight attached to a line for testing perpendicular surfaces for trueness; also, to test, or adjust, with a plumb line.

plumb cut: In roof framing, a cut made on a *rafter* parallel to the *ridge board,* at the point where the rafter and ridge board meet.

plumb rule: Same as *level*. See Fig. 56.

ply: One thickness of any material used for building up of several layers, as roofing felt or layers of wood, as in laminated woodwork.

363

Masonry Simplified

pneumatic: In building, a term applying to any machine or tool which is powered or driven by compressed air.

pneumatically applied concrete: See *shotcrete*.

pointed ashlar: In stonework, face markings on a stone made with a pointed tool.

pointing: A term used in masonry for finishing of joints in a brick, block, or stone wall.

pointing trowel: A small hand instrument used by stone masons or bricklayers for pointing up joints, or for removing old mortar from the face of a wall.

polygon: A figure bounded by straight lines. The boundary lines are called sides and the sum of the sides is called the perimeter. *Polygons* are classified according to the number of sides they have.

poor lime: In building, a lime containing more than 15 percent impurities; also called *meager lime*.

porch: A covered entrance to a building, projecting from the main wall with a separate roof; also, a type of veranda which often is partially enclosed.

portable: Anything which may be moved from place to place easily, such as portable power tools, concrete or mortar mixers, mobile homes, etc.

portable electric drill: A portable, electrically operated drill with a wide variety of bits for various on-the-job drilling operations. Has bits for wood, metal, and masonry drilling. See Fig. 67.

portable electric generator: A portable gasoline-powered generator of electricity, skid or trailer mounted, or provided with two-man handles, which can be carried from place to place, as electric power for various hand tools is needed.

portable electric plane: Electrically powered plane designed for on-the-job finishing purposes. Advantages include an ease of handling which results in highly accurate work. The power plane turns out neat, well-finished work in appreciably less time than planes which depend upon manual power. See Fig. 68.

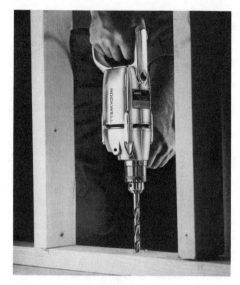

Fig. 67. Portable electric drill.

Fig. 68. Portable electric plane. (Porter Cable Machine Co.)

portal: An entranceway such as a door or gate, usually applied only to structures which are impressive in appearance or size as a massive church edifice.

portico: An open space covered with a roof supported by columns, often attached to a building as a porch but sometimes entirely detached from any structure.

Glossary

portland cement: A hydraulic cement, commonly used in the building trades, consisting of silica, lime, and alumina intimately mixed in the proper proportions, then burned in a kiln. The clinkers or vitrified product, when ground fine, form an extremely strong cement.

portland cement paint: A specially prepared paint made by mixing cement and water; used on concrete walls as a finish, and to protect the joints against water from rain and snow.

post: In building, an upright member in a frame; also a pillar or column.

post and beam construction: A system of construction currently used in one-story buildings in which post and beam-framing units are the basic load-bearing members. Fewer framing members are needed, leaving more open space for functional use, for easier installation of large windows, and more flexible placing of free standing walls and partitions. It is also adaptable for prefabricated modular panel installation. Wide roof overhangs, for sun protection and outdoor living areas, are simpler to construct when this framing system is used. Posts and beams may be of wood, structural steel or concrete. See *concrete bent construction*. Ceiling heights are higher for the same cubage, and it is reported that building is faster and cheaper. Roof deck can double as finished ceiling in the post, beam, and plank variation of the system. Problems include the necessity for extra insulation, difficulty in concealing wiring and duct work, and the necessity for extra care in the choice of materials and in planning. Also called *plank and beam framing*. See Fig 69.

post and pan: A term sometimes applied to half timbering formed of brickwork or of lath and plaster panels.

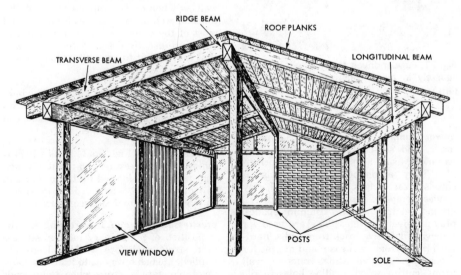

Fig. 69. Plank and beam framing requires careful placement of posts and beams. The roof planks must be strong enough to span the distance between beams.

Masonry Simplified

powder fastener: A special tool which drives metal pins by means of an explosive cartridge. Usually used to fasten wood or other materials to concrete. See Figs. 70 and 71.

Fig. 70. Powder-driven fastener: position and fire. (Ramset Fastening System.)

Fig. 71. Powder-driven fastener: pin is driven squarely in place. (Ramset Fastening System.)

power hammer: Portable electric, pneumatic, and self-contained gasoline-driven hammers, using a vibratory action principle. They accommodate such tools as chisels, frost wedges, solid drill steel, clay spades, tampers, diggers, asphalt cutters, ground rod drivers, offset trimming spades, plug and feathers, etc. They are used for removing defective brick from walls, hardened putty from steel sash, and mortar for repointing; for vibrating concrete wall forms, cutting wood, drilling holes in tile floors, digging holes for posts or sewers, and many other applications.

power troweler: A mechanical device, usually driven by a gasoline engine, used to perform the final finishing of extremely large areas of concrete.

Pratt truss: A special type of construction used in both roof and bridge building in which the vertical members are in compression and the diagonals are in tension.

precast concrete: Concrete building units which are cast separately at a central location and transported to their final positions. See Fig. 72.

prefabricate: To construct or fabricate all the parts, as of a house, at the factory in advance of selling so that the final construction of the building consists merely of assembling and uniting the standard parts.

prefabricated areawall: A corrugated steel wall which lines area space outside basement windows or crawl spaces. See Fig. 73.

prefabricated modular units: Units of construction which are prefabricated on a measurement variation base of 4 inches or its multiples and can be fitted together on the job with a minimum of adjustments. Modular units include complete window walls, kitchen units complete with installations, as well as masonry, wall panels, and most of the other components of a house. Units are usually designed in such a way that they will fit functionally into a variety of house sizes and plan types.

prefabricated skylight: A clear plastic bubble, set in an aluminum frame, often with built-in vent and exhaust fan, which can be set over a prepared opening.

pressed brick: A high-grade brick which is molded under pressure, as a result of which sharp edges are formed by the meeting of two surfaces and a smooth face, making it suitable for exposed surface work.

prestressed concrete: Prestressing is "the imposition of preliminary interval stresses in a structure before working loads are applied, in such a way as to lead to a more favorable state of stress when these loads come into action." Concrete is usually prestressed by means of high strength steel

Glossary

Fig. 72. Precast concrete wall sections. (Medusa Portland Cement Co.)

Fig. 73. Prefabricated areawall.

wire incorporated in it. If the wires are placed in tension, and held in this position before the concrete is placed, the process is called *pretensioning*. If the concrete is poured with pockets in it, where the wires can be placed and prestressed after the concrete is poured and cured, the process is called *post-tensioning*. In one type of construction, members are built up of units resembling concrete blocks, with adjacent faces ground smooth. Threaded reinforcing rods are placed in side splines of the units and extended through washers. Tension is applied to the rods by hydraulic jacks. Pretensioning makes it possible to use much less steel than is needed for structural steel or reinforced concrete buildings. The concrete members can be less bulky. It is also used in making crackless tanks for storing liquids.

priming: The first coat of paint put on for sizing and preserving wood.

profile: An outline drawing of a section, especially a vertical section through a structural part; that is, a contour drawing.

projecting belt course: A masonry term used when referring to an elaboration of a plain band course or cut-stone work projecting beyond the face of a wall for several inches.

projection: In architecture, a jutting out of any part or member of a building or other structure.

promenade tile: In masonry, unglazed machine-made tile; same as *quarry tile.*

puddle: In masonry, to agitate grout with a stick or rod after it has been poured into the cavity of a cavity wall to consolidate it and eliminate voids.

pugging: A coarse kind of mortar used for packing or covering, and laid between floor joists to prevent the passage of sound; mortar used to deaden sound; also called deadening.

pulley stile: In architecture, the upright pieces at the sides of a double-hung window frame on which the pulleys for the sash weights are fastened.

pulpit: A raised platform, as in a church, where the clergyman stands while preaching.

punch: A tool, usually a short steel rod, variously shaped at one end for different operations such as perforating.

purlins: Horizontal timbers supporting the common rafters in roofs, the timbers spanning from truss to truss.

putlog: A crosspiece in a scaffolding, one end of which rests in a hole in a wall; also, horizontal pieces which support the flooring of scaffolding, one end being inserted into *putlog* holes; that is, short timbers on which the flooring of a scaffolding is laid.

putty: A type of cement usually composed of a mixture of whiting and boiled linseed oil, which has been kneaded or beaten to the consistency of dough; used for securing

Masonry Simplified

panes of glass in sash, for filling crevices, and for other similar purposes.

putty in plastering: A cement consisting of lump lime slacked with water to the consistency of cream, and left to harden by evaporation until it becomes like soft putty. It is then mixed with sand, or plaster of paris, and used for a finishing coat.

puzzolano: A volcanic dust which has a hardening effect when mixed with mortar, producing a valuable hydraulic cement; first discovered at Pozzuoli, Italy.

Q

quadrangle: In architecture, an open court or space in the form of a parallelogram, usually rectangular in shape, partially or entirely surrounded by buildings, as on a college campus; also, the buildings surrounding the court.

quadrant: An instrument usually consisting of a graduated arc of 90 degrees, with an index or vernier; used primarily for measuring altitudes. Sometimes a spirit level or a plumb line is attached to the *quadrant* for determining the vertical or horizontal direction.

quarry-faced masonry: Squared stone as it comes from the quarry, with split face, only squared at joints; having the face left rough as when taken from the quarry, as building stone; masonry built of such stone.

quarry-stone bond: In masonry, a term applied to the arrangement of stones in rubble work.

quarry tile: In masonry, a name given to machine-made, unglazed tile. Also called *promenade tile.*

quarter bend: A bend, as of a pipe, through an arc of 90 degrees.

quarter sawing: The sawing of logs lengthwise into quarters, with the saw cuts parallel with the medullary rays, then cutting the quarters into boards, as in making quartered oak boards.

quatrefoil: In architecture, a single decorative feature consisting of an ornamental unit in the form of a four-leaved flower.

queen closure: A half brick made by cutting a whole brick in two lengthwise; also, a half brick used in a course of brick masonry to prevent vertical joints falling above one another. Sometimes spelled *closer.*

queen post: One of the two vertical tie posts in a roof truss or any similar framed truss.

queen truss: A truss framed with queen posts, that is, two vertical tie posts, distinguished from the king truss which has only one tie post.

quicklime: The solid product remaining after limestone has been heated to a high temperature. The process of producing lime is known as *lime-burning.*

quirk: A small groove, or channel, separating a bead or other molding from the adjoining members; also, an acute angle between moldings or beads.

quirk bead: A bead molding separated from the surface on one side by a channel or groove. A *double quirk bead* refers to a molding with a channel on each side of the beads.

quick molding: An architectural term usually applied to a molding which has a small groove, although sometimes the term is also used in reference to a molding with both a convex and a concave curve separated by a flat portion.

quoins: In architecture, large squared stones, such as buttresses, set at the angles of a building.

R

rabbet: In woodworking, a term used in referring to a groove cut in the surface or along the edge of a board, plank, or other timber, so as to receive another board or piece similarly cut.

rabbet joint: A joint which is formed by the fitting together of two pieces of timber which have been rabbeted.

racking: When workmen, in laying a stone or brick wall, approach a corner where two walls meet, they step back the end of each

Glossary

course so that it is shorter than the course below it. This procedure is known as *racking*. The workmen can then tie in their courses in the easiest manner, and the junction of the walls does not form a vertical line which might cause cracking, owing to uneven settlement of the different parts.

radial bar: A device made by attaching a point and pencil to a wooden bar which is then used for striking large curves.

radiating brick: In masonry, a brick which tapers in at least one direction, so as to be especially useful for curved work, as in building arches. Sometimes called *radius brick*. See *compass brick*.

radius tool: In masonry and cement work, a finishing tool used for shaping curved sections which must be smooth and true.

rafter plate: In building construction, the framing member upon which the rafters rest. See Figs. 15, 65, and 76.

rafters: The sloping members of a roof, as the ribs which extend from the ridge or from the hip of a roof to the eaves; used to support the shingles and roof boards. See Fig. 76.

rag work: In masonry, a term applied to any kind of rubble work made of small, thin stones.

rail: A horizontal bar of wood or metal used as a guard, as the top member of a balustrade; also, the horizontal member of a door or window. See Fig. 83.

raked joint: In brick masonry, a type of joint which has the mortar raked out to a specified depth while the mortar is still green.

rake molding: A gable molding with a longer face than that of the eaves molding. The face of the *rake molding* is worked out so that it will line up with the eaves molding.

rake or raking bond: In masonry, a method of laying the courses of brick in an angular or zigzag fashion, as is often seen in the end walls of Colonial houses. See *herringbone bond*.

rake out: In masonry, the removal of loose mortar by scraping, in preparation for pointing of the joints.

raking course: In masonry, a course of bricks laid diagonally between the face courses of a specially thick wall for the purpose of adding strength to the wall.

rammer: In building construction, a term applied to an instrument which is used for driving anything by force, as stones or piles, or for compacting earth; in concrete work, a kind of "stomper," used to pack concrete by removing the air bubbles.

ramp: A sloping roadway or passageway; also, a term used in architecture when referring to a short bend, slope, or curve usually in the vertical plane where a handrail, coping, or the like changes direction.

ramp and twist: In masonry, a term used when referring to work in which a surface both twisting and rising has to be, or is produced.

random ashlar: In masonry, a type of ashlar construction where the building blocks are laid apparently at random, but usually are placed in a definite pattern which is repeated again and again. See *ashlar*, Fig. 11.

random shingles: Shingles of different widths banded together; these often vary from 2½ inches to 12 inches or more in width.

random work: Any type of work done in irregular order, as a wall built up of odd-sized stones.

ranged rubble: Masonry built of rough fragments of broken stone or unsquared or rudely dressed stones, irregular in size and shape. See *rubble work*.

ranger: A horizontal bracing member used in form construction. Also called a *whaler* or *waler*. See Fig. 89.

rangework: Squared stone laid in horizontal courses of even height; same as *coursed ashlar;* also known as ranged masonry.

ratchet bit brace: A carpenter's tool consisting of a bit brace with a ratchet attachment which permits operation of the tool in close quarters.

ratchet drill: A hand drill which is rotated by a ratchet wheel moved by a pawl and lever.

ratchet wheel: A wheel with angular teeth on the edge into which a pawl drops or catches to prevent a reversal of motion.

ratio: The relation between two similar magnitudes in respect to the number of times the first contains the second, either integrally or fractionally, as the *ratio* of 3 to 4 may be written 3:4 or 3/4.

rat stop: A type of construction for a masonry wall which provides protection against rats by stopping them when they attempt to burrow down along the outside of the foundation.

Rawlplug: A fastener or holding device used in wood, glass, masonry, plaster, tile, brick, concrete, metal, or other materials. These devices are made of longitudinal strands of tough jute fiber compressed into a tubular form. See Fig. 7.

ready mix: Concrete mixtures ordered in advance which are delivered, ready to use, in special truck mounted mixers.

rebar: A contraction for *reinforcing bar* commonly used in building trades.

rebate: A woodworking term used when referring to a recess in or near the edge of one piece of timber to receive the edge of another piece cut to fit it; that is, a rabbet groove.

recess: An indentation in the surface of a room, as an alcove or bay window.

reeding: A general architectural term applied to various kinds of ornamental molding; for example, a small convex or semi-cylindrical molding resembling a reed; also, a set of such moldings as on a column; any ornamentation consisting of such moldings.

reflective insulation: Foil-surfaced insulation whose insulating power is determined by the number of its reflective surfaces, and which must be used in connection with an air space. This type of insulation also acts as a vapor barrier.

registers: In building construction, an arrangement of fixtures in floors or walls for admitting or removing hot or cold air for heating or ventilating purposes.

reinforced: To strengthen by the addition of new material, or extra material, for the reinforcement of concrete, iron or steel rods are embedded to give additional strength.

reinforced concrete: Concrete which has been strengthened by iron or steel bars embedded in it.

reinforced concrete construction: A type of building in which the principal structural members, such as floors, columns, and beams, are made of concrete, which is placed around isolated steel bars, or steel meshwork, in such a way that the two materials act together in resisting force. Fig. 74, left, shows reinforcement for a typical concrete floor. Fig. 74, right, shows the reinforcing bars in place for a concrete column.

reinforcing steel: Steel bars used in concrete construction for giving added strength; such bars are of various sizes and shapes. See *rebars*.

rendering: A term used in perspective drawing meaning to finish with ink or color to bring out the effect of the design.

retaining wall: Any wall erected to hold back or support a bank of earth; any wall subjected to lateral pressure other than wind pressure; also, an enclosing wall built to resist the lateral pressure of internal loads. See Fig. 75.

retarder: An admixture used to slow the setting process in concrete or mortar. Used occasionally during hot weather.

reticulated: In masonry, work in which the courses are arranged like the meshes of a net; work constructed or faced with diamond-shaped stones, or of stones arranged diagonally.

return nosing: In the building of stairs, the mitered, overhanging end of a tread outside the balusters. See Fig. 83. Same as *end nosing*.

reveal: In architecture, that part of a jamb or vertical face of an opening for a window

Glossary

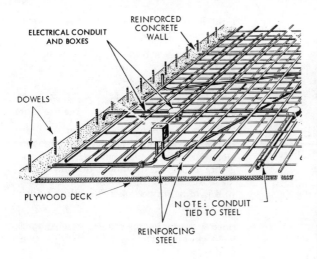

Fig. 74. Left: Horizontal steel reinforcement.
Right: Vertical steel reinforcement. (Adolphi Studios, Inc.)

or doorway between the frame and the outside surface of a wall; also, a term sometimes applied to the entire jamb or vertical face of an opening.

revolving door: A type of door commonly used in entrances to department stores or public buildings; a door with four vanes operating in a curved frame and mounted on a central vertical axis about which it revolves.

revolving shelf: Sometimes called a *lazy Susan,* this shelf revolves to provide easy access to the total shelf area. It is often placed in a closet, especially in the ordinarily unusable corners where two cabinets meet each other at right angles.

ribbon strip: A term used in building for a board which is nailed to studding for carrying floor joists.

ribbon windows: Two or more adjacent windows, each longer than it is high; usually placed with sills five feet or more above floor level.

rich lime: A quicklime which is free from impurities, used especially for plastering

371

Masonry Simplified

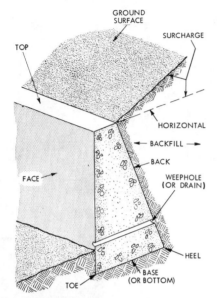

Fig. 75. Retaining wall.

and for masonry work. Also called *fat lime*.

ridge: The intersection of two surfaces forming an outward projecting angle, as at the top of a roof where two slopes meet. The highest point of a roof composed of sloping sides. See Fig. 76.

ridge capping: The covering of wood or metal which tops the ridge of a roof.

ridge pole: The horizontal member, or timber, at the top of a roof which receives the upper ends of the rafters.

ridge roof: A roof whose end view is a gable and whose rafters meet in an apex.

ridge tiles: Tiles used to cap the ridge of a roof.

ridge ventilator: A raised section on a roof ridge, provided with vents which admit air currents.

right angle: An angle formed by two lines which are perpendicular to each other; that is, the lines represent two radii that intercept a quarter of a circle; hence, a 90-degree angle.

right line: The shortest distance between two points; that is, a straight line.

ring shake: A separation of the wood between the annual growth rings of a tree. See *annual growth rings,* Fig. 9.

ripping: In woodworking, the sawing or splitting of wood lengthwise of the grain or fiber.

riprap: In masonry construction, broken stones or other similar material, thrown together loosely and without definite order, for a sustaining bed where a foundation wall is to be formed on soft earth or under deep water.

ripsaw: A saw having coarse, chisel-shaped teeth used in cutting wood in the direction of the grain.

rise: The distance through which anything rises, as the *rise* of a stair, the *rise* of a roof. Also, the vertical distance between the springing of an arch and the highest point of the *intrados.* See Fig. 42.

rise and run: A term used by carpenters to indicate the degree of incline.

riser: A vertical board under the tread of a stair step; that is, a board set on edge for connecting the treads of a stairway. See Fig. 83.

rod: A polelike stick of timber used by carpenters as a measuring device for determining the exact height of risers in a flight of stairs; sometimes called a *story rod*.

rolling partitions: This type of partition is made up of narrow slats, tongue-and-grooved with one another along the edges, which roll up on a shaft when passage is desired. They may be set either horizontally or vertically. See *folding door or partition*.

roll-up door: Constructed in horizontally hinged sections, and usually made of wood, these doors are equipped with springs, tracks, counterbalancers, and other hardware which pull the sections into an overhead position, clear of the opening. They are often motor-operated with manual, radio, or magnetic driver controls, and are commonly used for garages. See *tilt-up doors*.

Glossary

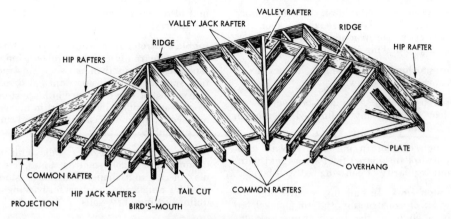

Fig. 76. Roof members.

roll-up screen: A type of metal or plastic window screen installed on the inside of a window frame, which may be rolled up on tracks along the window's sides for full daylight and view. It can be left in place all year round.

roman brick: A *solid masonry unit* whose nominal dimensions are 2″ x 4″ x 12″. The nominal dimensions vary from the specified dimensions by the addition of the thickness of the mortar joint with which the unit is designed to be laid but not more than ½ inch. The specified dimensions of roman brick are 1⅝″ x 3¾″ x 11¾″. Roman brick is sometimes made 16 inches or more in length to suit various construction needs.

roof members: In building construction, the various parts or members which compose a roof, as the framing members. The names of important *roof members* are given in Fig. 76.

roof truss: The structural support for a roof, consisting of braced timbers or structural iron fastened together for strengthening and stiffening this portion of a building.

room divider: A temporary curtain wall such as a *folding partition,* or permanent partition, which may or may not reach from floor to ceiling, as a bookcase or cabinet with *planter box.* These partitions serve to block off activity areas in a room, for various needs, while providing for flexibility of function.

rose window: A circular window decorated with ornamental designs similar to those found in the head of a Gothic window or in some ornate styles of vaulting; also, an ornamental circular window adorned with roselike tracery or mullions radiating from the center.

rosette: In architecture, any rounded ornament resembling a rose in the arrangement of its parts; any circular roselike unit of ornamentation with mullions or tracery radiating from the center; also, a decorative unit similar to a roundel filled with leaflike ornaments.

rostrum: A platform or stage on which a public speaker stands.

rotunda: A round-shaped building, or circular room, covered by a cupola or dome.

roughcast: A term used in the building trade for a kind of plastering made of lime mixed with shells or pebbles and applied to the outside of buildings.

rough floor: A subfloor serving as a base for the laying of the *finished floor* which may be of wood, linoleum, tile, or other suitable material. See Fig. 65.

roughing-in: In building, a term applied to doing the first or rough work on any part of the construction, as roughing plastering, plumbing, and stairs.

rough lumber: Undressed lumber as it comes from the saw.

rough opening: An unfinished window or door opening; any unfinished opening in a building.

rough rubble: In masonry, a wall composed of unsquared field stones laid without regularity of coursing but well bonded.

rough work: In building construction, the work of constructing the rough skeleton of a building; the rough framework, including the boxing and sheeting. The *rough work* may include making the rough frames for doors and windows and any similar work done in a factory and later moved to the building site ready for installing.

rout: A term in woodworking for cutting or gouging out material with a tool called a *router,* which is a special type of smoothing plane.

router: A two-handled woodworking tool used for smoothing the face of depressed surfaces, such as the bottom of grooves or any other depressions parallel with the surface of a piece of work.

routing: The cutting away of any unnecessary parts that would interfere with the usefulness or mar the appearance of a piece of millwork.

rowlock: In masonry, a term applied to a course of bricks laid on edge. Also, the end of a brick showing on the face of a brick wall in a vertical position. See Fig. 33.

rowlock-back wall: In masonry, a wall whose external face is formed of bricks laid flat in the ordinary manner, while the backing is formed of bricks laid on edge.

rubbing stone: In masonry, a stone used by bricklayers to smooth bricks which are designed for some particular purpose in a structure, as in a *gauged arch.*

rubble: Rough broken stones or bricks used to fill in courses of walls, or for other filling; also, rough broken stone direct from the quarry.

rubble concrete: In masonry work, a form of concrete reinforced by broken stones, especially that used in massive construction, such as solid masonry dams; also, masonry construction composed of large stone blocks set about six inches apart in fine cement concrete, and faced with squared rubble or ashlar.

rubble masonry: Masonry walls built of unsquared or rudely squared stones, irregular in size and shape; also, uncut stone used for rough work, such as for backing of unfinished masonry walls.

rubrication: The coloring, especially in red, of a background by use of enamel or paint.

rule joint: In woodworking, a pivoted joint where two flat strips are joined end to end so that each strip will turn or fold only in one direction; an example is the ordinary two-foot folding rule used by carpenters and other woodworkers.

running bond: In masonry, a form of bond used largely for partitions and veneer, in which every brick is laid as a stretcher, with each vertical joint lying between the centers of the stretchers above and below, making angle closers unnecessary. Same as *stretcher* or *stretching bond.* See, also *chimney bond.*

run of stairs: A term used when referring to the horizontal part of a stair step, without the nosing; that is, the horizontal distance between the faces of two risers, or the horizontal distance of a flight of stairs. This is found by multiplying the number of steps by the width of the treads. If there are 14 steps each 10 inches wide, then 14 x 10 equals 140 inches or 11 feet 8 inches, which is the *run of the stairs.*

run of work: A term used in reference to a steady run of jobs following one another in rapid succession; also, applied to a type of job which calls for the repeated production of a quantity of the same kind of article.

rustication: In building and masonry, the use of squared or hewn stone blocks with roughened surfaces and edges deeply bev-

Glossary

eled or grooved to make the joints conspicuous.

rustic beveled work: Masonry in which the face of the stones are smooth and parallel to the face of the wall. The angles are beveled to an angle of 135°, with the face of the stone so that, when two stones come together on the wall, the beveling forms an internal right angle.

rustic joint: In masonry, a sunken joint between building stones.

rustics: In masonry, bricks which have a rough-textured surface, often multicolored.

S

safe carrying capacity: In the building industry, a term used with reference to construction of any piece or part so it will carry the weight, or load, it is designed to support, without breaking down.

sag: To droop or settle downward, especially in the middle, because of weight or pressure; also, the departure from original shape, a dragging down by its own weight, as a sagging door.

salon: A large, more or less elegant, reception room where guests are entertained; also, more commonly, in the United States, an exhibit room where various works of art are displayed.

sandpaper: An abrasive paper, made by coating a heavy paper with fine sand or other abrasives held in place by some adhesive such as glue, used for polishing surfaces and finishing work.

sandstone: A building stone, usually quartz, composed of fine grains of sand cemented together with silica, oxide of iron, or carbonate of lime. *Grindstones* are made of sandstone in its natural state.

sap: In woody plants, the watery circulating fluid which is necessary to their growth.

sap streaks: Streaks showing through a finished wood surface which contains sapwood. Such streaks must be *toned out* in order to secure a uniform finish.

sapwood: The wood just beneath the bark of a tree; that is, the young, softwood consisting of living tissues outside the heartwood.

sash: A framework in which window panes are set.

sash bars: In building, the strips which separate the panes of glass in a window sash.

sashless window: Panes of glass which slide along parallel tracks in the window frame toward one another, to leave openings at the sides, are used as windows; also fixed pane sashless windows are often used for picture and clerestory windows as well as other purposes. Sliding panes lift out for cleaning; panes, tracks, and frame are designed for moisture proofing and scratchless sliding.

sash pin: A heavy gage barbed headless nail or pin used to fasten the mortise-and-tenon joints of window sash and doors.

saw arbor: The spindle, or shaft, on which a circular saw is mounted.

saw bench: A table or framework for carrying a circular saw.

saw gullet: The throat at the bottom of the teeth of a circular saw.

saw gumming: Shaping the teeth of a circular saw. Usually a grinding process.

sawhorse: A rack, or frame, for holding wood while it is being sawed; also, the ordinary trestle on which wood or boards are laid by carpenters for sawing by hand.

saw set: An instrument used for giving set to saw teeth.

saw-toothed skylight: In architecture, a term applied to a skylight roof with its profile shaped like the teeth of a saw.

saw trimmer: A machine used for sawing and trimming metal plates.

sawyer: One whose occupation is that of sawing wood or other material; sometimes used in a restricted sense meaning one who operates one of several saws.

S beam: Structural steel beam. Formerly called *I beam*.

scab: A short piece of timber fastened to two other timbers to splice them together.

scabble: The dressing down of the roughest irregularities and projections of stone which is to be used for rubble masonry. A stone ax or scabbling hammer is used for this work.

scaffold: An elevated, and usually temporary, platform for supporting workmen, their tools, and material while working on a building.

scagliola: An imitation of ornamental marble, consisting of a base of finely ground gypsum, mixed with an adhesive such as a hard cement, and variegated on the surface while in a plastic condition, with chips of marble or with colored graphite dust. When this mixture is hardened, it is finished with a high polish and used for floors, columns, and other interior work.

scale: An instrument with graduated spaces for measuring; also, a term applied to the outside covering, or coating, of a casing. In lumbering, estimating the amount of standing timber.

scaled drawing: A plan made according to a scale, smaller than the work which it represents but to a specified proportion which should be indicated on the drawing.

scaling: When the surface of a hardened concrete slab breaks away from the slab to a depth of about an $\frac{1}{8}$ in.

scantling: A piece of timber of comparatively small dimensions, as a 2x3 or 2x4, used for studding.

scarf joint: The joining of two pieces of timber by notching and lapping the ends; then fastening them with straps or bolts.

sconce: A decorative bracket projecting from a wall for holding candles.

scoring: To mark with lines, scratches, and grooves across the grain of a piece of wood with any kind of steel instrument, for the purpose of making the surface rough enough to make it a firmer joint when glued. In masonry, the first blows to the brickset or with the brick hammer that marks and starts the fracture of the masonry unit being cut. Also, decorative groove in the face of concrete block.

scotch: In masonry, a tool resembling a small pick with a flat cutting edge, used for trimming brick to a particular shape. Same as *scutch*.

scotia: A concave molding as at the base of a pillar or column; so called because of the dark shadow it casts. From the Greek word *skotia* meaning *darkness*.

scratch awl: A tool used by shopworkers for marking on metal or wood. It is made from a sharp-pointed piece of steel.

scratch coat: The first coat of plastering applied to a wall.

SCR brick: A *solid masonry unit* whose greater thickness permits the use of a single wythe in construction. Its nominal dimensions, which vary from the specified dimensions by the addition of the thickness of the mortar joint with which the unit is designed to be laid (but not more than $\frac{1}{2}$ inch), are 6″ x 2⅔″ x 12″. See Fig. 77.

screed: In concrete flatwork, the wood or metal straightedge used to *strike off* (level) freshly placed concrete. Also, the forms for concrete slabs.

screeding: *Striking off* (leveling) freshly placed concrete to the tops of the forms.

screw: A mechanical device used to unite wood or metal parts in structural work. It consists of a helix wound around a cylinder. Screws are of several varieties. See Fig. 78.

screw anchor: A metal shell, much like that used with an *expansion bolt,* which expands and wedges itself into the hole drilled for it. Also, like the expansion bolt, it is used to fasten light work to masonry construction. See *Rawplug anchor,* Fig. 7.

screw chuck: A contrivance for holding work in a wood-turning lathe, with a projecting screw as live center.

screw clamp: A woodworker's clamp consisting of two parallel jaws and two screws; the clamping action is obtained by means of the screws, one operating through each jaw.

screw driver: A woodworker's tool used for driving in or removing screws by turning them. The tool is made of a well-tempered

Glossary

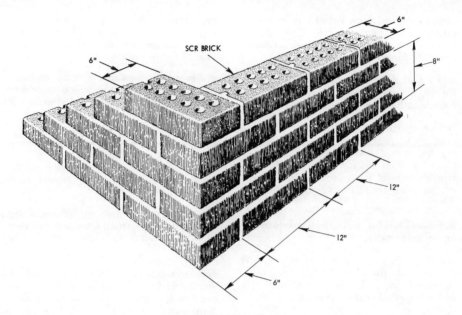

Fig. 77. SCR brick.

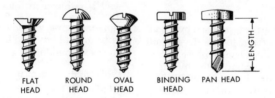

Fig. 78. Typical screws.

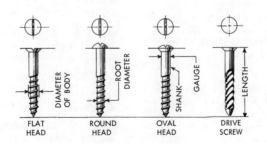

377

Masonry Simplified

steel bar or rod flattened at one end to fit into the slots of screw heads. The steel bar is then fitted into a handle made of tough wood reinforced to prevent splitting.

screw eye: A screw with the head shaped into a completely closed ring or circle, forming a loop or eye.

scriber: A carpentry tool, consisting of a compass of pressed steel with a pencil in one leg or end and a metal point in the other leg; used to draw a line to mark the irregularities of a surface in fitting cabinets or other trim members to the wall or floor.

scribing: Marking and fitting woodwork to an irregular surface.

scutch: In masonry, a bricklayer's cutting tool used for dressing and trimming brick to special shapes. It resembles a small pick and is sometimes called a *scotch*.

scutcheon or **escutcheon:** In carpentry, a term applied to a metal shield used to protect wood, as around the keyhole of a door; also, a metal plate of a decorative character.

seasoning of lumber: A term used by woodworkers when referring to the drying out of green lumber. The drying process may be accomplished either naturally, by allowing the lumber to dry in the air while sheltered from the weather under a shed, or the wood may be dried artificially in an oven, or kiln, specially prepared for that purpose.

sectional view: A drawing that shows the internal detail of a building, but supposes the building to be cut through in sections to exhibit certain features, such as wall thicknesses, sizes and designs of interior doors, fittings, and thickness of floors or other parts.

segment: Any portion of a whole which is divided into parts, as when an apple is cut into quarters each quarter is a *segment*. In geometry, a term specially applied to the portion of a circular plane bounded by a chord and an arc; for example, the diameter of a circular plane divides the plane into two equal segments.

segmental arch: A type of masonry construction where the curve of an arch, though an arc, or segment, or a circle, is always less than a semicircle.

self-faced: In masonry, a term applied to stone, such as flagstone, which splits along natural cleavage planes and does not require dressing.

semichord: One-half the length of any chord of an arc.

semicircle: A segment of a circle which is bound by the diameter and one-half of the circumference.

semicircular arch: In architecture, a type of masonry construction where the curve of an arch, that is the intrados, forms a half circle.

set: In woodworking, a term applied to a small tool used for setting nail heads below the surface of a piece of work; also, a term used in connection with the adjusting of some part of a tool, as to *set* saw teeth or to *set* a plane bit. See *nail set*.

set-off: A horizontal line shown where a wall is reduced in thickness.

settee: A long wooden seat or bench with arms and a high solid back, often having an enclosed foundation serving as a chest.

settlement: A term used in the building industry for an unequal sinking or lowering of any part of a structure which may be caused by the use of unseasoned lumber, by skimping in material, by the weakness of the foundation, or settlement of earth.

serration: A formation resembling the toothed edge of a saw.

shake: A defect in timber such as a fissure or split, causing a separation of the wood between the annual growth rings. See *windshake*, Fig. 94.

shakes: In the building industry, a term applied to a type of handmade shingles.

sheathing: In construction work, a term usually applied to the materials nailed to studding or roofing rafters, as a foundation for the covering of the outer surface of the side walls or roof of a building. See Fig. 65.

Glossary

shed: A one-story structure for shelter or storage, often open on one side. It may be attached to another building but frequently stands apart from other structures.

sheeting: In construction work, a term synonymous with *sheathing*.

shell construction: Construction in which the structure and enclosure are one, instead of consisting of a framework to carry a load and a covering to keep out the weather. It may be built of steel, masonry, wood, or other material, but reinforced concrete is most often used. A structure of this type derives its strength from the calculated distribution of tension, compression, and shear stresses over the entire thickness of the shell through shapes and balance. This method of load calculation cuts down or completely eliminates extra supports and bending of elements, as found in other construction, permitting the shell itself to be much thinner. Shells have a great capacity for carrying unbalanced loads, great reserve strength when damaged, and provide great economy in use of materials. They are particularly well adapted for roofs, which may be of different shapes. Some of the shapes used are: long barrels; short barrels; domes; umbrella in tension; and umbrella in compression. Related types of construction are geodesic, box frame, and *post and beam*.

shingles: Thin pieces of wood, or other material, oblong in shape and thinner at one end, used for covering roofs or walls. The standard thicknesses of wood shingles are described as 4/2, 5/2¼, and 5/2, meaning, respectively, 4 shingles to 2 inches of butt thickness, 5 shingles to 2¼ inches of butt thickness, and 5 shingles to 2 inches of butt thickness. Lengths may be 16, 18, or 24 inches. Wood shingles may be bought in random widths or dimensioned.

shiplap: In carpentry, a term applied to lumber that is edge dressed to make a close rabbeted or lapped joint.

shopwork: Any type of work performed mechanically in a shop.

shore: A piece of lumber placed in an oblique direction to support a building temporarily; also, to support as with a prop of stout timber.

shoring: The use of timbers to prevent the sliding of earth adjoining an excavation; also, the timbers used as bracing against a wall for temporary support.

short length: A term used by woodworkers when referring to lumber which measures less than 8 feet in length.

shotcreting: Pneumatic placing of concrete. The concrete is forced at high velocity through a nozzle onto a prepared surface; widely used for large structures with curved surfaces, such as swimming pools, reservoirs, architectural roofs, etc., where ordinary forming techniques would be difficult and expensive. Also used extensively in repairing damaged or deteriorated structures.

show rafter: An architectural term applied to a short rafter which may be seen below the cornice; often an ornamental rafter.

side cut: Both *hip* and *valley rafters* must have an angle cut, called a *side cut,* to fit against the *ridge* or *common rafter* at the top.

siding: In the building industry, a term applied to boards used for forming the outside walls of frame buildings. See *bevel,* also *drop siding.*

sieve: In masonry, a screen or open container with a mesh bottom; used for removing stones and large particles from sand.

silica brick: In building, a refractory material made from quartzite bonded with milk of lime; used where resistance to high temperature is desired.

sill: The lowest member beneath an opening, such as a window or door; also, the horizontal timbers which form the lowest members of a frame supporting the superstructure of a house, bridge, or other structure. See Figs. 6 and 65.

sill anchor: In building construction, a bolt embedded in the concrete or masonry foundation for the purpose of anchoring the sill to the foundation; sometimes called a *plate anchor.* Same as *anchor bolt.* See Fig. 6.

sill high: The distance from the ground level to the window sill. In masonry, the height from floor to sill.

single-pole switch: An electric device for making or breaking one side of an electric current.

site: The local position of a house or town in relation to its environment.

skewback: A sloping surface against which the end of an arch rests; that is, the course of masonry against which the end of the arch abuts.

skewback saw: A curved-back handsaw made to lessen its weight without sacrificing stiffness.

skew chisel: A woodworking tool with a straight cutting edge, sharpened at an angle, used in wood turning.

skew nailing: A carpentry term referring to the driving of nails on a slant, or obliquely. See *toenailing*.

skintled brickwork: In masonry, an irregular arrangement of bricks with respect to the normal face of a wall. The bricks are set in and out so as to produce an uneven effect on the surface of the wall; also, a rough effect caused by mortar being squeezed out of the joints.

skirting: The same as baseboard; that is, a finishing board which covers the plastering where it meets the floor of a room.

skylight: An opening in a roof or ceiling for admitting daylight; also, the window fitted into such an opening.

slab: A relatively thin slice of any material, such as stone, marble, or concrete; also, a term applied to the outside piece cut from a log.

slag cement: An artificial cement made by first chilling slag from blast furnaces in water, then mixing and grinding the granulated slag with lime, a process which produces a cement with hydraulic properties.

slag concrete: A concrete in which blast-furnace slag is used as an aggregate. Relatively light in weight, *slag concrete* can be used in almost every type of construction,

and is also valued because of its fire resistant properties, as well as for its insulating qualities against cold and sound.

slag sand: Any fine slag product, carefully graded, used as fine aggregate in mortar or concrete.

slag wool: A material made by blowing steam through fluid slag. The final product is similar to asbestos in appearance and is used for insulating purposes.

slaked lime: A crumbly mass of lime formed when quicklime is treated with water. Same as *hydrated lime*.

slaking: The process of hydrating quicklime into a putty by combining it with water.

slamming stile: A term used in carpentry when referring to the upright strip, at the side of a door opening, against which the door slams, or against which it abuts when closed; also, the strip into which the bolt on the door slips when the lock is turned.

sledge: A heavy hammer having a long handle, usually wielded with both hands and used for driving posts or other large stakes.

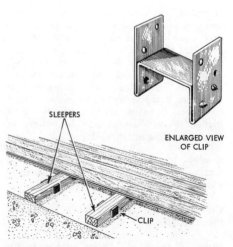

Fig. 79. Sleeper Clips.

Glossary

sleeper: A heavy beam or piece of timber laid on, or near, the ground for receiving floor joists and to support the superstructure; also, strips of wood, usually 2x2, laid over a rough concrete floor, to which the finished wood floor is nailed. See Fig. 79.

sleeper clips: Sheet-metal strips used to anchor wood flooring to concrete. See Fig. 79.

sleeve: In building, a tubular structure to provide openings for the installation of electric and plumbing services. Used particularly in solid concrete floors through which the services must pass. Also called *sleeve chase.* See Fig. 80.

sliding doors: Doors hung from an overhead track on which the panels are wheeled into a special recess at the sides to clear the opening. A related construction operates on a scissors type suspension attached to the narrow strip of wall at the inner end of the recess, and to that edge of the door which first enters the recess when it is pushed.

slip joint: In masonry, especially in brickwork, a type of joint made where a new wall is joined to an old wall by cutting a channel or groove in the old wall to receive the brick of the new wall. This method of joining the two walls forms a kind of telescopic, non-leaking joint.

slip sill: In the building trade, a term applied to a simple sill consisting of a stone slab just as long as the window is wide, and fitting into the walls between the window jambs. The *slip sill* differs from the *lug sill,* which is longer than the width of the window opening and is *let in* to the wall on each side. See *lug sill.*

slip stone: An oilstone used in wood turning and wood patternmaking for sharpening gouges. The small wedge-shaped stone has rounded edges and can be conveniently held in one hand while whetting a tool.

sloyd knife: A type of woodworker's knife used in the sloyd system of manual training. A special feature of this system, which originated in Sweden, is the use of wood carving as a means of acquiring skill in the use of woodworking tools. The Swedish sloyd system was a forerunner of the American manual-training system.

slump: In concrete work, the relative consistency or stiffness of the fresh concrete mix.

smoke shelf: In fireplaces, a projection at the bottom of the smoke chamber used to prevent down-rushing air currents from forcing smoke into the room. See Fig. 81.

smoothing plane: A small woodworking plane, usually not more than 9 inches in length, with a varying iron width measur-

Fig. 80. Sleeves.

Masonry Simplified

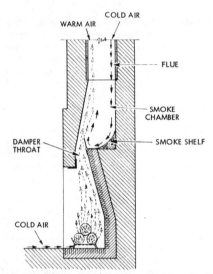

Fig. 81. The smoke shelf directs the cold air toward the warm rising air.

ing from 1¼ inches to 2¼ inches. It is used principally for smoothing and finishing surfaces.

sneck: A small stone used to fill in between larger stones, as in rubblework masonry; also, a term sometimes applied to the laying of a rubble-work wall, or snecked wall.

snecked masonry: A term applied to rubble walls in which the stones are roughly squared but of irregular sizes, and not arranged in courses.

socket chisel: A woodworker's tool of great strength and with sharp cutting edges on each side. Usually the upper end of the shank terminates in a socket into which the handle is driven. In the best quality tools of this type, the blade and socket are forged in one piece with no welded socket.

socle: An architectural term applied to a projecting member at the base of a supporting wall or pier, or at the bottom of a pedestal or column.

soffit: The underside of any subordinate member of a building, such as the under surface of an arch, cornice, or stairway.

soffit vent: An opening in the underside of a roof overhang which acts as a passageway into the house for air currents.

soil stack: In a plumbing system, the main vertical pipe which receives waste material from all fixtures.

soldier course: In masonry, a term applied to a course of bricks where they are laid so that they are all standing on end. See Fig. 33.

sole: In carpentry, a term applied to a horizontal foot piece on the bottom of a wall to which the studs are nailed.

solid masonry unit: A masonry unit whose cross-sectional area in every plane parallel to the bearing surface is 75 percent or more of its gross cross-sectional area measured in the same plane.

soot pocket: In chimney construction, an extension of a flue opening to a depth of 8 or 10 inches below the smoke-pipe entrance. The pocket thus formed prevents soot from collecting in the smoke pipe.

sound knot: A term used in woodworking when referring to any knot so firmly fixed in a piece of lumber that it will continue to hold its position even when the piece is worked; also, is solid across its face and hard as the wood encircling it.

soundproofing: The application of deadening material to walls, ceiling, and floors to prevent sound from passing through these structural members into other rooms.

spading: The compacting with a spade or shovel of newly placed concrete against the forms to prevent the formation of voids on the surface.

spall: A fragment or chip of stone or brick, especially bad or broken brick; in masonry, to reduce an irregular stone block to approximately the desired size by chipping with a stone hammer. Also spelled *spawl*. See *gallet*.

span: The distance between the abutments of an arch or the space between two adjoining arches (Fig. 41); also, the distance between the wall, or rafter, plates of a building.

Glossary

spandrel: The space between the exterior curve of an arch and the enclosing right angle; or the triangular space between either half of the extrados of an arch and the rectangular molding enclosing the arch.

specifications: Written instructions to the builder containing all the information pertaining to materials, style, workmanship, fabrication, dimensions, colors, and finishes supplementary to that appearing on the working drawings.

spike: In the building trade, a term commonly applied to a large-sized nail, usually made of iron or steel, used as a fastener for heavy lumber.

spike knot: In woodworking, a knot sawed lengthwise.

spire: In architecture, a tapering tower or roof; any elongated structural mass shaped like a cone or pyramid; also, the topmost feature of a steeple.

spirit level: An instrument used for testing horizontal or vertical accuracy of the position of any structural part of a building. The correct position is indicated by the movement of an air bubble in alcohol. Same as *plumb rule*. See *level*, Fig. 56.

splash block: A small masonry block laid with the top close to the ground surface to receive drainage of rain water from the roof and carry it away from the building.

splay: An inclined surface, as the slope or bevel at the sides of a door or window; also, to make a beveled surface, or to spread out, or make oblique; an angle greater than 90°.

splayed brick: A purpose-made brick having one side beveled off.

splay end: The end of a brick which is opposite the end laid squarely by rule.

split-level: A house in which two or more floors are usually located directly above one another, and one or more additional floors, adjacent to them, are placed at a different level. See Fig. 82.

spokeshave: A cutting tool, or plane, with a transverse blade set between two handles. This device is especially suitable for dressing rounded pieces of wood of small diameter, such as spokes or other similarly curved work.

spread footing: A footing whose sides slope gradually outward from the foundation to the base.

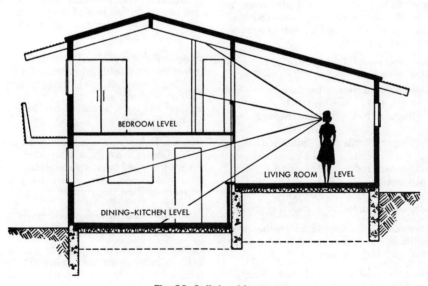

Fig. 82. Split-level house.

Masonry Simplified

sprig: In woodworking, a term applied to a small brad with no head; also, one of the small, triangular-shaped pieces of tin or zinc used for holding a pane of glass in a window sash.

springer: In architecture, a stone or other solid piece of masonry forming the impost of an arch; that is, the topmost member of a pillar or pier upon which the weight of the arch rests. See Fig. 54.

spring hinge: A joint with a spring built into it, used for self-closing doors, such as screen doors.

sprung molding: In carpentry, a term applied to a curved molding.

spur: A sharp-pointed carpenter's tool used for cutting veneer.

spur center: A term used by woodworkers when referring to the center used in the headstock of a woodturning lathe.

square: The multiplying of a number by itself; also, a plane figure of four equal sides, with opposite sides parallel, and all angles right angles. Shingles for the trade are put up in bundles so packed that 4 bundles of 16- or 18-inch shingles, when laid 5 inches to the weather, will cover a *square* 10 by 10, or 100 square feet, and three bundles of 24-inch shingles will also cover a *square*. An instrument for measuring and laying out work is called a *framing square*. See also *steel square*.

square measure: The measure of areas in square units.

144 square inches (sq. in.)	= 1 square foot (sq. ft.)
9 square feet	= 1 square yard (sq. yd.)
$30\frac{1}{4}$ square yards	= 1 square rod (sq. rd.)
160 square rods	= 1 acre (A.)
640 acres	= 1 square mile (sq. mi.)
36 square miles	= 1 township (twp.)

square root: A quantity of which the given quantity is the square, as 4 is the *square root* of 16, the given quantity.

squinch: A small arch built across an interior corner of a room for carrying the weight of a superimposed mass, such as the spire of a tower.

squint brick: In masonry, a brick which has been shaped or molded to a specially desired form; a purpose-made brick.

stabbing: In masonry a term used when referring to the process of making a brick surface rough in order to provide a key for plasterwork.

stack: In architecture, a large chimney usually of brick, stone, or sheet metal for carrying off smoke or fumes from a furnace, often a group of flues or chimneys embodied in one structure rising above a roof.

stack partition: A partition wall which carries the stack or soil pipe; sometimes constructed with 2″ x 6″ or 2″ x 8″ studs, and continuous from first floor to attic lines.

staff bead: In the building trades, a term applied to a strip of molding inserted between the masonry of a wall and a window frame, for protection against the weather.

staggered partition: In building, a type of construction used to soundproof walls. Such a partition is made by using two rows of studding, one row supporting the lath and plaster on one side of the wall and the other row supporting the lath and plaster on the other side of the wall.

staggered screeds: Wood screeds (2″ x 4″), imbedded in mastic, are staggered below flooring, instead of subflooring, to provide nailing surface for floor boards. This method has been approved by the F.H.A. for oak flooring.

staging: In building construction, the same as scaffolding, that is, a temporary structure of posts and boards on which the workmen stand when their work is too high to be reached from the ground.

stair: One step in a flight of stairs. Also called a stairstep. See Fig. 83.

staircase: A flight of steps leading from one floor or story to another above. The term includes landings, newel posts, handrails, and balustrades. See Fig. 83.

Glossary

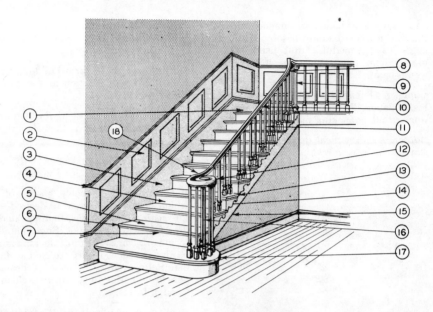

1. Landing
2. Raised-panel dado
3. Closed stringer
4. Riser
5. Tread
6. Tread housing
7. Cove molding under nosing
8. Goose neck
9. Landing newel post
10. Handrail
11. Baluster
12. Volute
13. End nosing
14. Bracket
15. Open stringer
16. Starting newel post
17. Bull-nose starting step
18. Concave easement

Fig 83. Staircase.

stairs: In building, a term applied to a complete flight of steps between two floors. *Straight run stairs* lead directly from one floor to another without a turn; *close string stairs* have a wall on each side; *open string stairs* have one side opening on a hallway or room; *doglegged* or *platform stairs* have a landing near the top or bottom, introduced to change direction.

stair treads: The upper horizontal boards of a flight of steps. See Fig. 83.

stair well: A compartment extending vertically through a building, and in which stairs are placed.

staking out: A term used for the laying out of a building plan by driving stakes into the ground showing the location of the foundation. To insure a clean edge when excavating, the stakes are connected with strong cord indicating the building lines. See *batter boards*, Fig. 17.

standard brick: In masonry, brick, size $2\frac{1}{4}''$ x $3\frac{3}{4}''$ x $8''$. Permissible variables are, plus or minus, $\frac{1}{16}''$ in depth, $\frac{1}{8}''$ in width, and $\frac{1}{4}''$ in length.

standard modular brick: Brick designed in accord with the standard $4''$ module. (See *modular dimension standards*) Modular

Masonry Simplified

dimensions are the *actual* dimensions plus the thickness of the mortar joint. For a 4" x 2⅔" x 8" modular unit using a ⅜" mortar joint, the actual size of the brick would be 3⅝" x 2¼" x 7⅝".

staple: A U-shaped metal fastener used to fasten paper, tiles, insulation, etc., to framework or other backing material.

stapler: A mechanical device for applying staples. Also see *air stapler*. See Fig. 84.

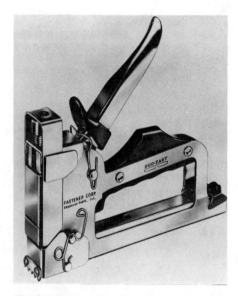

Fig. 84. Stapler. (Duo-Fast Fastener Corp.)

star drill: A tool with a star-shaped point used for drilling in stone or masonry.

Star Dryvin anchor: A type of *expansion anchor* used for securing wood structural parts to a masonry or concrete wall. See *anchors*, Fig. 7.

star expansion bolt: A bolt or screw having a shield of two semicircular parts which spread apart as the bolt is driven into the shield. Used for securing structural wood parts to a masonry wall.

starling: A protection around a bridge or pier made by driving piles close together to form an enclosure.

starting board: In form building, the first board nailed in position at the bottom of a foundation form.

starting newel: A post at the bottom of a staircase for supporting the balustrade. See Fig. 83.

starting step: The first step at the bottom of a flight of stairs. See Fig. 83.

steel forms: Removable pieces of steel which hold wet concrete in desired shapes for casting foundations, footings, and window frames on the spot. Some foundation formwork comes with interlocking modular hardware. Steel forms are said to last indefinitely, produce a clean, accurate face, and to be easier to set up and clean than wooden forms.

steel square: An instrument having at least one right angle and two or more straight edges, used for testing and laying out work for trueness. A term frequently applied to the large framing square used by carpenters.

steel wool: A mass of fine steel threads matted together and used principally for polishing and cleaning surfaces of wood or metal.

stepped footings: If a house is built on sloping ground, all the footings cannot be at the same depth; hence, they are stepped.

stile: In carpentry, one of the vertical members in a door or sash, into which secondary members are fitted.

stilt house: A house which is constructed on stilts above the ground; used mostly in hot, moist regions and on very uneven ground level sites; provides breeze passage underneath, protection from insects, and space for car.

stirrup or **hanger:** In building trades, a term applied to any stirrup-like drop support attached to a wall to carry the end of a beam or timber, such as the end of a joist. *Stirrups* or *hangers* may also be suspended from a girder as well as from a wall. See *hanger*.

Glossary

stonecutter's chisel: A stonemason's tool used for dressing soft stone. Also called *tooth chisel*.

stonemason: In building, one who builds foundations and walls of stone.

stool: In architecture, a term applied to the base or support of wood at the bottom of a window, as the shelflike piece inside and extending across the lower part of a window opening.

stoop: A raised entrance platform, with steps leading up to it, at the door of a building; sometimes the term is applied to a porch or veranda.

stop: In building, any device which will limit motion beyond a certain point, as a doorstop in a building, usually attached near the bottom of a door and operated by pressure from the foot. See *doorstop*.

storm door: An extra outside, or additional, door for protection against inclement winter weather. Such a door also serves the purpose of lessening the chill of the interior of a building, making it easier to heat, and also helps to avoid the effects of wind and rain at the entrance doorway during milder seasons.

storm sash: An additional sash placed at the outside of a window for protection against the severe weather of winter.

story rod: A rod or pole cut to the proposed clear height between finished floor and ceiling. The *story rod* is often marked with minor dimensions, as for door trims.

stove bolt: A special type of bolt with a nut. Formerly such bolts were provided with a coarser thread pitch than a machine bolt; however, the only difference now is, that without a nut a stove bolt is called a *machine screw*.

straightedge: A bar of wood or metal with the edges true and parallel, used for testing straight lines and surfaces; that is, gaging the accuracy of work. See Fig. 85.

stretcher: In masonry construction, a term applied to a course in which brick or stone lies lengthwise; that is, a brick or stone is laid with its length parallel to the face of the wall.

stretcher bond: In masonry, a bond which consists entirely of stretchers, with each vertical joint lying between the centers of the stretchers above and below, so that angle closers are not required. This type of

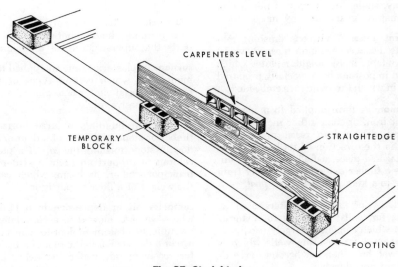

Fig. 85. Straightedge.

bond is used extensively for internal partition walls which have a thickness of a single tier of brick. See Fig. 22.

striking off: Same as *screeding*.

string: In building trades, a term applied to the inclined member which supports the treads and risers of a stair. Also called a *stringer*. See Fig. 83.

stringcourse or **sailing course:** In building, a horizontal band forming a part of the design, consisting of a course of brick or stone projecting from a wall for decorative purposes or to break the plain effect of a large expanse of wall surface.

stringer: A long, heavy, horizontal timber which connects upright posts in a structure and supports a floor; also, the inclined member which supports the treads and risers of a stair. See Fig. 83.

strip: In the building trades, a term applied to a narrow piece of wood, relatively long, and usually of a uniform width; also, used when referring to the breaking, tearing, or stripping off the threads of a bolt or nut.

stripping: In concrete work, the removal of forms from hardened concrete.

structural clay tile: A term applied to various sizes and kinds of hollow and practically solid building units; molded from surface clay, shale, fire clay, or a mixture of these materials, and laid by masons.

structural glass: A vitreous finishing material used as a covering for masonry walls. It is available in rectangular plates which are held in position by a specially prepared mastic in which the plates are embedded.

strut tenon: A term applied to a piece of wood or iron, or some other member of a structure, designed to resist pressure or weight in the direction of its length; used on a diagonal piece, usually on heavy timbers, as a timber extending obliquely from a rafter to a king post. See *king post*.

stucco: Any of various plasters used as covering for walls; a coating for exterior walls in which cement is largely used; any material used for covering walls which is put on wet, but when dry becomes exceedingly hard and durable.

stud: In building, an upright member, usually a piece of dimension lumber, 2" x 4" or 2" x 6", used in the framework of a wall. On an inside wall the lath are nailed to the studs. On the outside of a frame wall, the sheathing boards are nailed to the studs. The height of a ceiling is determined by the length or height of the studs. See Figs. 15 and 65.

stud walls: A stud wall consists of verticals usually spaced 16" apart between top and bottom plate. Stud walls include, or can include, window shapes and headers, and can be preassembled and moved into position.

subbase: In architecture, the lowest part of a structural base, which consists of two or more horizontal members, as the base of a column; also, a *baseboard*.

subfloor: In carpentry, a term applied to a flooring of rough boards laid directly on the joists and serving the purpose of a floor during the process of construction on the building. When all rough construction work is completed, the finish floor is laid over the subfloor. See Figs. 15 and 65.

subrail: In building a closed string stair, a molded member called a *subrail* or *shoe* is placed on the top edge of the stair string to receive and carry the lower end of the balusters.

substructure: The lower portion of a structure forming the foundation which supports the superstructure of a building.

suction: In plastering, a term applied to the manner in which certain types of plastering material *pull* when worked with a trowel; adhesive quality.

summer: In building, a large horizontal timber or stone; the lintel of a door or window; a stone forming the cap of a pier or column to support an arch; a girder; the principal timber, or beam, which carries the weight of a floor or partition.

sump: A pit or depression in a building where water is allowed to accumulate; for example, in a basement floor to collect seepage or a depression in the roof of a building for receiving rain water and delivering it to the downspout. A device used for remov-

Glossary

ing water from such a depression is known as a *sump pump*.

sunk panels: A term used in the building trade when referring to panels recessed below the surrounding surface.

superstructure: The framing above the main supporting level.

supplement of an angle: An angle which is equal to the difference between the given angle and 180 degrees. If the given angle is 165 degrees its supplement is 15 degrees.

surbase: In architecture, a molding above a base, as that immediately above the baseboard of a room; also, a molding, or series of moldings, which crown the base of a pedestal.

S4S: An abbreviation for the term *surfaced on four sides*.

surfacing of lumber: In woodworking, symbols are used to indicate how lumber has been surfaced, as *S1E*, surfaced on one edge; *S1S*, surfaced on one side; *S2S*, surfaced on two sides, and so on.

surveying: That branch of applied mathematics dealing with the science of measuring land, the unit of measure being the surveyor's chain, with 80 chains equal to 1 mile.

swing saw: A woodworker's tool, consisting of a circular saw mounted on a hinged frame suspended from above. When needed, the saw is pulled to the work which remains stationary. Also called *radial saw*.

symmetrical: Pertaining to any plane or solid body or figure which is well-proportioned, with corresponding parts properly balanced and harmonious in all details; anything which exhibits symmetry in size, form, and arrangement of its parts.

T

table: In carpentry, the insertion of one timber into another by alternate projections or scores from the middle; same as a coak. In architecture, a flat surface of a wall, usually raised, as a *stringcourse,* especially a projecting band of stone or brick where an offset is required. See *water table,* Fig. 91.

tabling: In masonry, the forming of a type of horizontal joint by arranging various stones in a course so they will run into the next course, hence preventing them from slipping; in carpentry, the shaping of a projection on a piece of timber, so it will fit into a recess prepared to receive it in another timber.

tackle: A construction of blocks and ropes, chains, or cables used for hoisting purposes in heavy construction work. Often spoken of as *block and tackle*.

tail beam or **tail joist:** Any timber or joist which fits against the header joist.

tailing: In building construction, any projecting part of a stone or brick inserted in a wall.

tail joist: Any building joist with one end fitted against a header joist.

take-up: In shopworking, any equipment or device provided to tighten or take up slack, or to remove looseness in parts due to wear or other causes.

tamp: To pound down, with repeated light strokes, the loose soil thrown in as filling around a wall.

tape: Any flexible narrow strip of linen, cotton, or steel marked off with measuring lines similar to the scale on a carpenter's rule. Usually the tape is contained in a circular case into which it can easily be rewound after using.

tarpaulin: Any material, usually either plastic or canvas, waterproofed with tar or paint and used for covering purposes, such as hatches of ships or boats, or anything exposed to the weather.

taut: Anything tightly drawn until it is tense and tight, with all sag eliminated, as a rope, wire, or cord pulled *taut*.

T bevel: A woodworker's tool used for testing the accuracy of work cut at an angle such as a beveled edge. See Fig. 86.

temperature stress rods: Steel rods placed horizontally in concrete slabs for preven-

389

Masonry Simplified

Fig. 86. T bevel.

tion of cracks, due to temperature changes, drying, etc., parallel to the reinforcing rods. The rods are the same physically as reinforcing rods, and are usually laid at right angles to, and almost in, the same plane as reinforcing rods.

template: A gage, commonly a thin board or light frame, used as a guide for forming work to be done.

tenon: In carpentry, a piece of lumber or timber cut with a projection, or tongue, on the end for fitting into a mortise. The joint formed by inserting a tenon into a mortise constitutes a so-called *mortise-and-tenon joint*.

tenon saw: In woodworking, any small backsaw used on the bench for cutting tenons.

tensile strength: The degree to which a structure can withstand *tension*.

tension: A stretching or pulling stress. Opposite of *compression*.

terminal: In carpentry, the extremity of any structural part, especially the finish of a newel post, or standard; also, a carving used for decorative purposes at the end of some structural member, such as a pedestal.

termite shield: A protective shield made of noncorrosive metal, placed in or on a foundation wall or other mass of masonry, or around pipes entering a building, to prevent passage of termites into the structure. See Fig. 87.

terrace: An elevated level surface of earth supported on one or more faces by a masonry wall, or by a sloping bank covered with turf.

terra cotta: A clay product used for ornamental work on the exterior of buildings; also, used extensively in making vases, and for decorations on statuettes. It is made of hard-baked clay in variable colors with a fine glazed surface.

terrazzo: A type of Venetian marble mosaic in which portland cement is used as a matrix. Though used in buildings for centuries, *terrazzo* is a modern floor finish, used also for bases, borders, and wainscoting, as well as on stair treads, partitions, and other wall surfaces.

terrazzo flooring: A term used in the building trades for a type of flooring made of small fragments of colored stone, or marble, embedded irregularly in cement. Finally, the surface is given a high polish.

tessellated: Formed of cubes of stone, marble, glass, or other suitable material (*tessera*), arranged in a checkered pattern, as in mosaic floors and pavements.

tessera: Any one of the small square pieces of marble, stone, tile, or glass used in mosaic work, such as in floors or pavements.

textured finish: In concrete work, a coarser final finish than that achieved by trowelling. It may be for decorative purposes or for safety, as for a skidproof floor or sidewalk.

thermal unit: Any unit chosen for the calculation of quantities of heat; that is, a unit of measurement used as a standard of comparison of other quantities of heat, such as Btu (British thermal unit).

thermostat: An automatic device for regulating the temperature of a room by opening or closing the damper of a heating furnace.

T hinge: A type of joint with an abutting piece set at right angles to a strap, thus forming a **T**-shaped hinge, used mainly on outside work, such as barn doors and gates.

three-ply: Anything composed of three distinct layers or thicknesses, as plywood used in building construction or in furniture making, in which the material used consists of three separate plies or layers.

Glossary

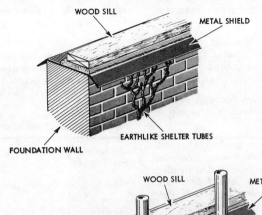
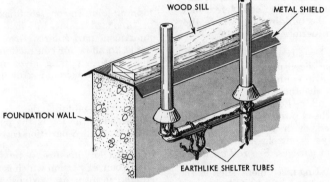

Fig. 87. Termite shields are placed over the concrete and masonry walls and also on pipes.

three-way switch: A switch used in house wiring when a light (or lights) is to be turned on or off from two places. A three-way switch must be used at each place.

threshold: In building construction, a term applied to the piece of timber, plank, or stone under a door.

through shake: A separation of wood between annual growth rings, extending between two faces of timber; similar to a *windshake*. See Fig. 94.

through stone: In stone masonry, a term applied to a stone which extends through a wall forming a bond. Also called *bondstone*.

thumb plane: In woodworking, a term sometimes applied to a small plane not more than 4 or 5 inches in length with a bit about 1 inch in width.

thumbscrew: A screw with its head so constructed that it can be turned easily with the thumb and finger.

tie: In architecture, anything used to hold two parts together, as a post, rod, or beam. See *anchors, metal ties*.

tie beam: A timber used for tying structural parts together, as in the roof of a building. Any beam which ties together or prevents the spreading apart of the lower ends of the rafters of a roof.

tile: A building material made of fired clay, stone, cement, or glass used for floors, roofing, and drains; also made in varied ornamental designs for decorative work.

tile hanging: A term applied to the hanging of tile on a vertical surface, such as a wall, to protect the wall against dampness. See *weather tiling.*

tile setting adhesives: Specially formulated glues or mastics, used instead of mortarbed, for tile setting. They are said to be clean, waterproof, less expensive, and faster.

tile shell: A construction tile unit consisting of a number of hollow cells separated by webs.

tilt-up construction: A method of constructing walls, and sometimes floors, by pouring concrete, or putting wooden walls together in flat panels, and, when completed, moving them to the building site where they are tilted into permanent place.

tilt-up doors: Usually consisting of a rigid panel of sheet steel, aluminum, or wood, these doors are equipped with springs, tracks, counterbalances, and other hardware, which pull the door clear of the opening to an overhead position. They are often motor-operated, with manual, radio, or magnetic driver controls, and are commonly used in garages. See *roll-up doors.*

tin snips: A cutting instrument, such as the ordinary hand shears, used by sheet-metal workers.

toeing: In carpentry, the driving of nails or brads obliquely; also, to clinch nails so driven. See Fig. 52.

toenailing: The driving of a nail, spike, or brad slantingly to the end of a piece of lumber to attach it to another piece, especially, as in laying a floor, to avoid having the heads of the nails show above the surface. See Fig. 52.

tongue: A projecting rib cut along the edge of a piece of timber so it can be fitted into a groove cut in another piece. Also, the shorter of the two extending arms of the *framing square,* usually 16 inches long and 1½ inches wide. The *blade* of the square forms a right angle with the *tongue.* The *octagon scale* and the *brace measure scale* appear on the faces of the tongue.

tooled joints: In masonry, mortar joints which are specially prepared by compressing and spreading the mortar after it has set slightly. *Tooled joints* present the best weathering properties.

tooth chisel: A chisel especially designed for cutting stone; same as *stonecutter's chisel.*

toothing: In masonry construction, the allowing of alternate courses of brick to project toothlike to provide for a good bond with any adjoining brickwork which may follow.

topping: A mixture of cement, sand, and water used in creating the finished surface of concrete work such as walks and floors. See Fig. 41.

topping joint: In concrete finishing, a space or break of about $\frac{1}{8}''$ made at regular intervals, particularly over expansion joints, to allow for contraction and expansion in the topping layer of sidewalks, driveways, and similar structures. See Fig. 41.

top plate: In building, the horizontal member nailed to the top of the partition studding. Also called partition cap. See Fig. 15.

torus: In architecture, a type of molding with a convex portion which is nearly semicircular in form, used extensively as a base molding.

tower: A structure whose height is proportionally much greater than its width, often surmounting a large building, such as a cathedral or church, usually less tapering than a steeple. A *tower* may stand alone entirely apart from any other building.

T plate: A **T**-shaped metal plate commonly used as a splice; also used for stiffening a joint where the end of one beam abuts against the side of another.

tracery: An architectural term applied to any delicate ornamental work consisting of interlacing lines such as the decorative designs carved on panels or screens; also, the intersecting of ribs and bars, as in rose windows, and the upper part of Gothic windows; any decorative design suggestive of network.

trammel: An instrument used for drawing arcs or radii too great for the capacity of the ordinary compass; a *beam compass* with adjustable points attached to the end of a bar of wood or metal used by draftsmen and shopworkers for describing unusually large circles or arcs.

transit: An instrument, commonly used by surveyors, consisting of four principal parts: (1) a telescope for sighting; (2) a spirit level; (3) a vernier or graduated arc for measuring vertical or horizontal angles; and (4) a tripod with leveling screws for adjusting the instrument.

transite: A fireproofing material used in walls, roofs, and for lining ovens. It is composed of asbestos and portland cement molded under high pressure and is sold under the trade name of *transite*.

transom: A term used in building for any small window over a door or another window.

transom bar: A crossbar of wood or stone which divides an opening horizontally into two parts.

trap door: A covering for an opening in a floor, ceiling, or roof; usually such a door is level, or practically so, with the surface of the opening which it covers.

trass: In plastering, a type of gray, yellow, or whitish earth common in volcanic districts, resembling puzzolano; used to give additional strength to lime mortars and plasters; also used in the making of hydraulic cement.

trass mortar: In masonry, a mortar made of lime, sand, and *trass* or brick dust; or a mortar composed of lime and *trass* without sand. The *trass* makes the mortar more suitable for use in structures exposed to water.

tread: In building, the upper horizontal part of a stair step; that portion of a step on which the foot is placed when mounting the stairs. See Fig. 83.

trefoil: In architecture, an ornamental three-lobed unit resembling in form the foliage of an herb whose leaf is divided into three distinct parts, such as the common varieties of clover.

treillage: A latticework erected for supporting vines, as in an arbor.

trellis: An ornamental structure of latticework over which vines are trained, such as a summerhouse, usually made of narrow strips of wood which cross each other at regular intervals.

tremie: A device for placing concrete under water consisting of a tube with removable sections and a funnel for receiving the concrete. The bottom of the tube is always immersed in the concrete, being gradually raised as the form is filled. See Fig. 88.

trestle: A braced framework, usually consisting of a horizontal beam supported at each end by a pair of spreading legs which serve as braces.

trestle table: A large drawing board supported by trestles.

triangle: A *polygon* enclosed by three straight lines called sides.

triangular scale: A draftsman's three-faced measuring device having six graduated edges. On one edge is shown a scale of full-size measurements, while on the other edges are shown various reductions in scale.

triangular truss: A popular type of truss used for short spans, especially for roof supports.

trim: In carpentry, a term applied to the visible finishing work of the interior of a building, including any ornamental parts of either wood or metal used for covering joints between jambs and plaster around windows and doors. The term may include also the locks, knobs, and hinges on doors.

trimmer: The beam or floor joist into which a header is framed.

trimmer arch: A comparatively flat arch, such as may be used in the construction of a fireplace.

trimmer beam: Usually two joists spiked together around a fireplace opening in floor framing.

trimming joist: A timber, or beam, which supports a header.

trowel: A flat steel tool, of various sizes and shapes, used to spread and smooth plaster, mortar, or concrete.

Masonry Simplified

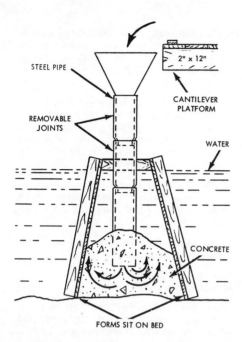

Fig. 88. Placing concrete under water with a tremie.

truss: A combination of members, such as beams, bars, and ties, usually arranged in triangular units to form a rigid framework for supporting loads over open spaces.

trussed beam: An architectural term applied to a beam stiffened by a truss rod.

try square: A tool used for laying out right angles and for testing work for squareness.

T square: A tool used by draftsmen. It consists of a ruler usually from 2 to 3 inches in width and from 1 to 5 feet in length, with a crosspiece attached to one end of the ruler or blade. The crosspiece, or head, is at least twice as thick as the blade.

tubular scaffolds: Scaffolds for interior and exterior construction work, made of tube steel. These scaffolds are lightweight, offer low wind resistance, and are easily dismantled. They are obtainable in several strengths for varying heights and types of work.

tuck pointing: In masonry, the repairing of worn or damaged mortar joints by raking out the damaged part of the mortar and replacing with fresh mortar.

turnbuckle: A type of coupling between the ends of two rods, used primarily for adjusting or regulating the tension in the rods which it connects. It consists of a loop or sleeve with a screw thread on one end and a swivel at the other, or with an internal screw thread at each end.

turning gouge: In woodworking, a tool used for roughing down woodwork in a lathe. The widths of gouges vary from $1/4$ to $1 1/2$ inches.

turrets: Small towers, often merely ornamental features at an angle of a large building. Turrets frequently begin at some distance above the ground, although they may rise from the ground or be built on corbels.

twelfth scale: A scale which divides the inch into 12 parts instead of 16; found usually on the back of the *framing square* along the outside edge. In this scale, one inch equals one foot. The *twelfth scale* makes it possible to reduce layouts to $1/12$ of their regular size and to solve basic right triangle problems.

twist bit: In woodwork, a tool used for boring holes in wood for screws. A tool similar to the twist drill used for drilling holes in metal, except that the cutting edge of the *twist bit* is ground at a greater angle.

twist drill: A drilling tool having helical grooves extending from the point to the smooth portion of the shank. This type of drill is made of round stock with a shank that may be either straight or tapering. It is used for drilling holes in metal. A similar drill used for wood is known as a *twist bit*.

T wrench: A tool for tightening a nut on a bolt. The T wrench consists of a handle, or lever, with a T-shaped socket which fits over and completely encircles a nut or bolthead. It may or may not have a ratchet but it sometimes has an extension to permit working in places not easily accessible.

U

U bolt: An iron bar bent into a U-shaped bolt, with screw threads and nuts on each end. Often called a clip, as a spring clip on an automobile.

umbrella house: A house with a parasol roof, which may be lattice work, covering the main structure and terraces. There is space for the passage of air currents between the covering roof and the house structure.

unit length rafter table: A table which appears on the *blade* of the *framing square*. It gives unit lengths of *common rafters* for seventeen different rises, ranging from 2 to 18 inches. It also gives the unit lengths for *hip* or *valley rafters,* difference in lengths of *jack rafters* set 16 inches on center, jack rafters set 24 inches on center, and the *side cuts* for jack and hip rafters.

unsound knot. A term used by woodworkers when referring to a *knot* which is not as solid as the wood in the board surrounding it.

upright: In building, a term applied to a piece of timber which stands upright or in a vertical position, as the vertical pieces at the sides of a doorway or window frame.

V

valley: In architecture, a term applied to a depressed angle formed by the meeting at the bottom of two inclined sides of a roof, as a gutter; also, the space, when viewed from above, between vault ridges.

valley rafter: A rafter disposed in the internal angle of a roof to form a *valley* or depression in the roof. See Fig. 76.

vane: Any flat piece of metal attached to an axis and placed in an elevated position where it can be readily moved by the force of the wind, such as a weathercock on a barn or steeple, indicating the direction of the wind.

vapor barrier: Material used to retard the passage of vapor or moisture into walls thus preventing condensation within them. There are different types of vapor barriers, such as membrane which comes in rolls and is applied as a unit in the wall or ceiling, and the paint type which is applied with a brush.

vault: In architecture, an arched structure of masonry usually forming a ceiling or roof; also, an arched passageway under ground, or any room or space covered by arches.

vaulting course: A course formed by the springers of a vault which usually are set with horizontal beds, either corbeled out or in projection.

veneer: Thin pieces of wood, or other material, used for finishing purposes to cover an inferior piece of material, thus giving a superior effect and greater strength with reduced cost.

veneer wall: A wall with a masonry facing which is anchored to the superstructure to provide architectural decoration, long life and low maintenance. The veneer is not load-bearing.

ventilating brick: See *air brick*.

ventilation: The free circulation of air in a room or building; a process of changing the air in a room by either natural or artificial means; any provision made for removing contaminated air or gases from a room and replacing the foul air by fresh air.

vent pipe: A flue, or pipe, connecting any interior space in a building with the outer air for purposes of ventilation; any small pipe extending from any of the various plumbing fixtures in a structure to the vent stack.

vent stack: A vertical pipe connected with all vent pipes carrying off foul air or gases from a building. It extends through the roof and provides an outlet for gases and contaminated air, and also aids in maintaining a water seal in the trap.

veranda: An open portico, usually roofed, attached to the exterior of a building. In the United States commonly called a *porch*.

verge: The edge of tiling, slate, or shingles projecting over the gable of a roof, that on the horizontal portion being called the *eaves*.

verge board: The board under the verge of a gable, sometimes molded. During the latter part of the nineteenth century, *verge boards* were often richly adorned with decorative carving, perforations, and cusps, frequently having pendants at the apex. The term *verge board* is often corrupted into *bargeboard*.

vermiculated: Pertaining to stone or other material with designs worked on the surface, giving it the appearance of being worm-eaten.

vertical: Pertaining to anything, such as a structural member, which is upright in position, perpendicular to a horizontal member, and exactly plumb.

vestibule: A small entrance room at the outer door of a building; an anteroom sometimes used as a waiting room.

V & C V: An abbreviation for the term meaning **V** *grooved* and *center* **V** *grooved;* that is, the board is **V** grooved along the edge and also center **V** grooved, on the surface.

viaduct: Any elevated roadway, especially a bridge of narrow arches of masonry, or reinforced concrete, supporting high piers which carry a roadway or railroad tracks over a ravine or gorge. In the United States, viaducts are sometimes of steel construction consisting of short spans carried on high steel towers.

vibrator: In concrete work, a special tool for compacting freshly placed concrete within the forms. The vibrating *head,* which is either pneumatically or mechanically driven, is immersed in the concrete as it is being placed, causing rapid and thorough consolidation.

vista: A view, especially one seen through a long narrow passage as between rows of houses facing on an avenue.

vitrified tile: In building construction, pipes made of clay, baked hard and then glazed, so they are impervious to water; used especially for underground drainage.

volute with easement: The spiral portion of a handrail which sometimes supplants a newel post in stair building. See Fig. 83.

voussoir: In architecture, any one of the wedge-shaped pieces, or stones, used in forming an arch. See Fig. 54. The middle one is called the *key*.

voussoir brick: Building brick made especially for constructing arches. Such brick are so formed that the face joints radiate from a common center.

W

wainscot: The wooden lining of the lower part of an interior wall. Originally, only a superior quality of oak was used for this purpose, but now the term includes other materials.

wainscoting: The materials used in lining the interior of walls; also, the process of applying such materials to walls.

wainscoting cap: A molding at the top of a wainscoting.

wales: In concrete formwork, the horizontal timbers on the outside of the form to which the ties are fastened, and which hold the forms in line. Also called *walers* and *whalers*. See Fig. 89.

wall beam: In masonry, a metal member or type of anchor fastened to a floor joist to tie the wall firmly to the floor. The anchor extends into the masonry wall and, at the end of the anchor, there is some kind of bolt or wall hook, which may be either **L**-shaped or **T**-shaped, for holding the anchor in the wall.

wall bed: In building, any one of the various types of beds which fold or swing into a wall or closet when not in use. A type of bed commonly used where the conserving of space is important, such as in city apartment buildings.

wallboard: An artificially prepared sheet material, or board, used for covering walls and ceilings as a substitute for wooden boards or plaster. There are many different types of *wallboard* on the market.

wall coping: The covering course on top of a brick or stone wall; also referred to as capping. Where porches or other similar spaces are enclosed with solid walls to the

Glossary

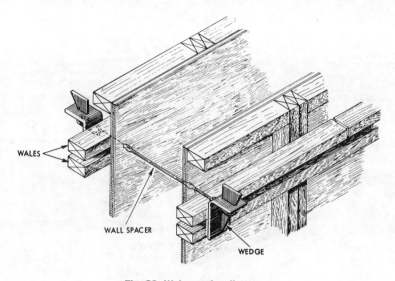

Fig. 89. Wales and wall spacers.

height of the porch railing, the material which is used as a finish is called *coping*.

wall cornice: A kind of coping with a cornicelike finish at the top of a masonry wall; also, a finish for the top of a wall.

wall plates: Horizontal pieces of timber placed on the top of a brick or stone wall under the ends of girders, joists, and other timbers to distribute the weight of the load or pressure of the superstructure, especially the roof.

wall spacers: In concrete work, a type of tie for holding the forms in position while the concrete is being poured, and until it has set. For typical *wall spacers,* see Fig. 89.

wall tie: A device, in any of various shapes, formed of $\frac{1}{4}''$ diameter steel wire, the purpose of which is to bind together the tiers of a masonry wall, particularly those in hollow wall construction. See Fig. 58. Also, a contrivance, usually a metal strip, employed to attach or secure a brick veneer wall to a frame building. See Fig. 60.

wane: Bark, or lack of bark or wood, from any cause, on edge or corner of a piece of lumber.

warped: In woodworking, a term applied to any piece of timber which has been twisted out of shape and permanently distorted during the process of seasoning.

water bar: A strip of material inserted in a joint between wood and stone of a window sill to prevent or bar the passage of water from rain or snow.

water joint: In stone work, a joint protected from rain and snow by sloping the surface of the stone away from the wall, so it will shed water easily.

water lime: a lime which will harden under water.

waterproofing walls: In concrete work, the making of walls impervious to water, or dampness, by mixing a compound with concrete, or by applying a compound to the surface of the wall. A method of waterproofing the foundations of walls is shown in Fig. 90.

water putty: In woodworking, a powder

Masonry Simplified

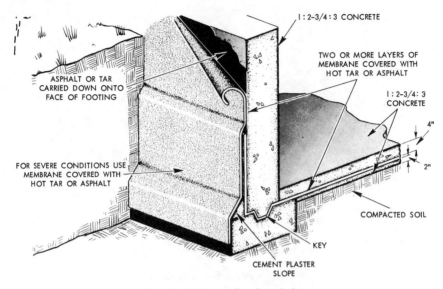

Fig. 90. Waterproofing foundation.

which, when mixed with water, makes an excellent filler for cracks and nail holes. It is not suitable for glazing purposes.

water table: A ledge or slight projection of the masonry or wood construction on the outside of a foundation wall, or just above, to protect the foundation from rain by throwing the water away from the wall. See Fig. 91.

W beam: A wide-flanged steel beam. See Fig. 92.

weatherboards: Boards shaped so as to be specially adaptable for overlapping at the joints to prevent rain or other moisture from passing through the wall. Also, called *siding*.

weathered: In masonry, stonework which has been cut with sloped surfaces so it will shed water from rain or snow. See *water joint*. In carpentry, a term applied to lumber which has been seasoned in the open air.

weathering: A slope given to the top of cornices, window sills, and various moldings to throw off rain water.

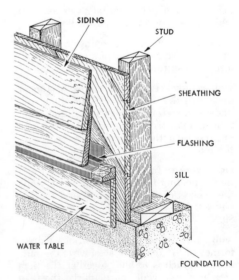

Fig. 91. Water table.

weather strip: A piece of metal, wood, or other material used to cover joints around

Glossary

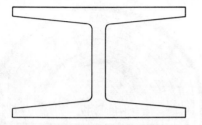

Fig. 92. W beam.

doors and windows to prevent drafts and to keep out rain and snow.

weather tiling: Tiles hung on battens to form a weatherproof covering for a vertical surface such as a wall. See *tile hanging*.

weep hole: In masonry cavity or veneer walls, an opening through the mortar joints at regular intervals to allow for drainage of condensed moisture. Also, a drain through retaining walls.

welded wire fabric: Heavy steel wire welded together in a grid pattern; used for reinforcing concrete slabs. See Fig. 93.

wellhole: An open space such as a shaft or well in a building, as for a staircase; also, the open space about which a circular stairs turns.

western framing: See *platform framing*.

wet rot: A term used by woodworkers for decay of lumber or wood, due particularly to moisture and warmth.

white oak: An American oak of the eastern part of the United States. It is the hardest of American oaks, characterized by its heavy, close grain. Used extensively where strength and durability are required.

wicket: A small door set within a larger door; also, a window or similar opening closed by a grating through which communication takes place, as a cashier's window.

winders: Treads of steps used in a winding staircase, or where stairs are carried around curves or angles. *Winders* are cut wider at one end than at the other so they can be arranged in a circular form.

winding stair: A circular staircase which changes directions by means of winders or

Fig. 93. Welded wire fabric. (United States Steel)

a landing and winders. The wellhole is relatively wide and the balustrade follows the curve with only a newel post at the bottom.

windlass: A device for hoisting weights, consisting usually of a horizontal cylinder turned by a lever or crank. A cable, attached to the weight to be lifted, winds around the cylinder as the crank is turned, thus raising the load to whatever position is desired.

window: An opening in an outside wall, other than a door, which provides for natural light and ventilation. Such an opening is covered by transparent material inserted in a frame conveniently located for admitting sunlight and constructed so that it can be opened to admit air.

window head: In architecture, a term applied to the upper portion of a window frame.

window jack: A portable platform which fits over a window sill projecting outward beyond the sill; used principally by painters.

window seat: A seat built in the recess of a window, or in front of a window.

window wall: An outside wall, of which a large portion is glass. Glass area may consist of one or more windows. A window wall may be made up entirely of windows.

windshake: A defect in wood, so-called because of the belief that it is caused by wrenching of the growing tree by the wind. Since there is a separation of the annual rings from each other around the trunk of the tree, this defect is cuplike in appearance and is sometimes known as *cupshake*. See Fig. 94.

wing: In building, a term applied to a section, or addition, extending out from the main part of a structure.

wire-cut brick: Brick formed by forcing plastic clay through a rectangular opening designed for the purpose, and shaping clay into bars. Before burning, wires pressed through the plastic mass cut the bars into uniform brick lengths.

wire glass: In building construction, a type of window glass in which wire with a coarse

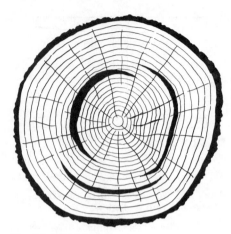

Fig. 94. Windshake.

mesh is embedded to prevent shattering of the glass in case it is broken; also, to protect a building against intruders. *Wire glass* is used principally in windows of buildings where valuables are kept, such as a United States mint; also used as a safety measure in case of fire.

wire ties: Short lengths of wire in various shapes and gages for reinforcing the bond between two members. They may be embedded in mortar; nailed; twisted around and between masonry, wood, or metal. Wire ties are usually of cement-coated steel or galvanized metal.

withe: In masonry, a single vertical tier of brick. Also spelled *wythe*.

wood brick: A wooden block, the size and shape of a brick; built into brickwork to provide a hold for nailing finish material. A nailing block.

wood screws: Wood fasteners of various types and sizes, ranging from No. 0 to 30, and in length from $\frac{1}{4}$ inch to 6 inches. Length is measured from largest bearing diameter of head to the point of the screw. Threads extend for seven-tenths of the length, beginning at the point. Screws are made in oval-, round-, and flat-headed types, while gimlet points are standard. See Fig. 78.

Glossary

wood turning: The process of shaping pieces of wood or blocks into various forms and fashions by means of a lathe.

woodwork: Interior fittings of wood, such as moldings and staircases; also, work done in or with wood objects or parts made of wood.

working drawing: In architecture, a drawing or sketch which contains all dimensions and instructions necessary for carrying a job through to a successful completion.

wreath: The curved section of a stair rail, curved in both the vertical and horizontal planes; used to join the side of a newel post to the ascending run of the handrail.

wreath piece. In stair building, the curved section of the handrail string of a curved or winding stair. Any ornamental design intertwined into a circular form.

wrecking bar: A steel bar about ¾ of an inch in diameter and 24 to 30 inches in length, used for prying and pulling nails. One end of the bar is slightly bent with a chisel-shaped tip and the other end **U**-shaped with a claw tip for pulling nails.

Y

year ring: One of the clearly defined rings in a cross section of a tree trunk, showing the amount of annual growth of the tree. Each ring represents one year of growth; also called *growth ring* and *annual ring*. The rings are made up of cells or tubes which convey sap through the tree. Year rings are shown in Fig. 9.

yoke: In architecture, a term applied to the horizontal top member of a window frame.

Z

Z bar: A heavy wire fabrication shaped in the form of the letter **Z**, usually about 4" x 6" in size. See Fig. 58. Such ties are used to bind together the two separate sections of a cavity wall, the ends of each tie being embedded in the horizontal mortar joint of both tiers at intervals of 24".

zeprex: A lightweight mineral composed of siliceous material, cement, chemicals, and water; can be used for roof decking, walls, ceilings, and floors. Stronger than concrete of similar density, this material can be nailed, drilled, and hand-sawed. It is incombustible, termite proof, has minimum shrinkage and swelling, and high thermal insulation value.

zoning: A term applied to the division of a certain municipal subdivision into districts which may have different types of regulation. Such a condition is brought about by local ordinance under the police power of a state, granted by specific legislation commonly called an *enabling act*. Zoning laws pertain to the use of land in a particular area.

Index

Numerals in **bold type** refer to illustrations.

A

Accelerants, 54
Admixtures,
 concrete, 200
 mortar, 54, 55, 98, 99
Aggregate, concrete, 186-191, **188, 190**
Aggregate, concrete block, 123
Aggregate proporations for concrete, 193, 194
Air entraining cement, 49
Air entrainment, concrete, 199, 200
American Society for Testing and Materials (ASTM), 49, 51, 53, 56, 62, 68, 70, 126, 127, 191
Anchor bolts, 114, **115**, 223, **223**
Anti-freeze admixtures, 98, 99
Apprentice, 1, 7
Apprenticeship standards, 7, **8**, 9
Architect, 6, 254
Asphaltum, 177, **177**, 178, **178**, 228, **228, 229**, 230
Asphaltum board, 225
Autoclaves, 124, 125, **125**

B

Backing, concrete block, 148, 149, **149**
Bars, steel reinforcing, 91, **92**, 92, 151, 153, 221, **221**
Beam block, **129**, 130
Beam support, 157, 159, **159**
Bearing, joist, 104, **105**
Blast furnace slag, 190
Bleeding in concrete, 235
Block planning, 139, 140, 141, **141**, 142
Blueprints, 9, 10, 255-273
 details, 261, 264, **264, 265, 266**
 elevations, 261, **263**
 floor plans, 260, 261, **262**
 longitudinal sections, 265
 scale, 256
 schedules, 265, **266**
 sections, 261, **264, 265**
 symbols, 265, **267, 268, 269, 270, 271**, 272, **272**
 types of lines, 257, **259**, 260
Bolts, anchor, 114, **115**, 223, **223**
Bond stone, 180, **180**
Bonding face brick, 89, **89**, 90
Bonds, brick
 common, 78, **79**, 82, **83**
 Dutch, 78, **79**
 English, **79**, 80
 English cross, 78, **79**
 Flemish, **79**, 80
 running- 78, **79**
 stack, **79**, 80
 stretcher, 78, **79**
Brick classification, 67-70
Brick, history of, 61, 62
Brick masonry, 61-121
Brick, color & surface finishes, 67
Brick courses, 71-80
Brick, face, 68, 69
Brick grades 70, **70**
Brick hammer, 29, 33, **34**
Brick manufacture, 62-67
Brick masonry,
 construction details, 99-106
 tools for, 29-39
Brick, methods of molding, 63, 64, **64**
Brick, non-modular, 67
Brick properties, 70
Brick, raw materials, 62, 63, **63**
Brick set, 29, 35, **35**
Brick, sizes and shapes, 67, 68, **69**

402

Index

Brick veneer, 95, 97, 98, **98**
Brick, weight of, 66
Buck anchor, 117, **117**
Builder's level, 230, **231**
Building codes, 54, 57, 58, 72, 98, 99, 200
Building enclosure, 25, **27**, 28
Building paper, 164, **164**
Built-in-place formwork, 208, 209, **209**, 210, **210**
Bulking in sand, 194
Bull float, 41, 42, **42**, 234, **236**
Bullnose block, 128, **129**
Bureau of Apprenticeship and Training, 7

C

Calcium chloride, 54
Calk, 163, **163**, 164, **164**
Carbon black, 55
Cavity walls, brick, 90, **90**, 91, **91**
Cement
  damaged, 192
  distribution, 58, 192
  paste, 185
  storage, 192
Chemicals, safety practices, 20
Chimneys, brick, 105, **105**, 106
Chimneys, concrete block, 173, 174, **174**, 175
Cleaning brickwork
  cleaning compound method, 120
  sandblasting, 120
  steam and water-jet process, 120

water-jet process, 120
Codes, building, 54, 57, 58, 72, 98, 99, 200
Cold weather practices, bricklaying, 98, 99
Colorimetric test, 52, **52**
Columns, brick, 102, **102**
Columns, concrete, 221, **221**
Columns, concrete block, 157, **158**
Common bond, 78, **79**
Common bond brick wall, 82-90
Concrete
  admixtures for, 200
  coarse aggregates for, 188-190, **190**
  colored, 244, 245
  fine aggregate, 187, 188, **188**
  finishing, 233-241
  mixers, 44, **44**, 200, 201, **201**
  mixing, 201, 202, **202**
  portland cement for, 191, 192
  proporting mixtures for, 192-199
  ready mixed, 200, 201, **201**
  special uses, 249-252, **250-252**
  trial proportions, 196-199, **198**
  water for, 185, 191, 193, 198
Concrete block
  dimensions, **129**, 130, 131, 139, 140
  grades, 126, 127
  hollow load-bearing, 127
  maintenance, 182
  manufacture, 123-126, **125, 126**
  materials for, 123

  painting, 182-184
  sizes and shapes, 128, **129**, 130, 131
  special, 131-137
  three cored, 138, **138**, 139
  two cored, 138, **138**, 139
  types, 127
  unit classification, 127
Concrete block construction
  beam support, 157, 159, **159**
  block planning, 139-142
  bond beams, 153, **153**, 154, **154**
  brick and block walls, 148, 149, **149**
  cavity walls, 140-151, **151**
  chimneys, 173, 174, **174**, 175
  concrete floor support, 160, **160**
  control joints, 162-166, **162-164**
  cornice details, 175, **176**
  details, 148-177
  duct details, 176, **176**
  joint finishing, 147, **147**, 148, **148**
  joist floors, 160-162, **161**
  joist support, 155-157, **156**
  laying a concrete block wall, 139-148
  layout, 142, **142**, 143, **143**
  lintels, 153-155, **154, 155**
  pilasters and columns, 157, **158**
  preparation of materials, 139
  reinforcement, 151-155, **152-155**

403

steel bar reinforcing, 153-155, **153-155**
wall intersections, 165, 166, **165**
waterproofing foundations, 177-179, **177-179**
Concrete bricks, **129,** 130
Concrete finishing, 233-241
Concrete flatwork,
 brooming, 240, **240**
 bull floating, 234, **234,** 236
 curing, 241, 242, **242**
 darbying, 234, 235, **236**
 driveways, **243,** 246
 edging, **235,** 238
 floating, 238, **239**
 jointing, 235, **237**
 screeding, 233-235, **234**
 site preparation, 230, 231, **231**
 tools for, 39-44
 troweling, 238-240, **239, 240**
Concrete footings and foundations
 formwork for, 207-216, **207-216**
 slab-at-grade, 225, **225,** 226, **226**
Concrete, precast, 249, **250**
Concrete steps, forms for, 215, 218, **219, 220,** 221
Concrete steps, thickness of, 248, **249**
Concrete vibrators, 222, **222,** 223
Concrete walls, 213-224, **214, 217, 224**
Construction Specifications Institute (CSI) Format, 11
Contracting practice, 5-7
Contractor, 2, 5-7

Control joint blocks, 163, **164**
Control joints, concrete block, 162-164, 166
Corbelling, 106, **106**
Core blocks, 137, **137**
Corner block, 128, **129**
Corners, brick wall, 82-89
Corners, concrete block wall, 139-146
Cornices in concrete block walls, 175, 176, **176**
Coursed rubble masonry, 180, **180**
Courses, header, 71, **71**
Courses, lengths & heights, 80, **81,** 82
Courses, rowlock, 71, **72**
Courses, soldier, 71, **72**
Courses, stretcher, 71, **71,** 84, **85,** 86
Crushed stone, 188, 189
CSI Format, 11
Curves, concrete sidewalk, 243, **243,** 244, **245**

D

Damaged cement, 192
Darby, 41, **41,** 234, 235, **236**
Decimals, 276-279
Decimals, changing to fractions, 279
Decimals, operations with, 277, 278
Deformed steel bars, 92, **92**
Department of Labor, 7
Dimensions, brick 67, 68, **69**
Dimensions, concrete block, 128, **129,** 139, 140

Dimensions, concrete block actual and nominal 130, 131, 139, 140
Dimensions, scaling and, **259**
Door schedule, 265, **266**
Doors in concrete block wall, 169, 172, **172**
Doors, openings for, 93, **94,** 95, 169-173
Double corner block, 182, **192**
Drain tile, 179, **179, 229,** 230
Driveways, concrete, 246, **243**
Dry-press process, 64
Ductwork in concrete block walls, **176,** 177
Dutch bond, 78, **79,** 80

E

Edger, 42, **42**
Electricity, safety practices, 19, 20
Electric saw, 37, **38**
Elevation drawings, 255, **258,** 261, **263**
English bond, **79,** 80
English cross bond, 78, **79,** 80
English to metric conversions, 279-281
Estimating, 2, 5, 284, 285
Estimating areas, volumes and materials, 284-285
Estimator, 2
Extruded mortar, 136, 137

F

Face block, 131, **132**

Index

Face brick, types, 68, 69
Filler block, **129,** 130
Firebrick, 69, 70
Firebrick, mortar for, 70
Fire clay, 69, 70
Fireplaces, 107, **107,** 108, **108**
First Aid, 15-17
Flashing, 102, 103, **103,** 104, 149, 150
Flashing, materials, 102, 103
Flemish bond, **79,** 80
Floors, concrete, 246-248, **247**
Flue linings, 105, **105,** 106
Footings & foundations, brick, 99, **100**
Footings, forms for, 206, **206,** 207, **207**
Footings, keys, in, 207, **208**
Footings, T type, 207, **207,** 208, **208**
Foreman, 2
Formwork, concrete, 205-221
Formwork, removal of, 224
Foundation formwork, 208-217, **209-217**
Fractions, 274-276
Fractions, changing to decimals, 278, 279
Fractions, operations with, 275, 276
Furring clip, SCR brick, 112, **112**
Furring strips with SCR brick, 111, 112, **112,** 115, **115**

G

Geometry, elementary, 281-285

Glazed block, 131, **132**
Grade beam & pier construction, 226, **227**
Grading crushed stone, 189
Grading gravel, 189, 190
Grading sand, 53, **53,** 54
Gravel, grading, 189, 190
Groover, 42, **42,** 43
Grout, 92, **93,** 215, 224

H

Head joints, 74, **74**
Head plate, **114,** 176
Header, 71, **71**
Header block, 128, **129**
Header course, 71, **71**
High early strength cement, 48, 55, 192, 200
Hollow load-bearing block, 127, 138, **138**
Horizontal reinforcement, concrete block, 150, 151, **151,** 152, **152** 153, **153**
Housekeeping, 28
Hydrated lime, 46, 49-51
Hydration, 241
Hydrochloric acid, 120

I

Indenture, 1, 2, **3,** 4
Insurance, 5, 6
Intersections, concrete block walls, 165, **165,** 166
Isolation joints, 233, **233**

J

Jamb blocks, 93, **94,** 172, 173, **173**
Jitterbugs, 40, **41**
Joint Apprenticeship Committee, 2, 7, 8, 9
Joint finishing, 77, 78, **78**
Jointers, 29, **30,** 31, 32, **32**
Joints, bed, 73, **73,** 74, **74**
Joints, closure, 75, **75,** 76, **76,** 86
Joints, concave, 77, **78,** 147, **147**
Joints, cross, 75, **75**
Joints, head, 74, **74,** 75
Joints, isolation, 233, **233**
Joints, raked, 77, **78**
Joints, thickness, 76, 77
Joints, rodded, 77, **78**
Joints, V, 77, **78,** 147
Joist anchors, 104
Joist bearings, 104, **105**
Joists, 104, **105,** 113, 114, **114, 115,** 155, 156, **156,** 157, 160, **160,** 161, **161**
Journeyman, 2

K

Keys, 207, **208**
Kiln burning, 64-66
Kilns, 64-66, **65-66**

L

Labor, Department of, 7
Lath, metal, 166, 175, 176, **165, 174, 176**
Layout, concrete block, 142, **142**

405

Level, builder's, 230, **231**
Leveling instruments, 230, **231**
Levels, 29, **30**, 33, **33**, **34**
Lien laws, 5
Lift, 224
Lightweight concrete, 191
Lime, manufacture, 49-51
Line, masons, 29, **30**, 35, **36**
Lintel block, **129**, 130
Lintels, steel, 95, **96**
Longitudinal sections, 265
Lug sill, 169

M

Maintenance, brick, 118-120
Masonry as a trade, 1-28
Metal lath, 166, 175, 176, **165**, **174**, **176**
Metal sash block, **129**, 130
Metal ties, 89-91, 98, 149, **89**, **98**, **150**
Metric to English conversions, 279-281
Mineral oxide pigments, 55, 244, 245
Modular coordination, 140
Modular sizes, brick, 67, 68, **69**
Modular sizes, concrete block, 130, 131, 138, 139, **129**, **138**
Moisture in sand, 194, **194**
Molding bricks
  dry-press process, 64
  soft-mud process, 64
  stiff-mud process, 63, **64**
Mortar
  admixtures, 54
  colors, 55
  durability, 55, 56
  ingredients, 46-55, **47**
  lime for, 49-51, **50**
  measurements, 58
  mixer, 18, 37, 38, **18**, **38**
  mixing, 58, 59, **59**
  strength, 55, 56
  water for, 54
  water rententivity, 56
  workability, 56
Mortar, extruded, 136, **137**
Mortar joints, 72-78, 146-148, 181, 182
Muriatic acid, 120

N

National Concrete Masonry Association, 177
Nominal dimensions, brick, 68, **69**
Nominal dimensions, concrete block, 130, 131, 138, 139, **129**, **138**

O

Operations with decimals, 277
Operations with fractions, 275
Organic matter in sand, 52, **52**, 187

P

Parapet walls, flashing for, 103, **103**
Parging, 76, **77**
Partition block, **129**, 130
Periodic kilns, 65, 66
Pier block, 128, **129**
Pigments, mineral oxide, 55, 244, 245
Pilasters brick, 99, 100, 101, 102, **101**
Pilasters, concrete, 221
Pilasters, concrete block, 157, **158**
Plan view, 261, **262**
Plasticity, 49, 56, 57
Plate, head, 114, **176**
Plot plan, 261
Plumb rule, 29, **30**, 33, **33**, **34**, 100
Pointing mortar, 119
Portland cement
  history of, 46-48
  ingredients of, 48
  types for concrete, 191, 192
  types for mortar, 49
Portland Cement Association, 177, 244
Power joint cutter, 44, **44**
Power saws, 18, 19, **19**
Power tools and equipment, 37, 39, 43, 44, 201, 231, 234, 239, **38**, **43**, **44**, **231**, **235**, **237**, **240**
Power trowel, 43, **43**
Precast concrete, 249, **250**
Precut stone, 181, **181**
Proportioning concrete mixtures, 192-199

Q

Quicklime, 51
Quoins, 182, **182**

R

Random rubble masonry, 180, **180**

Index

Rate of suction, 72, 73
Ready mixed concrete, 200, 201, **201**
Reinforcement, concrete, 221
Reinforcing steel bars, 92, **92**
Reinforcing, brick wall, 91, 92, 93, **93**
Repointing brickwork, 118, **118**, 119
Retardants, 54
Risers of steps, 248
Rolok-bak wall, 91, **91**
Rough buck, 215, **218**
Rubble, 190, 191
Rubble concrete, 190
Rules, measuring, 29, **30**, 36, **37**

S

Safety, 13-28
  building enclosure, 25, **27**
  clothing, 17
  chemicals, 20
  electricity, 19, 20
  scaffolds, 20-25, **21**, **23-25**
  tools and equipment, 17, 18, **18**, 19, **19**
Salamanders, 25
Salmon brick, 65
Sand,
  cleanness of, 51, 52
  grading, 53, 54, **54**
  organic matter in, 52, **52**
  sieve analysis, 53
  silt content, 51, 52, **52**
Sandblasting, 121

Sash windows, metal, 116, **116**, 169, **171**
Sash windows, wood, **172**
Scaffolds,
  built up, 20
  hanging, 21, 24
  rolling, 20
  steel, 22, 23, **25**
  swing stage, 21, **21**, 24
  trestle, 20
  wood, 21, **22, 23, 24**
Scored block, 131, **131**
SCR brick, 109-118
  door and window installation, 116-118, **116, 117**
  foundations for, 112-114, **113, 114**
  furring clip for, 112, **112**
  head plate anchorage, 114, 115, **115**
  wall construction, 111, **111**, 112, **112**
Screeding, 233, 234, **234, 235**
Screeds, 40, 231-235, **231, 234, 235**
Segregation in concrete, 40
Shadow block, 136, **136**
Shotcreting, 249, 251, 252, **251, 252**
Sidewalks, 242-246
Sidewalk intersections, 243, **243**, 244
Sieve analysis, sand, 53, 54
Sill high, 93-95
Sill, lug, 169
Sill, slip, 169, 171
Silt content in sand, 51, 52, **52**
Slab-at-grade floors, 225-227, **226, 227**
Slag, blast furnace, 190
Sled runner, 32, 147, **32, 148**
Slip sill, 169

Slope, concrete floors, 247, **247**
Slump, 195
Slump block, 131, **133**
Slump test for concrete, 195, **195**
Small Homes Council, 225
Snap ties, 213, 214, **214**
Sodium hydroxide, 52
Soldier courses, 71, **72**
Solid brick, **129**, 130
Specifications, 9, 10, **10**, 11, 260, 272, 273
Split block, 131-135, **133, 135**
Stack bond, **79**, 80
Steel reinforcement, concrete, 221, **221**
Steel square, 29, 35, **30, 35**
Steps, concrete, 248, 249, **249**
Steps, self-supporting, 218, **220**
Stiff mud process, 63, 64, **64**
Stone, crushed, 188, 189
Stone masonry,
  bond stone, 180, **180**
  coursed rubble, 180, **180**
  mortar for, 181
  precut stone, 180, 181
  random rubble, 180, **180**
Stone, precut, 180, **181**
Story pole, 104, 105, 144, **145**
Stretcher block, 128, **129**
Stretcher bond, 78, **79**
Structural Clay Products Research Foundation, 109
Sulfate attack, 191, 192
Superintendent, 2
Symbols, 265, 272, **267-272**

407

Masonry Simplified

T

Tampers, 40, **41**
Test, colorimetric, 52, 53
Textured block, 131, **132**
Tile, drain, 179, **179, 229,** 230
Tilt-up construction, 249, **250**
Tools, 29-44
Trade unions, 6
Transverse reactions, 265
Treads of steps, 248
Trial proportions for concrete, 196-199, **198**
Trowels,
 brick, 29-31, **30, 31**
 concrete, 43, **43**
T type footing, 207, **207,** 208, **208**
Tuckpointing, 118, **118,** 119
Tunnel kilns, 65, 66, **68**

U

U.S. Department of Labor, 7

V

V joint, 77, 78, **78, 79**
Vapor barrier, 225, **225**
Veneer, brick 95, **97,** 98, **98**
Veneer, stone, 179
Ventilation, brick wall, 103, **104**

W

Wales, 205, **212,** 213
Wall patterns, concrete block, 133, 137
Wall ventilation, 103, 104, **104**
Walls, concrete, 223, 224, **224**
Water,
 for concrete, 185, 191, 193, 198
 for mortar, 54
Water-cement ratio, 190
Water retentivity, 56
Waterproofing
 foundations, concrete, 228, **228, 229,** 230

Waterproofing
 foundations, concrete block, 177-179, **177-179**
Weep holes, 90, 91, 103, 149, 150
Welded wire fabric, 221, **221,** 232
Window, double hung, 169, 172
Window, metal sash, 169, **171**
Windows, openings for, 93-95, 166-169, **94, 167**
Wire ties, **210,** 211, 213, **213**
Wood sash jamb block, 128, **129**
Working drawings, 260-272

Z

Z tie bars, **89,** 163, **163, 164**